DATA SCIENCE AND INNOVATIONS FOR INTELLIGENT SYSTEMS

Demystifying Technologies for Computational Excellence: Moving Towards Society 5.0

Series Editors: Vikram Bali and Vishal Bhatnagar

This series encompasses research work in the field of Data Science, Edge Computing, Deep Learning, Distributed Ledger Technology, Extended Reality, Quantum Computing, Artificial Intelligence, and various other related areas, such as natural-language processing and technologies, high-level computer vision, cognitive robotics, automated reasoning, multivalent systems, symbolic learning theories and practice, knowledge representation and the semantic web, intelligent tutoring systems, AI and education.

The prime reason for developing and growing out this new book series is to focus on the latest technological advancements-their impact on the society, the challenges faced in implementation, and the drawbacks or reverse impact on the society due to technological innovations. With the technological advancements, every individual has personalized access to all the services, all devices connected with each other communicating amongst themselves, thanks to the technology for making our life simpler and easier. These aspects will help us to overcome the drawbacks of the existing systems and help in building new systems with latest technologies that will help the society in various ways proving Society 5.0 as one of the biggest revolutions in this era.

Data Science and Innovations for Intelligent Systems
Computational Excellence and Society 5.0
Edited by Kavita Taneja, Harmunish Taneja, Kuldeep Kumar, Arvind Selwal, and Ouh Lieh

Artificial Intelligence, Machine Learning, and Data Science Technologies
Future Impact and Well-Being for Society 5.0
Edited by Neeraj Mohan, Ruchi Singla, Priyanka Kaushal, and Seifedine Kadry

Transforming Higher Education Through Digitalization
Insights, Tools, and Techniques
Edited by S. L. Gupta, Nawal Kishor, Niraj Mishra, Sonali Mathur, and Utkarsh Gupta

A Step Towards Society 5.0
Research, Innovations, and Developments in Cloud-Based Computing Technologies
Edited by Shahnawaz Khan, Thirunavukkarasu K., Ayman AlDmour, and Salam Salameh Shreem

Computing Technologies and Applications
Paving Path Towards Society 5.0
Edited by Latesh Malik, Sandhya Arora, Urmila Shrawankar, Maya Ingle, Indu Bhagat

For more information on this series, please visit: https://www.routledge.com/Demystifying-Technologies-for-Computational-Excellence-Moving-Towards-Society-5.0/book-series/CRCDTCEMTS

DATA SCIENCE AND INNOVATIONS FOR INTELLIGENT SYSTEMS

Computational Excellence and Society 5.0

Edited by
Kavita Taneja, Harmunish Taneja
Kuldeep Kumar, Arvind Selwal, and
Eng Lieh Ouh

CRC Press
Taylor & Francis Group
Boca Raton London New York

CRC Press is an imprint of the
Taylor & Francis Group, an **informa** business

MATLAB® is a trademark of The MathWorks, Inc. and is used with permission. The MathWorks does not warrant the accuracy of the text or exercises in this book. This book's use or discussion of MATLAB® software or related products does not constitute endorsement or sponsorship by The MathWorks of a particular pedagogical approach or particular use of the MATLAB® software.

First edition published 2022

by CRC Press
6000 Broken Sound Parkway NW, Suite 300, Boca Raton, FL 33487-2742

and by CRC Press
2 Park Square, Milton Park, Abingdon, Oxon, OX14 4RN

ISBN: 978-0-367-67627-8 (hbk)
ISBN: 978-0-367-67628-5 (pbk)
ISBN: 978-1-003-13208-0 (ebk)

DOI: 10.1201/9781003132080

Typeset in Times
by MPS Limited, Dehradun

Contents

Editors

Kuldeep Kumar works at the Department of Computer Science and Engineering, National Institute of Technology Jalandhar, India. He received his Ph.D. degree in computer science from the National University of Singapore in 2016. Prior to joining the institute, he worked for two years in the Birla Institute of Technology and Science, Pilani, India. He received his Ph.D. in computer science from the School of Computing, National University of Singapore (NUS SoC), Singapore in 2016. He has several publications in reputed international journals/conferences. https://orcid.org/0000-0003-1160-9092

Eng Lieh Ouh received his Ph.D. in computer science (software engineering) from the National University of Singapore. He is involved in several large-scale information technology industry projects for a decade at IBM Singapore and Sun Microsystems before joining academia as an educator. His research areas are software reuse, software architecture design, design thinking, and software analytics. He has experiences delivering courses for the postgraduate students, undergraduate students, and industry participants in the software engineering areas including design thinking, practical software architecture design, security engineering, and mobile development. He received multiple teaching excellence awards and industry projects recognition awards throughout his career. https://orcid.org/0000-0001-7759-348X

Arvind Selwal works as an assistant professor in the Department of Computer Science and Information Technology, Central University of Jammu, India. His research interests include machine learning, biometric security, digital image processing, and soft computing. He has contributed more 25 research articles in reputed international journals and he has authored a book titled *Fundamentals of Automata Theory and Computation*. He is an active member of the Computer Society of India (CSI) and he is undertaking two research projects from DRDO, New Delhi, India. https://orcid.org/0000-0002-1075-6966

Harmunish Taneja is an assistant professor at DAV College, Sector 10, Panjab University, Chandigarh. He received his Ph.D. in computer science and applications from Kurukshetra University, Kurukshetra. He has more than 21 years of teaching and research experience. He has guided 05 M.Phil. students and more than 90 PG students of various universities. Four students are pursuing their Ph.D. under his guidance. He is a reviewer of many reputed journals and has been a review committee member of many conferences. His research interests are information computing, mobile ad hoc networks, image processing, data science, recommender systems, and system simulation. He has published and presented over 55 papers in international journals/conferences of repute. He has also authored books of computer science and applications. https://orcid.org/0000-0003-1745-0748

Kavita Taneja is an assistant professor at Panjab University, Chandigarh. She received her Ph.D. in computer science and applications from Kurukshetra University, Kurukshetra, India. She has published and presented over 60 papers in national/international journals/conferences and has had best paper awards in many conferences including IEEE, Springer, Elsevier, ACM, and many more. She is a reviewer of many reputed journals and has been a technical program committee member of many conferences. She has also authored and edited computer books. She has more than 18 years of teaching experience in various technical institutions and universities. She is also member of BoM, Academic Council, and Board of Studies of many universities and institutions. She has guided scholars of Ph.D./M.Phils., more than 100 PG students of various universities, and currently four students are pursuing a Ph.D. under her guidance at Panjab University. Her teaching and research activities include mobile ad hoc networks, simulation and modeling, and wireless communications. https://orcid.org/0000-0002-9348-3587

Contributors

Amit Kumar Bindal is an associate professor in the Department of Computer Science & Engineering, M. M. Engineering College, M. M. (Deemed to be University) Mullana, Ambala, Haryana, India. He received Ph.D. from Maharishi Markandeshwar University, M. Tech (Computer Engineering) from Kurukshetra University and a B. Tech. in computer engineering from Kurukshetra University Kurukshetra. He has been in teaching and research and development since 2005. He has published about 60 research papers in international, national journals, and refereed international conferences. His current research interests are in wireless sensor networks, underwater wireless sensor networks, sensors, and IoT etc. https://orcid.org/0000-0002-2624-8077

Michael Boguslavsky is presently the head of AI and is leading the AI team developing ML credit analytics for a start-up trade finance platform. The team builds predictive models for SME credit risk and develops trade credit event prediction models based on supply chain graph flows analysis. Prior to this he was an advisor for Blackstone Alternative Asset Management, advising the leading alternative asset manager on fund structuring, portfolio optimization, risk modelling, and placement for alternative, private, and illiquid credit, hedge funds, infrastructure, and emerging market debt funds. He has also worked in banks and asset management companies over 15 years as acting head of EMEA Equity Derivatives Structuring and head of ALM and Quantitative Analytics, as well as being a trader and quantitative analyst. He holds a Ph.D. in mathematics from the University of Amsterdam; a Ph.D. in computer sciences from the Russian Academy of Sciences, Moscow; and a master's degree in mathematics and applied mathematics. https://orcid.org/0000-0001-8423-9403

Amit Chadgal completed the M.Tech. in 2020 at the Department of Computer Science and Technology, Central University of Jammu. Prior to M.Tech., he completed a B. Tech. in 2018 at the Department of Information Technology, National Institute of Technology Srinagar, India. The research topic of his M. Tech. was "light-weight cryptography techniques in IoT" at the Department of Computer Science and Technology, Central University of Jammu. https://orcid.org/0000-0002-9013-308X

Ashutosh Deshwal is currently pursuing a Ph.D. in computer science and engineering at Thapar Institute of Engineering, Patiala, Punjab, India, since July 2019. His areas of interest include blockchain, machine learning, Internet of Things, embedded system, and product analysis. https://orcid.org/0000-0001-8444-1572

Paul R. Griffin is a member of the faculty of Singapore Management University (SMU) teaching postgraduate and undergraduate students in IT and FinTech as an associate professor of information systems. He gained a Ph.D. at the Imperial College London in 1997 on quantum well solar cells and thermophotovoltaics and is now researching disruptive technologies applications and impact. In particular, his research covers decentralized solutions and the application of quantum computing focusing on financial applications. With a number of projects ongoing for consensus, trade finance, and portfolio optimization, he has been advising companies since 2014 and presenting at events, judging hackathons and moderating panel discussions on FinTech. Prior to SMU he was leading application development on global, regional, and local projects for over 15 years in the United Kingdom and Asia in the finance industry. During this time, as well as leading internal IT development teams, he worked on outsourcing, off-shoring projects, and IT support. With a Black Belt in Six Sigma, he is keen on quantifying quality and ensuring efficiency in current processes and during change. https://orcid.org/0000-0003-2294-5980

Anu Gupta, has been working as a professor in computer science and applications at the Panjab University, Chandigarh since July 2015, where she has been working as a faculty member since 1998. She has the experience of working on several platforms using a variety of development tools and technologies. Her research interests include open-source software, software engineering, cloud computing, and data mining. She is a life member of the "Computer Society of India" and "Indian Academy of Science." She has published more than 25 research papers in various journals and conferences. https://orcid.org/0000-0002-1403-5023

Keshav Gupta is a pre-final year student at the Indian Institute of Information technology, pursuing an integrated postgraduate in information technology. He has been associated with machine learning and data science for the better part of the last three years. He also has experience working with various machine learning based start-ups. He has worked as a research scholar in his institute for research project related multimodal AI and has formidable ranks in various data science challenges and hackathons. His research interests include computer vision, natural language processing, and time series forecasting analysis. https://orcid.org/0000-0002-5763-0690

Afshan Hassan is a Ph.D. research scholar in Chitkara University, Punjab, India. She has completed her B.Tech. (Computer Science & Engineering) degree from Islamic University of Science & Technology, J&K, India in 2014 and M.Tech. (Computer Science & Engineering) degree from Chandigarh Group of Colleges, Landran, India in 2019. She is a gold medalist in B.Tech. (CSE). She has several research publications in Scopus – Indexed journals of high repute. Her research interests include wireless sensor networks, ad hoc networks, machine learning, and IoT security. https://orcid.org/0000-0001-5133-2927

Junye Huang is a quantum developer advocate at IBM. He is part of the Qiskit community team whose mission is building an open, diverse, and inclusive quantum community. He is focusing on promoting quantum education in the Asia Pacific region. He organized the first Qiskit university hackathon in the world and is a guest lecturer for two quantum computing courses at National University of Singapore. In addition, he is supporting regional and global educational initiatives such as the IBM Quantum Challenge and the Qiskit Global Summer School. His passion for quantum computers drives him to create educational games for quantum computers such as QPong, a quantum version of Pong which he created at the first Qiskit camp. QPong was subsequently ported to a physical Quantum arcade machine and toured around Europe, including the EU Quantum Flagship Event in Helsinki in October 2019. He is a Ph.D. candidate in experimental low temperature physics at the National University of Singapore. https://orcid.org/0000-0001-7896-0595

Stan Jarzabek received a M.Sc. and Ph.D. from Warsaw University. He has been a professor at Bialystok University of Technology since 2015; in 1992–2015 he was an associate professor at the Department of Computer Science, National University of Singapore; in 1990–92 he was a research manager of CSA Research Ltd in Singapore. Before, Stan taught at McMaster University, Canada, and worked for an industrial research institute in Warsaw. Stan's research interest is software engineering (software reuse and maintenance), and in recent years mHealth – the use of mobile technology for psychotherapy support, patient monitoring, and data collection/analysis. https://orcid.org/0000-0002-7532-3985

S. Jaya is current Ph.D. research scholar (Full Time) in the Department of Computer Science at Sri Sarada college for Women (Autonomous), Salem-16, Tamilnadu, India. She has published three papers and one chapter in IGI Global. Her area of interest is digital image processing. https://orcid.org/0000-0003-0361-8052

A. Kamaraj received his B.E. degree in Electronics and Communication Engineering from Bharathiar University, Coimbatore, Tamil Nadu, India in 2003. He completed his post graduation from Anna Universiy, Chennai in the field of VLSI Design in 2006. He completed his Ph.D. at Anna University, Chennai in 2020. Currently he is an associate professor in the Department of Electronics and Communication Engineering, MepcoSchlenk Engineering College, Sivakasi, India. His research interests include digital circuits and logic design, reversible logic and synthesis, and advanced computing techniques. During his 14 years of teaching, he has published 21 papers in international journals and 23 papers in national and international conferences. He has filed two patents and was granted one copyright. He has been member of IETE and ISTE. https://orcid.org/0000-0001-6952-2374

Robert J. Kauffman holds the Endowed Chair in Digitalization at the Copenhagen Business School, sponsored by Danske Bank, Copenhagen Airport A/S, and the Danish Society for Education and Business. He earlier served at Singapore Management University's School of IS as professor of IS Management and Associate Dean (Research, Faculty). He was the W.P. Carey Chair in IS at Arizona State until 2011, and professor and director of the MIS Research Center, Carlson School of Management, University of Minnesota until 2007. His graduate degrees are from Cornell and Carnegie Mellon. His research focuses on senior management issues in strategy, information, technology, economics and society, digital commerce, the FinTech Revolution, and resource and energy sustainability. His past employment was in international lending, payment networks, trade services, investments and trading, technology innovation, and other financial services. His funded research has spanned these and hospitality services, digital entertainment, IT consulting, airlines, and agribusiness. https://orcid.org/0000-0002-3757-0010

Balwinder Kaur, is a research scholar in computer science and applications, Panjab University, Chandigarh where she has been working as an assistant professor since 2011. She has working experience of more than 14 years. She received her MCA degree from Kurukshetra University. Her research interest is in educational data mining. https://orcid.org/0000-0001-2222-5903

Aman Khera received his Ph.D. (Business Laws) from Panjab University, Chandigarh in the year 2014 and Ph.D. (Business Management & Commerce) from Panjab University, Chandigarh in the year 2018. He did his MBA (HRM) at Punjab Technical University in the year 2004 and LLB at Panjab University Chandigarh in the year 2009 and LLM at Kurukshetra University, Kurukshetra in the year 2011. He has seven years of corporate experience in HR. He has published 23 research papers in both national and international journals and has presented various papers at national and international conferences. He has guided one Ph.D. student and is guiding two Ph.D. students. Presently he is working as an assistant professor in University Institute of Applied Management Sciences (UIAMS), Panjab University, Chandigarh since 2011. His area of specialization is HRM and business law. https://orcid.org/0000-0003-4909-2332

Meenu Khurana has a Ph.D. in computer science and engineering at Chitkara University Institute of Engineering and Technology, Chitkara University, Punjab, India. She has won scores of academic awards during her graduate and post-graduate studies. She is a certified programmer for Java 2 Platform from Sun Systems, USA and also a certified programmer from Brain bench, USA. She has over 26 years of experience in industry, academics, research, and administration to her credit. Her areas of expertise are mobile ad-hoc networks, vehicular ad-hoc networks, wireless technologies, network security, sensor networks, machine learning, cloud computing, curriculum designing and development, and pedagogical innovation. Her areas of interest include artificial intelligence, machine learning, and algorithms. She has guided research candidates in areas of expertise. https://orcid.org/0000-0001-6515-7939

Ashok Kumar is an ex-professor in the Department of Computer Science & Engineering, M. M. Engineering College, M. M. (Deemed to be University) Mullana, Ambala, Haryana, India and former professor of Kurukshetra University. He was in teaching and research and development for more than 40 years. He has published many research papers in international, national journals, and refereed international conferences. His current research interests are in software engineering and digital image processing. https://orcid.org/0000-0003-1240-7054

Pawan Kumar is pursuing Ph.D. from Lovely Professional University, Punjab, India. He is serving as an assistant professor in the School of Computer Applications, Lovely Professional University, Punjab, India. He has an experience of more than 14 years in academics. He has qualified UGC-NET and GATE in 2012. His research interests include machine learning and data analytics. He is a lifetime member of Computer Society of India (CSI) and Indian Science Congress Association (ISCA). https://orcid.org/0000-0003-1698-3286

J. Senthil Kumar received his B.E. degree in electronics and communication engineering from M.S. University, Madurai, Tamil Nadu, India in 2003. He has completed his post graduation from Anna Universiy, Chennai in the field of Communication Systems in 2005. He has completed his Ph.D. at Anna University, Chennai in 2017. Currently he is an associate professor in the Department of Electronics and Communication Engineering, MepcoSchlenk Engineering College, Sivakasi, India. His research interests include robotics, Internet of Things, and embedded systems. During his 15 years of teaching career, he has published 24 papers in international journals and nine papers in national and international conferences. He has published five patents and granted one copyright. He has completed three sponsored research projects worth of 78 Lakhs. He has been member of IETE and ISTE. http://orcid.org/0000-0002-9516-0327

M. Latha is currently working as associate professor of computer science in Sri Sarada College for Women (Autonomous), Salem-16, Tamilnadu, India. She has 24 years of teaching experience. Her area of research includes software engineering and digital image processing. Her h-index is 5 and i10 index is 4. https://orcid.org/0000-0002-2648-765X

Umesh Kumar Lilhore is an associate professor in Chitkara University Institute of Engineering and Technology, Chitkara University, Punjab, India. He is Ph.D. in computer science engineering and M. Tech. in computer science and engineering. He has research publications in SCI-Indexed international journals of high repute. His research area includes AI, machine learning, computer security, computational intelligence, and information science. https://orcid.org/0000-0001-6073-3773

Kamakshi Malik is working as an assistant professor in the Department of Management and Commerce in DAV College, Chandigarh. She has a total teaching experience of more than a decade. Her research interests include employee engagement, organizational behavior, and human resource management and is currently pursuing a Ph.D. from IKG, Punjab Technical University, Jalandhar. She has published 16 research and review articles. She has also presented her work at various national and international conferences.

Palvinder Singh Mann received his bachelor's degree (B.Tech.) with honors (Institute Gold Medal) in information technology from Kurukshetra University, Kurukshetra, Haryana, India, M.Tech. in computer science and engineering from IKG Punjab Technical University, Jalandhar, Punjab, India, and Ph.D. in computer science and engineering from IKG Punjab Technical University, Jalandhar, Punjab, India. Currently, he is working as an assistant professor at DAV Institute of Engineering and Technology, Jalandhar, Punjab, India. He has published more than 50 research papers in various international journals, international conferences, and national conferences. His research interests include wireless sensor networks, computational intelligence, and digital image processing. https://orcid.org/0000-0002-9859-6193

Ogawa Masayoshi graduated from Singapore Management University with degrees in economics and information systems. He is passionate about developing greater understanding of the world and never passes up the opportunity to learn. Currently, he is a pricing actuarial executive in the life insurance industry. https://orcid.org/0000-0002-1605-6477

Gulab Kumar Patel received his master's in mathematics and computing from Indian Institute of Technology, Guwahati, Assam, India, in 2017. Currently, he is pursuing his doctoral work on queuing theory in the faculty of Mathematical Sciences from Indian Institute of Technology (Banaras Hindu University), Varanasi, Uttar Pradesh, India. His area of research interests fall into realms of deep learning, statistics, and queuing theory. https://orcid.org/0000-0002-5641-8532

Devendra Prasad is a professor in Chitkara University Institute of Engineering and Technology, Chitkara University, Punjab, India. He has received his M.Tech. (Computer Science and Engineering) and Ph.D. (Computer Science and Engineering). He has supervised several M.Tech. and Ph.D. students. He has research publications in SCI-Indexed international journals of high repute. His research interest includes network security and fault-tolerant mobile ad-hoc, wireless sensor networks, and machine learning. https://orcid.org/0000-0002-1771-4670

Asha Rani is an assistant professor at Gujranwala Guru Nanak Khalsa College, Ludhiana (Punjab). She is a research scholar in the Department of Computer Science and Applications Panjab University, Chandigarh, India. She has published and presented over 12 papers in national/international journals/conferences. She has more than seven years of teaching experience in various institutions. Her teaching and research activities include recommendation systems, multi-criteria decision making methods, and data mining techniques. https://orcid.org/0000-0001-5481-2141

Arpana Rawal is chairman, Board of Studies; faculty of computer science and engineering; former head, Department of Information Technology; professor, Department of Computer Science and Engineering at Bhilai Institute of Technology; Durg affiliated to Chhattisgarh Swami Vivekanand Technical University, Bhilai, India. Her graduation is in the discipline of computer technology from Nagpur University (1994) and post-graduation in the same discipline from Pt. Ravi Shankar Shukla University, Raipur (2003). She was awarded a doctoral degree in the faculty of computer science and engineering in 2013. She has teaching experience of more than 20 years and research experience of 15 years with current areas of research interest in innovative machine learning techniques to combat high obfuscation levels of plagiarism detection, natural language processing, text mining, temporal sequence mining, educational data mining, machine-assisted recommender systems, recognition of text document images, universal image steganalysis, and database forensics. https://orcid.org/0000-0002-6819-6498

Guramritpal Singh Saggu is pursuing an integrated M.Tech. in information technology from Indian Institute of Information Technology, Gwalior. He has been working and exploring the domains of data science and machine learning for the past few years and has prior experience of working with multiple organizations in developing their machine learning and backend pipelines. He has worked as a research scholar in his institute for research project related multimodal AI and has had formidable ranks in various data science challenges and hackathons. His research interests include computer vision, natural language processing, and time series forecasting analysis. https://orcid.org/0000-0002-761 9-3553

Lim Geok Shan graduated summa cum laude from Singapore Management University in 2020 with a bachelor of science (Information Systems) and a second major in advanced business technology (Information Security and Assurance). During her time as an undergraduate, she conducted a systematic literature review of software multitenancy architecture. She also worked as an undergraduate teaching assistant for software engineering and database management classes. She has a keen interest in the field of software engineering and security. https://orcid.org/0000-0003-1965-4648

Deepika Sharma received her master's degree in computer applications (MCA) from the Department of Computer Science and Information Technology, Central University of Jammu, Jammu and Kashmir, India. Presently, she is pursuing her Ph.D. from Central University of Jammu in the area of biometrics security. Her research interests include biometrics, machine learning, and deep learning. https://orcid.org/0000-0003-1376538X

Manmohan Sharma is serving as an associate professor in school of computer applications, Lovely Professional University, Punjab, India. He has a vast experience of more than 20 years in the field of academics, research, and administration with different universities and institutions of repute such as Dr. B.R. Ambedkar University and Mangalayatan University. He has been awarded his doctorate degree from Dr. B.R. Ambedkar University, Agra in 2014 in the field of wireless mobile networks. His areas of interest include wireless mobile networks, ad hoc networks, mobile cloud computing, recommender systems, data science, and machine learning, etc. A large number of research papers authored and co-authored, published in international or national journals of repute and conference proceedings come under his credits. He is currently supervising six doctoral theses. Three M.Phil. degrees have already awarded under his supervision. He has guided more than 1,000 PG and UG projects during his service period under the aegis of various universities and institutions. He worked as a reviewer of many conference papers and member of the technical program committees for several technical conferences. He is a member of various professional/technical societies including Computer Society of India (CSI), Association of Computing Machines (ACM), Cloud Computing Community of IEEE, Network Professional Association (NPA), International Association of Computer Science and Information Technology (IACSIT), and Computer Science Teachers Association (CSTA). https://orcid.org/ 0000-0001-9445-5898

Sarita Simaiya, is an associate professor at Chitkara University Institute of Engineering and Technology, Chitkara University, Punjab, India. She is Ph.D. in computer science engineering and an M.Tech. in computer science and engineering. She has research publications in SCI-Indexed international journals of high repute. Her research areas include artificial intelligence, machine learning, computer security, and wireless sensor networks. https://orcid.org/0000-0001-7686-8496

Ravinder K. Singla has been a professor of computer science and applications at Panjab University, Chandigarh since July 2004 where he has been a faculty member since 1988. He has held many positions including dean of university instructions, dean research, director, computer center, chairperson of the Department of Computer Science and Applications; coordinator, TIFAC-DST; project leader; and programmer/analyst. His research interests include scientific computing, Linux networking, mobile computing, open-source software, and software cost estimation. He is also a member of the editorial board of the Panjab University Research Journal (Science)-New Series. He is a life member of the Computer Society of India. He has published more than 70 research papers in various journals and conferences. https://orcid.org/0000-0001-9120-0267

Raghuraj Singh is currently pursuing a Ph.D. in computer science and engineering at Dr. Br. Ambedkar National Institute of Technology, Jalandhar, Punjab, India, since July 2019. His areas of interest include software fault prediction, machine learning, and soft computing and its applications in modern devices. https://orcid.org/0000-0002-0634-1485

Randeep Singh is a research scholar in the Department of Computer Science and Engineering, M. M. Engineering College, M. M. (Deemed to be University) Mullana, Ambala, Haryana, India. Randeep Singh received an M.Tech. from Kurukshetra University. Randeep Singh has been in teaching and research and development since 2008. He has published about 15 research papers in international and national journals and refereed international conferences. His current research interest is in software engineering. https://orcid.org/0000-0003-2371-4148

Neelam Swarnkar received her diploma in engineering in information technology in 2003 from Government Polytechnic College, Durg, Chhatisgarh. She received her B.E. degree in information technology in 2006 from Bhilai Institute of Technology, Durg, Chhattisgarh, India and received and M.E. degree in computer technology and applications in 2010 from Shri Shankaracharya College of Engineering and Technology, Durg, Chhattisgarh, India. Currently, she is working towards a Ph.D. at the Bhilai Institute of Technology Research Centre, Durg under Chhattisgarh Swami Vivekananda Technical University, Bhilai, Chhattisgarh, India. Her research interests are information security, steganography, and steganalysis. https://orcid.org/0000-0002-9940-4034

Brian R. Tan focuses on the study of value-creation in his research. He obtained his Ph.D. from the University of Seattle, Washington, and has served on the faculty at the Nanyang Business School at Nanyang Technological University (NTU) and the National University of Singapore (NUS) Business School. His industry background comes from management consulting at the Boston Consulting Group, and B.R.I.T. Management Consulting. He also serves as an advisor with a start-up marketing agency in Vietnam called K2, Kreative Kommunications. His efforts at integrating academia with industry have led to the development of a thinking framework called the "Wisdom Approach." It utilizes quantum mechanics principles with broad neuroscience foundations and supports teaching the cognitive aspect of wisdom. The application component of the "Wisdom Approach" has been used to teach MSc and MBA students at NUS. Brian is currently applying it to create better thinking machines and artificial wisdom. https://orcid.org/0000-0002-0275-846X

Rakesh Wats, is a professor and head of the Media Engineering Department at NITTTR, Chandigarh. He has a master's in civil engineering and doctorate in quality management. He holds his expertise in education, educational management, curriculum development, and media design and development. He has three modules, one MOOC in addition to more than 75 research publications to his credit. He has working experience of more than three decades in academics, research, and industry. https://orcid.org/0000-0001-8985-6777

1 Quantum Computing: Computational Excellence for Society 5.0

Paul R. Griffin[1], Michael Boguslavsky[2], Junye Huang[3], Robert J. Kauffman[4], and Brian R. Tan[5]

[1]School of Information Systems Singapore Management University, Singapore
[2]Tradeteq, London, United Kingdom
[3]IBM Quantum, Singapore
[4]Copenhagen Business School Denmark, and School of Information Systems, Singapore Management University, Singapore
[5]B.R.I.T. Management Consulting, Singapore

CONTENTS

DOI: 10.1201/9781003132080-1

1.1   INTRODUCTION

Quantum computing is a key part of building an intelligent systems infrastructure for Society 5.0 and can be used in the future across the main pillars of fintech, healthcare, logistics, and artificial intelligence (AI). Intelligent systems based on data science are machines that are sufficiently advanced to be able to perceive and react to external events. Quantum computers offer various avenues to go beyond systems using classical computers and extend computational excellence beyond its current state. Digital innovation underpins the concept of Society 5.0 for a better future with an inclusive, sustainable, and knowledge-intensive society that uses information computing. A key to realizing this society is to utilize gargantuan volumes of data in real-time in intelligent systems. The sharing of information in Society 4.0 has been insufficient and integration of data problematic, whereas Society 5.0 integrates cyberspace and physical space. For example, in Society 5.0, the huge amount of data from physical Internet of Things (IoT) devices are required to be analyzed, processed, and fed back to robotic devices interacting with people in various forms. In Japan alone, the next 15 years is expected to see a growth in IoT and robotics of US$20 bn and US$70 bn, respectively (JapanGov News, 2019). However, the aim of Society 5.0 is to balance economic development and solutions for social issues to bring about a human-centered society. This chapter shows how quantum computing can be applied to many current challenges and open up new opportunities with innovative ways that align better with human thinking.

Classical computing has brought society great benefits over many years from the abacus 3,300 years ago through to modern computing from Alan Turing in the 20th century – to the latest smartphones we are now familiar with. Computers have enabled products and services that humans cannot provide alone such as increasing productivity, enhancing communication, storing vast amounts of data, sorting, organizing, and searching through information amongst many more. However, many problems still exist such as data security, scalability, manageability, and interoperability. Furthermore, Moore's Law increases in computing power is now beginning to fail (Loeffler, 2018). Stefan Filipp, a quantum scientist at IBM Research, has stated that to "*continue the pace of progress, we need to augment the classical approach with a new platform, one that follows its own set of rules. That is quantum computing*" (Singh, 2019). Using the advantages of quantum over classical computing, it is possible to increase computing capacity beyond anything that classical computers can achieve.

Quantum computing was suggested in the 1980s by Manin (1980) and Feynman (1982). In the past few years, it has become a reality and accessible to everyone, with IBM putting the first quantum computer on the cloud in 2016. Now, in 2021, there are dozens of quantum computers online with processing capabilities much better than the first one. While there is little doubt that quantum computers can outperform classical computers for some processes, such as unstructured search problems (Grover, 1996), it is not clear whether and how quantum computing will be advantageous for a particular business need or, indeed, worth the effort to investigate further.

This chapter is aimed at providing a framework to assess the likelihood that quantum computing will be an area that is worthwhile to get involved in for particular business opportunities and challenges. The main differences for quantum computing are *superposition* and *entanglement*. Traditional computers use bits of either 0 or 1. In contrast, quantum computers use qubits existing in a state that is best described as the probability of being either 0 or 1. This is called the *superposition of states* (Nielsen & Chuang, 2010).[1] Qubits also exhibit entanglement, whereby they may be spatially nearby or far apart, may interact with one another at certain times, and yet are not able to be characterized as being independent of one another. The result is that two qubits may work together as if they were one larger qubit. This is fundamentally different from bits that are always kept separate in classical computing. However, current quantum computers are noisy and have an insufficient number of qubits to be able to show provable advantages. And even the widely publicized Google experiment (Arute et al., 2019) is still held to be contentious (Pednault, 2020). Even more contentious are *annealer-type quantum computers* (Rønnow et al., 2014). These will not be covered in this chapter as a result, and we will focus on *gate-type quantum computers* instead.

Considering the Society 5.0 issues of data volumes, real-time processing and linking data, we present a framework to assess what business needs may potentially be addressed by quantum computing and how quantum computing is different to classical systems. There are four areas of concern: the data, the processing, the infrastructure, and the environment. (See Figure 1.1.) First, data may be complex,

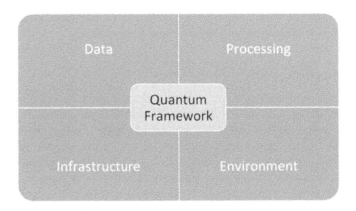

FIGURE 1.1   The main areas of concern in the quantum application framework.

natively quantum or probabilistic, for example, chemical reactions involving quantum particles and financial derivatives pricing including many predictor variables. Second, processing may involve a range of different data analytics approaches, such as simulation, machine learning (ML), optimization, and AI. Third, current quantum computing infrastructure solutions vary in their types and specifications, with the overall marketplace in the throes of rapid technological innovation. Last, the environment may have legal regulations in terms of what can be done, how data can be used, and how physical and human resources can be used while protecting private information.

The remainder of this chapter is laid out as follows. Section 1.2 reviews the main differences between quantum computing and classical computing. It introduces the concepts of qubits, quantum states, and quantum operations. It also gives overviews of different approaches to quantum hardware, as well as the end-to-end process of running a quantum algorithm. Available software and debugging challenges are also discussed, along with some common quantum algorithms. Section 1.3 looks at current managerial concerns and related business problems that quantum computing can potentially address. Section 1.4 offers a framework showing the areas that need to be addressed when considering whether to build a quantum computing solution. Using this framework, Section 1.5 provides an example of its application for a *quantum neural network* (QNN) solution for the credit rating of *small and medium-size enterprises* (SMEs). Section 1.6 looks at the future of quantum computing, covering improvement areas for theoretical development, hardware, and integration with analytics software.

While there is no expectation that quantum computers will outperform current classical models in the near term, exploring the requirements and limitations of current quantum computers can be useful for thinking through how to develop a future system to meet business objectives when quantum computing reaches a suitable level of maturity.

1.2 QUANTUM COMPUTING FUNDAMENTALS

This section reviews the fundamental properties of quantum computers, and the end-to-end workflow from data preparation to analyzing the quantum circuit output. We further discuss the available software, with a focus on Qiskit as an example. Hardware and available algorithms are also assessed.

1.2.1 KEY CONCEPTS

The fundamental concepts of superposition, entanglement, interference, qubits, quantum gates, the concept of Bloch's sphere, and adiabatic annealers are key to understanding quantum computers. We will explain these concepts here.

Superposition. The fundamental processing and storage unit of quantum computers is the quantum bit, also called a *qubit*. A qubit can represent 0 or 1 like a classical bit. However, it can also represent 0 and 1 in a superposition state. A classical bit is either in the state 0 or 1. In contrast, a qubit is in a state which can be characterized by a general formula, $|\psi\rangle = a|0\rangle + b|1\rangle$, where a and b are probability amplitudes represented by complex numbers. Although a qubit can be in a superposition, when it is measured or read out in a computer language, the result can only be 0 or 1 and not both at the same time. However, by measuring a qubit multiple times, the probability of obtaining 0 or 1 is obtained. The probability of 0 is $|a|^2$ and 1 is $|b|^2$, and due to the conservation of the total probability being equal to 1, $|a|^2 + |b|^2 = 1$.

Entanglement. Another strange phenomenon related to qubits is entanglement. It refers to the correlation of different qubits in a system. If we extend the general formula for a qubit state to two qubits, we have: $|\psi\rangle = a|00\rangle + b|01\rangle + c|10\rangle + d|11\rangle$. Qubits are entangled if the state of one qubit is correlated with by the other. For example, when $b = c = 0$, the state becomes $a|00\rangle + d|11\rangle$. In this state, the two qubits' states are strongly correlated so that when the first qubit is $|0\rangle$ the second qubit is always also $|0\rangle$, and when the first qubit is $|1\rangle$ the second qubit is always $|1\rangle$.

Interference. Quantum states can interfere with each other due to the *phase* difference between the probability amplitudes. *Quantum interference* is similar to wave interference and can be understood in the same way. For example, when two waves act together in phase, their probability amplitudes sum to create a higher probability, and when the probability amplitudes are out of phase, their amplitudes cancel out. (See Figure 1.2.) This phenomenon can be utilized by quantum algorithms to speed up their calculations. For example, the Grover (1996) search algorithm uses interference to increase the probability of finding the right answer and reduce the probability of the wrong ones.

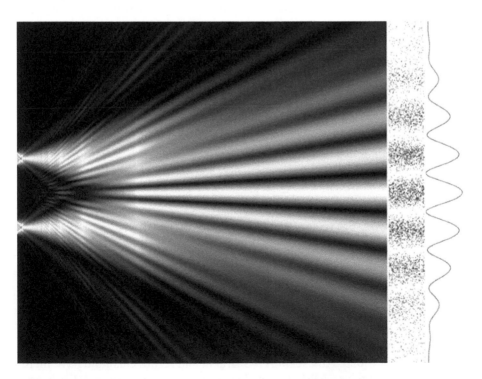

FIGURE 1.2 Interference pattern from a simulated double-slit experiment with electrons. *Source:* Alexandre Gondran, distributed under a CC-BY-SA-4.0 license.

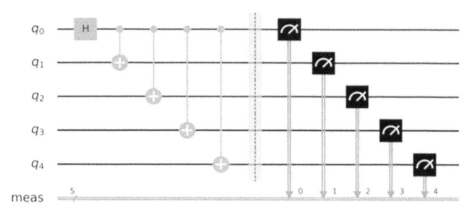

FIGURE 1.3 An example of a quantum circuit.

Quantum gates. Quantum algorithms can be represented by quantum circuits. Quantum circuits consist of quantum gates like the AND gate and the XOR gate from classical logic circuits. A common set of universal gates which can perform any quantum operation includes the Clifford gates and the T gates. This set of universal gates is composed of CNOT, H, S, and T gates.

Hadamard (H) gates map |0⟩ to |0⟩ + |1⟩ and |1⟩ to |0⟩ − |1⟩. H gates are ubiquitous in quantum algorithms for generating superpositions. S and T gates are part of the phase shift gates family. They leave |0⟩ unchanged and map |1⟩ to $e^{i\phi}$ |1⟩ (with $\phi = \pi/2$ and $\pi/4$ for the S and T gates, respectively). Phase gates are important for manipulating phases of quantum states to achieve intended interference.

Controlled NOT (CNOT) gates act on two qubits, unlike the H, S, and T gates, which act on a single qubit only. CNOT gates perform a NOT operation on the second qubit whenever the first qubit is in the state |1⟩ and can be described as mapping the state |a, b⟩ to the state |a, a ⊕ b⟩. CNOT is important for generating entanglement between qubits.

An example of a quantum circuit is shown next. (See Figure 1.3.) This circuit has an H gate on qubit-0 and four CNOT gates between qubit-0 (the control) and each of the remaining qubits (the target). Finally, the qubits are measured, and the outcomes are stored in classical registers.

1.2.2 Hardware, Software, Alorithms, and Workflow

We introduce a range of available software and hardware, followed by a general workflow for a quantum program's execution. The IBM Quantum Experience and Qiskit are used together as an example to illustrate a typical workflow.

Available hardware. As of January 2021, there were a handful of quantum computing hardware providers. They include: IBM Quantum Experience, Rigetti Quantum Cloud Services, AWS Braket, Microsoft Azure Quantum, Xanadu Quantum Cloud, D-Wave Leap, Quantum Inspire (from QuTech), and Origin Quantum Cloud. All except AWS Braket and Azure Quantum have their own hardware systems. AWS Braket has three different external hardware providers: D-Wave (annealer), IonQ (trapped ions), and Rigetti (super-conducting qubits), while Azure Quantum has IonQ (trapped ions), Honeywell (trapped ions), and Quantum Circuits (superconducting qubits). IBM Quantum Experience, Quantum Inspire, and Origin Quantum Cloud have systems that are open for the public free of charge, while the others charge for access.

IBM Quantum Experience (IQX) is the leading provider of quantum computing services. In 2016, IQX put the first quantum computer on the cloud. Since then, the platform has grown tremendously and now has more than 10 quantum systems (of up to 15 qubits) with free access and an additional more than 10 premium access quantum systems (with up to 65 qubits) for the IBM Quantum Network's partners. IQX has more than 250,000 users who collectively run more than 1 billion quantum circuits each day, and have published more than 250 related research papers (IBM.com, 2020). The IBM Quantum Network has more than 200 partners in industry and academia. IBM also released its quantum hardware roadmap in the annual Quantum Summit in September 2020, and laid out the steps toward building quantum systems with more than a million qubits. IBM aims to release processors with 127 qubits in 2021, 433 qubits in 2022, and 1,121 qubits in 2023 over the next three years (Gambetta, 2020). The quantum systems will be large enough to investigate the implementation of *quantum error correction* (QEC), to open the door to the practical implementation of many quantum algorithms.

Available software. There is a range of software development kits (SDKs) available for writing and running quantum programs, including Qiskit from IBM, Cirq from Google, QDK from Microsoft, Forest from Rigetti, and ProjectQ from ETH Zurich (LaRose, 2019). All of these SDKs, except QDK, are based on Python, which allows easy integration of the Python ecosystem's capabilities for scientific computing and ML. Many companies also work on quantum software packages for specific domain applications to interface with the SDKs mentioned previously. Notable examples include: Qiskit Aqua for chemistry, ML and optimization, from IBM; PennyLane for ML from Xanadu; OpenFermion for chemistry, from Google; and TensorFlow Quantum for ML, from Google.

Available algorithms. The most famous quantum algorithm is attributable to Shor (1997), who showed an exponential quantum speed-up related to the best-known classical algorithms for factorization. It is likely to threaten the existing cryptography infrastructure, however, it also requires a large number of qubits and gate operations, which also require QEC capabilities. The best estimate is that it will require on the order of millions of physical qubits (Gidney & Eker, 2019). Other famous textbook algorithms such as Grover's algorithm also require QEC capabilities, which will be difficult to implement in NISQ devices within the next three to five years.

There is a class of quantum-classical hybrid algorithms called *variational quantum eigensolver algorithms* (VQE, a kind of the more general variational quantum algorithms, VGAs). These are suitable for implementation on NISQ devices (Peruzzo et al., 2014). Such algorithms contain both quantum circuits and classical procedures that are invoked iteratively. The quantum circuits in each iteration have a small number (< 100) of qubits and a small number (also < 100) of quantum gates, and these can be run with current NISQ devices. The

results of the quantum circuits are then fed to classical procedures for calculation and optimization to determine the parameters of a quantum circuit in the
next iteration of the algorithm. This hybrid approach enables harnessing the
power of quantum computers in the NISQ era before QEC becomes widely
available.

Quantum computing workflow stages and examples. A typical quantum program
workflow includes three stages: data loading, data processing execution, and data
extraction.

- In the *data loading stage*, data are loaded into the memory of classical
  computer and converted to the states represented by qubits.
- In the *data processing execution stage*, a qubit's state is changed by the
  application of quantum gates. A compiler converts a logical circuit to a circuit
  that can be executed on a physical quantum processor, considering physical
  qubit connections and native gate sets.
- In the *data extraction stage*, all qubit states are read out by measurements.
  Due to the probabilistic nature of quantum states, multiple measurements are
  usually needed to sample the probability distribution in order to obtain a
  meaningful understanding of the solutions.

We next will look at Qiskit. It consists of four different modules: Terra for
writing and running quantum circuits; Aer for high-performance simulation;
Ignis for analyzing and mitigating noise; and Aqua for quantum algorithms and
applications. Using Qiskit and the IQX, we illustrate an example of how a
quantum program is executed on a quantum computer via the cloud. (See
Figure 1.4.)

For this circuit, we don't need to load any data as the qubits are initialize to |0⟩ by
default. In the circuit execution stage, the program written in Python first needs to
go through a compiler to convert it to a circuit that can run on a five-qubit quantum
processor, such as the *ibmq_vigo* quantum processor. The H gate in Figure 1.3 is
converted to a U2 gate that is native to IBM quantum processors. Such processors
are superconducting devices that are usually not connected to all of the other qubits.
However, the circuit written in Qiskit does not take this into account. So, it's the
compiler's job to convert the logical circuit in Figure 1.3 to a circuit that can be run
on the actual hardware. (See Figure 1.4 again.)

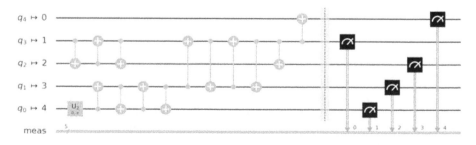

FIGURE 1.4 An example of a compiled quantum circuit.

If a CNOT gate is applied to qubits that are not directly connected, the compiler will need to apply a SWAP gate to move the quantum states of the qubits to two qubits that are connected. SWAP gates are implemented by using three CNOT gates, each of which generally has a ten times higher error rate than single qubit gates have. In the NISQ era, quantum processors do not have QEC and any errors may make a quantum program fail completely.

In Qiskit, this optimization procedure is implemented by a *transpiler* (i.e., a source-to-source compiler or transcompiler) consisting of a series of passes that optimize circuits based on different algorithms. The transpiled circuit is converted to a *quantum object* (Qobj) *file* and then sent to the IQX system (McKay et al., 2018). The Qobj file's contents control the electronics necessary to convert the program into microwave pulses transmitted to the quantum processor inside a cryogenic dilution refrigerator, which ensures that a low temperature is maintained. At the end of a quantum program, all the qubits will be read out, and the circuits and measurements will have run multiple times to sample the measurement outcome instead of only capturing a single-point measure.

1.3 QUANTUM COMPUTING NEEDS AND SERVICE INDUSTRY APPLICATIONS

We next consider a variety of business needs, and a new perspective on operational, managerial, and strategic problem-solving and decision-making that will benefit from the application of the quantum computing approach in the services industries. In addition to sharing our perspective on the reasons for the upwelling of interest in the new computational methods, we also provide some characteristics of appropriate business applications in operations, and the travel and hospitality services. We further offer a deeper reading of several prominent problem areas in contemporary financial services.

1.3.1 Business Needs and Concerns

Organizations operate in complex and dynamic economies driven by increasing social connectivity, the free flow of decentralized information, and technological advances that allow tracking of individuals using a network perspective. To deal with this increase in complexity and making good decisions (Nijs, 2015), managers in different kinds of organizations should be encouraged to move from a linear transaction logic and toward a networked logic that emphasizes value.

From the perspective of a *networked logic of value* (i.e., business value across business units, processes, product families, business partners, and smart networks in global value chains), and to make good decisions effectively, there is a need for businesses and organizations to capture and understand information from different sources. Information from the *internal environment*, such as employee-based or patent-related data, can help an organization to orchestrate its resources so it can develop higher-level capabilities and competencies. Likewise, information from the *external environment*, such as customer preference data, competitor information, or macro-environment trends, can aid an organization in decisions related to pricing, market segmentation, customer targeting, and product positioning. Understanding how to utilize both internal and external information types is critical for organizations to achieve sustainable advantage, driven by a networked logic of value.

To obtain relevant and meaningful information for decisions, organizations must consider four areas of managerial concern and related business problems. (See Table 1.1.)

Relevant information can only be obtained if the data used is accurate and captures what is being measured in a meaningful way. Data pertaining to real-life situations, such as customer preferences, competitor actions or employee data are inherently unstable, complex and dynamic. They often occur as range estimates (rather than just as point estimates); they also exhibit stochastic variation or are not easily measured or estimated. They interact with each other in complex probabilistic, interdependent and constrained ways (Ménard, Ostolic, Patel, & Volz, 2020).

1. *Data and information.* It is necessary to use well-specified and reliable data inputs to obtain good information for decision-making, lest such a decision process will devolve to GIGO thinking – or "garbage in, garbage out." This acronym implies that bad inputs will result in bad output, and this principle applies more generally to all analysis and logic that cannot support meaningful evidence-based conclusions.
2. *Explanatory power and processing.* Considerations about respondent bias (Furnham, 1986) and the explanatory power of the solution methodology used should also be considered. In addition, the nature of processing is often linked with the types of data to be analyzed, as well as the computational activities that are required to achieve a solution for a complex problem.

TABLE 1.1

Concerns and Related Business Problems to Consider

Managerial Concerns	Examples of Related Business Problems
Data and information	Conversion of data with regard to numerous forms to information/knowledge; text, sentiments, natural language, images/videos, map/spatial representations; and firm/competition/market sources
Explanatory power and processing	Power of varied solutions/approaches; robustness/vulnerability to data problems; effectiveness of handling different kinds of complex problems; trying to match human-related conditions
Infrastructure and capabilities	Tech infrastructure; computational/staff support; human limitations in working on NP-hard vs. soft organizational problems; and limit to value in firm-level data gathering/problem models
Environment and regulation	Business sustainability, fair allocation of resources; appropriate use of physical/human resources; ensure compliance; and protect personal info (customers/partners/employees)

3. *Infrastructure and capabilities.* Organizations also must make judgment calls on the type of resources to utilize to develop capabilities and competencies in information gathering and interpretation. Resource types under consideration include the organization's choice of IT systems, its processes, systems, and staff capabilities.
4. *Environmental and regulatory constraints.* Since organizations operate within a larger societal context, they must develop systems within environmental sustainability and other constraints. Thus, governmental regulations on personal information protection and privacy laws, financial constraints, and security considerations are all important.

1.3.2 Decision Problem Framing and Computation

Quantum computing is relevant in such situations, not just because of its potential computing supremacy or its advantage for some kinds of problems, but because of the way in which it achieves solutions via optimization, simulation, and unique distillation approaches (Arute et al., 2019). To understand how quantum computing can achieve supremacy, we should understand its source of power. The power from quantum computing comes from the qubit, exhibiting the characteristics of a range estimate and with multiple qubits, it is possible to express additional levels of correlation and stochasticity, to characterize problems of higher complexity (Arora & Barak, 2009). To create computational power, computer scientists have developed processes that leverage the nature of the qubit, through the integration of its stochastic range and point estimates, in a manner that makes it possible (with additional hardware developments) to obtain problem solutions with faster parallelism and speed. In contrast, classical computers operate in a linear manner, which slows them down. Both quantum computing and classical computing are competitive on problems of small to modest size.

There are three implications based on the source of power and differences in computing methods:

1. Quantum computing, like traditional computing, supports parallelism. A key contrast is that classical computers use more linear approaches to processes of computation though.
2. Suitably large *quantum machine learning* (QML) systems may lead to radical advances in the creation of thinking machines and AI that approaches the capabilities of humans.
3. Quantum computers are suitable for problems that are complex, combinatorial, and stochastic, where judgment in decision making is important. A wisdom-oriented approach towards thinking, by considering range-estimate inputs and requiring judgment becomes more relevant when compared to an intelligence-based approach (Jeste et al., 2010).

With these characteristics in mind, we consider some business problems in different service industries to which quantum computing seems uniquely suited.

1.3.3   THE RANGE OF BUSINESS PROBLEM AREAS THAT CAN BE ADDRESSED

The business issues to which quantum computing methods apply possess characteristics that often are observed. (See Table 1.2.) We caution the reader to recognize, however, that quantum computing in late 2020 is at the apex of its hype cycle, but there are not yet "killer applications" or completed commercial tools that will permit applications of extreme complexity.

First, most problems to which the methods are appropriate involve considerable computational complexity for traditional computers to handle – especially problems that are non-deterministic and of *non-polynomial time-complexity* (NP-hard). In

---

TABLE 1.2

Problems that Quantum Computing Addresses: Optimization of Complexity, Stochastic Drivers, Real-Time Computation, Intelligent Simulation, and Rugged Landscape Analytics

Problem Characteristics	Explanation	Examples
Complexity in optimization and simulation.	Solves NP-hard models for shortest-times, shortest-routes, lowest costs.	Traveling salesman problem (TSP). Traffic flow optimization. Airline route scheduling.
Stochastic modeling and solution considerations are prominent.	Leverages quantum (qubit) entanglement and probabilistic superposition of 0/1 bits.	Product delivery networks. Taxi routing in congested traffic. Service system control with measure errors.
Real-time solutions are required by business.	Applies quantum computing speed for app-specific improvements vs. Moore's Law limits.	Perishable goods revenue yield management. Real-time financial portfolio risk management.
Problems conceptualized for rugged landscape quantum computing.	Landscapes with uplifting mountains and settling low points for optimization.	Terrorist network member identification. Voice, speech, and facial recognition. Autonomous and driverless vehicle routing.
Intelligence needed for obtaining solutions appropriate for individuals.	Wisdom-based problem representation applied to multi-factor and soft model choices.	Smart mobility platform apps for people. Pollution mitigation, sustainability controls. Genomic data analytics for personal care.

addition, stochastic modeling is required in many problem settings, such as taxi routing in congested city streets, and the control of complex systems when volatility and measurement errors are present. Third, an increasingly important problem characteristic is business solutions that must produce a computational result in real-time or near real-time in order to support a business process. Intraday revenue yield pricing for perishable goods (e.g., airline seats, hotel rooms and rental cars) is an exemplar. Fourth, other problems require consideration of "rugged landscapes," which make it hard for ML solutions that do hill-climbing optimization to succeed. Finally, a fifth class requires machine intelligence to find solutions that are tailored to individuals, such as healthcare genomics-based treatment decisions. Ideally, what is required is a more wisdom-based problem representation, so it is possible to find good solutions in the presence of hard and soft constraints, and objective and subjective goals.

In the routing optimization domain of NP-hard problems, the well-known *traveling salesman problem* (TSP) stands out as one to which heuristic methods have been successful, for example, to address the 1.9 million-city World TSP Tour (http://www.math.uwaterloo.ca/tsp/world/index.html). More value-laden quantum computing use cases with strong industrial relevance are related to the airline industry. Othmani, Ettl, and Guonaris (2020) have identified areas in the progression of quantum computing since the 1900s development of quantum science, then quantum-ready proofs-of-concept, and commercial advantage for solving real-world problem in the industry. The applications they point to include: contextually enhanced service personalization; untangling operational disruptions (irregular ops management) in air travel, that requires dynamic updating and optimization of crew, take-off slots, and equipment, all while trying to address customer satisfaction and cost over-run concerns. Quantum computing offers promise for solving the vexing matter of overcoming the fragmentation of the overall problem. According to the authors, quantum computing is able to address data and variable range-uncertainty, while re-aggregating the fragmented optimization sub-problems that will produce a more efficient and competitive advantage.

In the past 10 years, applied research and corporate practices with data analytics and problems that can be studied from the large-scale availability of the digital traces of people as consumers, social media users, users of mobile phones, and the tracking of people who exercise has given rise to new directions in predictive modeling and "living analytics" (as with our work at the Living Analytics Research Centre (LARC, https://larc.smu.edu.sg/) of Singapore Management University). It stands to reason that new directions in data analytics are now able to support innovative work in other areas that are open to innovations with quantum computing methods. Many business and social problems are naturally modeled in ways that play to the strengths of the quantum paradigm. Examples are the modeling and recognition of complex voice, facial expressions and emotions, and other image recognition problems. Even fast computation by traditional computers has not been sufficient due to limitations in how such problems have been approached in computational terms.

Another direction lies in creating the computational basis for doing things such as terrorist network member identification and driverless, autonomous vehicle routing (Burkacky, Pautasso & Mohr, 2020). These involve the conceptualization of related problems with *rugged landscape optimization* (i.e., discontinuous modeling surfaces), so that the typical methods of *hill climbing* in optimization and ML are reduced in their power, especially with dynamic changes in their content and environment. We note the innovations based on the "changing landscape" analogy for problem identification and quantum computational methods. Finally, there are many such problems that require a smart approach, for "wicked" operational, social, and healthcare problems, like smart mobility platform design for changing urban traffic and transportation opportunities (Akrout, 2020), and policy analytics for pollution and sustainability, as well as genomic data analytics.

1.3.4   THE UNIQUE ROLE OF QUANTUM COMPUTING IN FINANCIAL SERVICES APPLICATIONS

Among the various settings noted above, it's beneficial to consider problem areas that can potentially be treated in the financial services space by quantum computing approaches. They include offering enhanced power, flexibility, and representational authenticity to achieve smart, nearly real-time solutions for complex optimization, stochastic modeling, and intelligent choice problems. Financial markets are essentially complex systems that exhibit a high degree of *randomness*, for example, in the movement of equities' and derivatives' market prices, as well as foreign exchange rate pairs and interest rates. Their *volatilities* have been observed, and numerous applications and problem contexts have been analyzed in university and industry research. They have used portfolio construction models and optimization, neural networks (NNs), and Monte Carlo simulation, and ML and AI approaches. These further leverage configural patterns that are hard for human analysts to identify in big data, due to the presence of stochastic variation and the distributional mechanics of asset returns. In all these cases, the large size of the data sets used to calibrate changing risk levels in day-to-day operations typically require a huge amount of computational power for understanding the necessary relationships in the data so meaningful forecasts can be made (Lee, 2020).

Quantum computing is attractive for financial applications, like many-asset, multi-market portfolio construction and risk management controls. It also is relevant for establishing intraday trading paths to find ways to successfully exploit cross-asset or cross-market arbitrage opportunities. Other domains include near real-time, irrevocable settlement of securities trades, and the voluminous and volatile payments in continuous matching systems that maintain adequate settlement system participants' liquidity. These exhibit knowledge about the degree of the cross-state independence and correlations of asset price realizations from distributions that are the basis for the evolving lattice of assets prices in a market, which could utilize the property of *quantum entanglement*, such that it is present.

Finance practitioners know that a measurement of the state of such systems at any time is inherently random, though there is underlying cross-correlations, which result in the cointegrated evolution of asset prices and value-at-risk, a popular loss estimate at some level of statistical certainty over time. These states can be de-scribed in mathematical finance as *wave functions*, which quantum computing hardware has the potential to speed up the simulations in comparison to more speed-limited classical computer hardware and algorithms. Another consideration going forward is that the extent to which computational speed and power are likely to become available with lower electricity costs, and beneficial implications for more sustainable electricity consumption by financial institutions.

An application to *securities trading settlement transactions*, involving the ex-change of a delivered financial instrument for an irrevocable cash payment, has been reported by Barclays Bank and IBM Research (Braine, Egger, Glick, & Woerner, 2019). The authors utilize quantum algorithms to make a mixed-binary math programming model's optimization faster. This is done for value maximiza-tion of continuously submitted batches of securities trades, with settlement pro-cessing that considers counter-party credit, collateral facilities, and regulatory compliance rules as objective function constraints. Traditional computation uses problem representations that support simulated annealing, which mimics a physical process of heating some material and lowering its temperature gradually. This re-duces material defects, and the overall energy of the system. The analogy for fi-nance is related to how computational algorithms approximate globally optimal solutions in a complex and constrained system, based on the weighted sum of settled transactions. The basic math programming formulation is transformed into an unconstrained model with a lambda-penalty function (like Kuhn-Tucker quad-ratic optimization) to support a solution with one qubit for each of the securities transactions to be settled, and a wave function for each possible settlement trans-action that together form a batch for actual settlement. Quantum computation le-verages unique features of the stochastic optimization, while reducing the liquidity risk for a participant that cannot settle its net funds position, and credit risk for system participants, should there be a fall in the value of traded securities.[2]

Another finance problem that deserves comment was presented by Woerner & Egger (2019). It is on *quantum risk analytics*, in which the quantum *amplitude estimation* (AE) algorithm is estimated for an unknown parameter. The problem the authors applied their approach to is the estimation of the uncertain price evolution for a U.S. Treasury bill, with a daily trading value of about US$500 bn and ag-gregate government debt as of 2016 of approximately US$14.8 tr. T-bill prices are subject to yield curve risk over time (i.e., as a fixed income security, from changing interest rates across the bills' maturities). For their analysis, the authors used constant maturity treasury (CMT) rates, to calculate the daily risk of a one-bill portfolio toy problem.[3] They did this for one-day changes in CMT based on the distributions of three correlated, underlying explanatory principal components called *shift, twist,* and *butterfly*. They are known to account for 96% of CMT's daily variations. Among them, only the first two were retained due to the low correlation (quantum entanglement) of the third component with them (i.e., due to its in-dependence and thus lack of suitability for quantum analysis).

Three qubits were used to represent shift uncertainty, and two were used to represent twist uncertainty. The authors' approach was intended to deliver a quadratic speed-boost in comparison to traditional Monte Carlo simulation for

T-bill diffusion in two-asset portfolios, which adds to forecasting problem complexity. They admitted that their sample simulation was not able to be to scaled up to more qubits for estimation, since they lacked the requisite quantum computing hardware to demonstrate the power-gain beyond massively-parallel, traditional computer hardware.[4]

We next turn to a discussion of a new framework for the implementation of quantum computing solutions, with special attention given to more technical aspects than have been discussed so far.

1.4 APPLICATION FRAMEWORK

In this section, we provide a framework describing the areas to consider when applying quantum computing to a business need. This is based on the assumption that advantage may be gained from using quantum computing, and a deeper look into what that entails is required. Taking into account what has been discussed in Section 1.2 on the basics of quantum computing and in Section 1.3 on current business needs, we now describe a framework for applying quantum computing to particular business problems. We will cover the four areas of concern presented in Figure 1.1 (data, processing, technical, and governance issues) by looking at the data preparation, algorithm, quantum circuit design, and some issues around integration. (See Figure 1.5.) These implementation areas are likely to be iterative as development continues towards the final solution and the state-of-the-art continues to evolve in all areas of quantum computing.

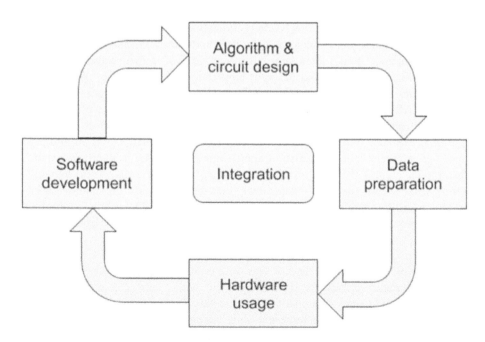

FIGURE 1.5 The iterative development cycle.

1.4.1 Algorithm Design

Once the business needs have been defined, we must consider designing a suitable algorithm for the problem(s) involved in the business need. This can be one of the hardest steps as it involves both a knowledge of the business and a grasp of quantum algorithms. However, there are a number of algorithms already available that have been well established and used on a range of problems. Some of the algorithms can be found in online tutorials with sample code, and there are also consultancies available to advise potential users further. If there is nothing available that is suitable, then a new algorithm can be designed. It is most likely that even well-described algorithms will require modification for the business need. There are also companies working on embedding quantum algorithms into common data analytics packages, such as MATLAB, to utilize the power of quantum computing while hiding the complexity of quantum algorithms (http://horizonquantum.com/).

We next list a number of common problems that could exploit quantum properties and provide some algorithms examples that are suitable to explore. (See Table 1.3.)

TABLE 1.3

Problem Type, Quantum Properties, and Sample Algorithms

Problem Type	Quantum Property	Sources of Sample Algos
Black-box and oracle problems, discrete logarithm problems, integer factorization, boson sampling problems.	Interference between qubits provides an efficient quantum Fourier transform operation.	Deutsch and Jozsa (1992). Bernstein and Vazirani (1997). Quantum phase estimation (Simon, 1997). Shor's algorithm (Shor, 1997).
Searching an unstructured data set, BQP-complete problems, decision making, modelling and simulation, linear systems.	Amplitude amplification using conversion between probability amplitude and phase.	Grover (1996).
Element distinctness, triangle-finding, rugged landscape.	Quantum walks display exponential speed-up.	Ambainis (2007).
Graph theory problems, minimize the energy expectation to find the ground state energy.	Hybrid quantum/classical algorithms.	Quantum approximation optimization algo (QAOA) (Farhi, Goldstone, & Gutmann, 2014). Variational quantum eigensolver VQE) (Peruzzo et al., 2014).
ML and optimzation including: least squares fitting, semidefinite programming (SDP), NNs, and combinatorial optimization.	Uses quantum superposition.	QAOA (Farhi et al., 2014). VQE (Peruzzo et al., 2014).

The entire scope of the software that needs to be developed can be broken into various components: preparing and loading the data with classical code; creating a quantum circuit using qubits and quantum gates; executing the quantum circuit with classical code using quantum libraries; and processing the output of the quantum computer with classical code.

1.4.2 SOFTWARE DEVELOPMENT

With an algorithm decided, we now must develop the classical application that runs the algorithm with the quantum circuit. The classical code may just run the quantum circuit once or it may run it many times, and then rebuild the quantum circuit depending on the output. An example is when a hybrid neural network (NN) uses classical gradient descent to optimize the quantum weights in a hidden layer, or via a variational design where the quantum gates are rebuilt after each run.

Data preparation. The data to be input to the algorithm will need to be prepared. This will depend on the type of data and the type of quantum circuit. Depending on the quantum circuit, the data will have to be loaded in different ways. Data can be loaded either as binary data, as qubit rotations, or as quantum states.

The first way, *binary data loading*, is similar to a classical bit, where each one is encoded directly onto the pure state of the qubit. A classical 0 is encoded as a $|0\rangle$ state and 1 as a $|1\rangle$ state. This is simple to understand and, assuming that the default state of a qubit is $|0\rangle$, then the circuit will have either a NOT gate immediately applied to the input qubit for a $|1\rangle$ state or nothing, so the qubit stays in a $|0\rangle$ state. However, this method has the lowest data density of information encoding and is not utilizing the potential range of probability amplitude and phase amplitude superposition of the qubit.

Encoding classical values directly onto qubit probability rotation angles is possible using rotation gates, either on the probability amplitude or the phase amplitude after scaling the values between 0 and π radians. Going beyond π actually decreases the amplitude, for example $0 = 2\pi$. Further note also that the probabilities have an angular dependence on the rotation angle. To apply a probability of P to the qubit will need an angle θ, where θ (radians) = arcos($2P - 1$), using an Ry gate for amplitude probability and an Rx gate for phase probability. This method is simple to apply and is easy to use with simulators and real backends. However, it still does not utilize the possibility of applying values to all quantum states.

Classical values can also be applied onto the qubit quantum states. In IQX, this is implemented using the *statevector* object and the *initialize* function of the quantum circuit object. This can utilize the full availability of states and map complex data directly onto the qubits. However, there is a need to normalize the *statevector* data for real computers before initialization. Some suggestions for how to handle the different data types include: *binary* – use binary; *integers* – scale use rotation or state; *floats* – scale and use rotation or state; *complex* – use state; text – convert to integer, scale and use rotation or state; *image* – convert to integer, scale and use rotation or state; and *objects* – serialize and convert to integer, and also scale and use rotation or state.

Creating the quantum circuit. The code to create a quantum circuit defines the registers, adds single and multiple qubits gates and measurement instructions. Registers are of two types: quantum and classical. Quantum registers contain qubits and classical registers contain classical bits. Classical registers will be used to collect the results of the quantum circuit execution and may be used to interact with the quantum circuit during the execution on simulators, such as the IF statement in Qiskit (Foy, 2019).

Qubits are initialized in the ground $|0\rangle$ state. The first operation is the a rotation gate to encode a classical value or a Hadamard gate to put the qubit into a 50-50 probability superposition. Single- and multi-qubit gates are then used to operate on qubits affecting the whole state of the quantum register. So, measurements are performed on the qubits in the quantum registers. The measured qubits may be the same as the input qubits or they may be ancilla used to collect the final circuit output.

Executing the quantum circuit. Two architectures are now considered: (1) a static quantum circuit that is executed once and has all the logic in it, and (2) a variational approach where the quantum circuit is modified after each run most likely to converge upon a result.

- *Static quantum circuit*: The quantum circuit is designed and executed on a quantum computer. Data are prepared in the classical code and initialized onto the input qubits of the quantum circuit. The code then defines a quantum backend which executes the quantum circuit, and then waits until a result is returned. (See Figure 1.6.) Before attempting to run the circuit, the circuit depth should be checked to ensure that it does not exceed the *coherence time* of the quantum computer, the amount of time a quantum state can continue in

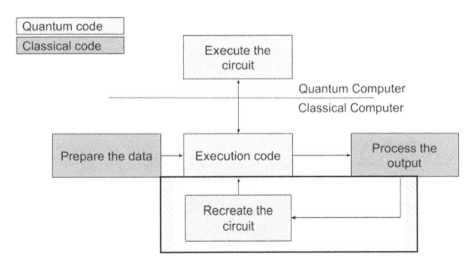

FIGURE 1.6   The combination of classical and quantum code to execute a quantum circuit. The shaded area is the extra step for variational quantum circuits.

the same stable form. (See the glossary in Appendix Table A1 for additional details.) It is likely that the high-level quantum circuit gate operations will need to be decomposed into the native gates of the quantum computer to understand the actual circuit depth that will be executed on the backend. Also, it is necessary to check the scaling of the algorithm's quantum depth with the quantum width, since algorithms can scale significantly.

- *Variational quantum circuits:* The entire system couples the quantum and classical code. The classical code recreates the quantum circuit after each time it is executed, depending on the results from the quantum circuit. (See Figure 1.6.)

The execution of a quantum circuit takes milliseconds, but the whole time for the call to the quantum computer can take time to come back, especially if the circuit jobs are queued and there is a significant queue depth. In this case, the calling code needs to have a wait or callback function to continue the processing after the call is complete.

Processing the output of the quantum computer. When a quantum computer provides an output, it needs to be retrieved by a classical computer and processed to be used by downstream systems. Measurement destroys the quantum state on the qubit and is usually performed at the end of a quantum circuit's execution. Measurements are only performed for the probability amplitude on one axis (the basis) of the Bloch sphere, which on IBM machines is the z-axis. The results are acquired from executing the quantum circuit a number of times, called shots, which default to 1,024. Each time a quantum circuit is executed it only returns a bit (0/1), depending on the probability amplitude for each state in the quantum register. For example, in Table 1.4 there are two qubits that have four states in total. The counts are how many times the circuit gave a 1 in that state. These counts can be calculated to obtain the probability of each state occurring in the output. (See Table 1.4.)

To find the probability of each qubit being in the $|1\rangle$ state sum the probabilities for that qubit across all the states. In the above example, with the probability of q_0 being $|1\rangle$, and $0.33 + 0.36 = 0.69$, whereas q_1 is 0.31. The probability outputs can then be analyzed by classical code.

TABLE 1.4

The Count and Probability Example for a 2-Qubit Circuit

State	Count of 1	Probability
00	160	16%
01	336	33%
10	158	15%
11	370	36%

1.4.3 HARDWARE FOR QUANTUM COMPUTING

The quantum hardware used is also important to the success of the solution. As well as the quantity of the input data that needs to be processed, the fidelity, qubit coupling and coherence time of the qubits are also important. Noise in the qubits and gates will also need to be considered however, noise may be a useful for some business cases such as decision making.

Circuit width and number of physical qubits. Inspecting an algorithm will give a good idea about the quantity of logical qubits required. From the number of logical qubits plus the number of qubits needed for correcting errors, the total number of physical qubits can be determined. For example, if the problem approach is to do optimization, and there are 10 features to optimize on and another 3 qubits are needed for error correction, then at least 13 physical qubits are needed.

Circuit depth and coupling. The algorithm that has been designed will give an idea of the circuit depth. But the qubit coupling of the quantum device will also affect the circuit depth by potentially adding more gates to move information around. For example, on the IQX, with the *ibmq_vigo* processor device, there is no direct coupling between qubit 0 and 4. So, if qubit 0 is entangled with qubit 4, there has to be more gates (SWAP gates) used to move the quantum information between the qubits. (See Figure 1.7.) Each gate adds to the circuit depth and time spent by the circuit that could be used elsewhere. Matching the circuit design to the coupling can increase performance. As the circuit's mismatch to the coupling increases, the number of gates increases and the coherence times can run over what is desired.

Device adjustments. Finally, on some backends such as IQX, it is possible to adjust the qubit interactions at the lowest level of the physical interactions. In IQX, for example, it is possible to adjust the microwave pulse that controls the qubit rotation. This can also help with reducing errors. Finally, it also is worth noting that

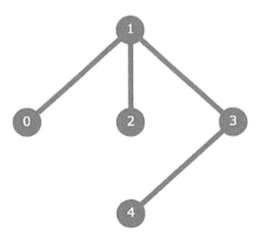

FIGURE 1.7 Coupling map for ibmq_vigo v1.2.1.

qubits can have more than two states. While the fidelity of even the third state is not sufficient for computation, this property would increase the computing power further than now with m^n states, where m is the number of states in a qubit and n is the number of qubits, if a quantum register were possible.

1.4.4 INTEGRATION WITH OTHER IT SYSTEMS IN THE FIRM

Currently, as quantum computing is still being used for research purposes, there is no integration with companies' internal IT systems. However, when a significant quantum advantage has been shown for business processes, then the quantum computer must be integrated into business flows so its value can be realized. It will then be important to consider data privacy and proprietary data processing issues. In light of this, current cloud-based quantum computers need to be assessed on how they transmit, store and delete data and quantum circuits that are used. Simulations can likely be performed locally for developing and testing quantum circuits, but to gain quantum advantage, real quantum computers will need to be used. Though it is possible to buy a real quantum computer now, the initial and maintenance costs are known to be quite high.

1.5 CASE: IMPLEMENTING A QUANTUM NEURAL NETWORK FOR CREDIT RISK

We next return to the quantum computing application framework described earlier, to illustrate a current QML project that the first two coauthors are currently working on.

1.5.1 CREDIT RISK ASSESSMENT

The business problem of quantitative credit risk assessment is not new. The first systematic data-driven approach to this problem dates back at least to 1968 when Altman (1968) published his pioneering Z-score formula for predicting bankruptcy. That work used linear discriminant analysis to predict the likelihood of bankruptcy from four observed financial ratios of a company. Since then, this formula has been refined many times with thousands of academic papers and hundreds of business applications introducing new features more sophisticated prediction models using larger company data sets. The vast majority of these modeling improvements are still focused on accounting data. This means that the model inputs are changing very slowly (e.g., once a year for a typical non-listed company). In addition, the input feature set is quite narrow – with no more than a dozen accounting data fields that are available for most corporates, and the input features are observed with a significant delay. These constraints limit the complexity of sensible models and the use of powerful modern ML approaches would result in heavily over-engineered solutions.

Our interest in credit scoring comes from the need for risk assessment in trade finance. Trade financing often exposes lenders to the risks of SMEs, as many of them operate in emerging markets with little timely and reliable information available. This means that classical accounting-based scoring approaches have limited utility in many cases. However, rapid digitization of trade finance means that a wealth of new information is available to lenders. The credit decisions can be made based on the information available at the level of individual transactions. Many trade finance users are engaged in high-intensity flows of relatively homogenous transactions. Many lenders have access to transaction data of dozens or even hundreds of thousands of borrowers. Moreover, these observed transactions are often inter-linked, either geographically or along industry verticals. Transaction data credit scoring makes the credit assessment problem with too few features available into a problem with a thousand or even millions of feature observations per day. Tradeteq uses sophisticated graph ML algorithms running on cloud GPU farms to calibrate and optimize these models.

The ongoing introduction of blockchains and IoT into the supply chains means that potentially relevant data flows are about to intensify by another three or four orders of magnitude, with real-time shipment location and conditions tracking available for more and more goods in transit. Thus, transaction credit scoring will soon become more taxing for existing computing systems.

Of course, classical computing and ML, in particular, are developing quite quickly, with rapid progress at all levels of the technological stack – from hardware to algorithms. However, to be ready for the increasing intensity of the data flows, we need to explore all available options for future ML systems, including the new QML systems.

Credit risk assessment problems are similar to certain problems that were already explored on quantum systems, for example, classification and regression ML and optimization problems. However, one still needs to ascertain to which extent peculiarities of credit risk problems are well suited for quantum or hybrid quantum and classical ML architectures. Credit risk ML problems often exhibit strongly imbalanced data sets with asymmetric noise, where one class is much less frequent and noisier than others. Supply chain graph topology matters a lot for the risk and needs to be reflected in a transaction risk model. It remains to be seen how well quantum systems will be able to handle these problem features though. The current state of quantum systems and QML algorithms means that they are quite far behind the capabilities of modern classical ML architectures. Even for relatively simple company credit scoring, our current models use over 300 features per company. On current quantum systems with limited number of qubits, it is not possible to process this number of features. and reaching these levels will take years. Theoretically, quantum systems may offer large advantages as they can process a very large number of possibilities simultaneously. In the future it is expected that quantum systems will reach a scale to be useful for these problems. To be ready for that, we need to build expertise and community knowledge about QML techniques and their capabilities.

We hope that quantum computing may become a good fit for these problems because, as explained earlier, the problems are dealing with high and growing volumes of complex data. So, model recalibration needs to be done quickly. Quantum algorithms have already been developed and tested for a number of similar problems, including classification ML, optimization, and graph ML. Our data and models are already residing in computing clouds, and we have already solved the accompanying security and access problems. Thus, connectivity to quantum systems is not going to add massive architectural complexity. Overall, *quantum neural networks* (QNNs) alongside classical processing seem to be a good match for the business needs.

1.5.2 Algorithm Design for a Quantum Neural Network (QNN)

The literature for QNNs contains two basic designs, one for a simple 2-qubit perceptron similar to what is offered by Entropica Labs (2020) and another involving a hybrid quantum/classical model that uses PyTorch and Qiskit (Jupyter Book Community, 2020).[5]

2-qubit perceptron. Here, q_0 and q_1 are the input qubits and q_2 is the output qubit from a Toffoli gate. R_{y1} and R_{y2} are the input values as rotations, and R_{y3} and R_{y4} are the neural network weights to be trained. (See Figure 1.8.) The circuit is re-created with new input and weight rotations and re-run for each data row. The output of the circuit is compared to the expected 0/1 output (binary in this case for "defaulted or not"). The best set of weights is the optimum accuracy from this routine. The main steps of the algorithm are: Data is prepared and loaded; a quantum circuit is created; the quantum circuit is executed; the output of the quantum circuit is processed.

The hybrid approach for this routine involved these steps. (See Figure 1.9.) Data is prepared and loaded; the data is input to the classical first layer; a quantum circuit is created (see Figure 1.10) based on the output of the first layer; the quantum circuit is executed; the output from the quantum circuit is used as input to the classical last layer.

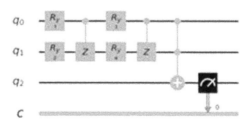

FIGURE 1.8   The 2-qubit perceptron quantum circuit.

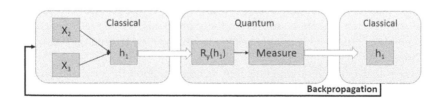

FIGURE 1.9   Hybrid algorithm using a quantum circuit as a hidden layer in a back-propagation NN.

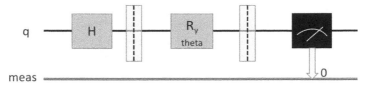

FIGURE 1.10   The quantum circuit used in the hidden layer.

Finally, the output of the last layer is fed back to the input layer adjusted for back-propagation.

Our approach was to first assess two features of credit data using these simple (1- or 2-qubit) designs and then to expand them to more features. There are other algorithms written, including variational approaches (e.g., Tacchino et al., 2019) in which the states hold weight information, as well as approaches that can also capture the training accuracy in the phase (e.g., Liao et al., 2019). As the implementation of these models is not easy, it is important to start small and increase complexity after getting meaningful results on smaller problems.

1.5.3   SOFTWARE DESIGN FOR QNN

With the algorithm decided, the application needed to be coded for data preparation and loading, executing the quantum circuit, and performing output processing. Qiskit was chosen as the we already had some experience and access to the IBM Q machines. First, the sample code was executed to ensure that all worked as expected, and then the code was modified to use the credit rating data.

Data preparation. Before the data is encoded onto the qubits, it is important to cap outliers. The cut-off points are determined empirically, and the step is performed classically. This step enables consistent scaling of feature values between 0 and π radians. Additionally, this step ensures that when the model is deployed for prediction, new data points will not take on values out of the usual range.

The data are encoded as amplitude rotations on the input qubits. These are coded into the circuit that is re-built at each data row. Rotation was used so far for the perceptron and hybrid models as it is straightforward and maps relatively easily to data features. State was not used due to the normalization issue with the scaling needing to be done across the data set not at each data row. With other models, such as Liao et al. (2019) with the whole data set applied in one shot, then this would be the best approach allowing for enormous data inputs and processing. The data set is going to be huge in the end state, but for now a smaller set of 2,000 rows of data with 40 features can be prepared.

- *For the 2-qubit perceptron* (see Figure 1.8), classical data preparation is used at first. The values for each row of data are put onto a *z*-rotation of the input qubits, and two more rotations are made after an entanglement for the neuron weights. A range of weight values are tested for each row of data, the result is stored outside the quantum circuit, and the best set of weights chosen as the final best solution.

- *For the hybrid approach*, there is 1 qubit in a hidden layer between two classical layers (Jupyter Book Community, 2020). The output from the first classical layer is applied as a rotation to the qubit in the hidden layer. The output of the hidden layer then is used as the input to the second classical layer.

The quantum circuit's design can be further modified:

- *For the simple perceptron*, different single and multiple gates can be selected, and then the ROC (receiver operating characteristic) and AUC (area under the curve) metrics can be tested for each design.
- *For the hybrid approach*, the circuit starts in a simple way with 1 qubit and is then expanded with multiple qubits and entanglement operations.

Executing the quantum circuit. We next consider how the quantum circuit will be run.

- *For the simple perceptron*, the same quantum circuit is executed for each combination of weight values for each data row and the accuracy of the output is stored. The most accurate set of weights is then chosen as the best NN model and the test data set is measured for the ROC and AUC metrics.
- *For the hybrid approach*, the quantum circuit is embedded with the hidden layer of a classical NN model. The quantum-related code is coded into a QuantumClass with Qiskit and a QuantumClassicalClass with PyTorch.

Processing the output. The output of the both models is a single probability of classification, and a threshold is applied to give a final binary result for the prediction of whether a company will default or not. For the simple perceptron model the output qubit probability of 1 is used. For the hybrid model the output of the output classical layer provides the classification output. In both instances, the outputs are analyzed and visualized using standard statistics packages, such as Sklearn (sclkit-learn.org).

1.5.4 HARDWARE FOR QUANTUM CREDIT SCORING

For the credit-scoring use case, we expect to need a large number of qubits to be available to enable good use of a range of features. IQX was chosen given the availability of machines with a relatively large number of qubits for the quantum circuits, effective software development with Qiskit, and good support. One issue with these models is if there is the use of queuing on the backends compared to having a dedicated time slot. Having a queue means that executing the quantum circuit multiple times can significantly increase the overall time if there is a long queue. This is not such a big problem for a proof-of-concept, but would not be feasible for production.

Coupling, meanwhile, is not a problem for small circuits. But, as the size of the NN grows in terms of features, weights and entanglement, the circuit will need to align more with the coupling of the device and the design will need to be optimized for the topology.

Further, coherence time has not been an issue so far with small circuits but moving to more sophisticated designs using variational circuits and storing accuracy in a state will require more qubits and much deeper circuits. The whole quantum circuit must then execute and optimize within the coherence time. Qubit and gate noise has also not been such a big issue with the smaller circuits, but again, with more sophisticated circuits errors are expected and will need to be dealt with. A potential trade-off could be made between circuit complexity and quantum speed-up, as the hybrid model can alternate between classical and quantum layers to avoid noise problems due to circuit depth, at the expense of pure-quantum speed-ups.

1.5.5 ISSUES FOR MOVING FROM A STAND-ALONE TO AN INTEGRATED SYSTEM

Our work on credit scoring is currently a stand-alone project, and it is not yet integrated into a production process. Thus, there is no integration downstream as yet, but a data pipeline was developed for upstream use, before the data are sent to the models. This involved cleaning, scaling, and linearizing the data using standard packages. The output of the models has been compared to a benchmark using classical ML, and ranking coefficients have been used to compare how the classical and quantum solutions differed. The data on this project are publicly available, with some proprietary processing required so that using cloud computing for this research is fine. Also, with the hybrid model, the quantum circuit model is less likely to be required to be protected.

1.6 CONCLUSION

Quantum computing is positioned to be key for the future and for Society 5.0. There are many business needs that could be greatly helped by the use of quantum algorithms. However, the quantum advantage has not yet been demonstrated beyond any doubt and exploring quantum computing has a steep learning curve. In this chapter, we have given some background for the key features of quantum computing, mapped them to current business needs, and shown how to begin to apply them to actual business problems. In the future, we are looking forward to improved quantum hardware, better software development environments, a greater understanding of the physics of quantum phenomena, and greater networked thinking to better utilize quantum computing. Quantum hardware is consistently improving for qubit fidelity and coherence times. In addition to the rapid improvements in the qubits of quantum computers in the coming years (Hackett, 2020), software development environments also will improve in usability and visualization as well as integration with current data analytics packages, such as the recent IBM Circuit composer upgrade and Horizon Quantum's integration with MATLAB®(http://horizonquantum.com/). Though advances are being made in understanding quantum phenomena, it is not an easy path and there still is no guarantee of success. But as Johann Wolfgang von Goethe is credited with saying:

Whatever you can do, or dream you can, begin it. Boldness has genius, power and magic in it.

ACKNOWLEDGMENTS

We benefited from the book editors' guidance and suggestions related to our revision of this chapter. Paul Griffin thanks Singapore Management University, the Monetary Authority of Singapore, Tradeteq, and OneConnect for their generous support. Michael Boguslavsky also acknowledges the Monetary Authority of Singapore. Junye Huang thanks IBM Corporation for its support. Robert J. Kauffman acknowledges Danske Bank, Copenhagen Airport A/S, Think Tank DEA, and the Endowed Chair in Digitalization at Copenhagen Business School for financial assistance. Finally, Brian R. Tan would like to thank Kreative Kommunication, Vietnam, a marketing agency, for its support. All errors and omissions are the sole responsibility of the authors.

NOTES

1 For a glossary with terms and definitions, see Appendix Table A1 at the end of this chapter.
2 In fact, Barclays has been testing the operation of quantum computing with quantum computing hardware that only uses 16 qubits at a maximum (Crosman, 2019). In other words, inadequate representational and computational power for a fully scaled version of its securities trade settlement process, which was reported as having been run on a 7-qubit quantum processor and a depth of 3 gates, with 6 participants and 7 security trades to be settled.
3 This used the T-bill CMT historical distribution of 1-day prices for maturity at time t as a random variable.
4 For more background, see Egger et al. (2020) and Orús, Enrique and Lizaso (2019) for finance domain quantum computing application areas, with consideration of future research, hardware, and applications.
5 See Wittek (2014) for QML fundamentals and Biamonte et al. (2017) for a broader review of QML methods.

REFERENCES

Akrout, M. (March 2020). *Will quantum computing be the next generation of automotive technology?* Retrieved from https://www.prescouter.com/2020/03/will-quantum-computing-be-the-next-generation-of-automotive-technology/
Altman, E. I. (1968). Financial ratios, discriminant analysis and the prediction of corporate bankruptcy. *Journal of Finance, 23*(4), 189–209.
Ambainis, A. (2007). Quantum walk algorithm for element distinctness. *SIAM Journal of Computing, 37*(1), 210–239.
Arora, S., & Barak, B. (2009). *Computational complexity: A modern approach.* Cambridge: Cambridge University Press.
Arute, F., Arya, K., Babush, R., Bacon, D., Bardin, J. C., Barends, R., ... Martinis, J. M. (2019). Quantum supremacy using a programmable superconducting processor. *Nature, 574,* 505–510.
Bernstein, E., Vazirani, U. (1997). Quantum complexity theory. *SIAM Journal of Computing, 26*(5), 1411–1473.

Biamonte, J., Wittek, P., Pancotti, N., Rebentrost, P., Wiebe, N., & Lloyd, S. (2017). Quantum machine learning. *Nature, 549*(7671), 195–202.

Braine, L., Egger, D. J., Glick, J., & Woerner, S. (2019). *Quantum algorithms for mixed binary optimization applied to transaction settlement.* Retrieved from https://arxiv.org/abs/1910.05788

Burkacky, O., Pautasso, L., & Mohr, N. (2020, September 2). *Will quantum computing drive the automotive future?* Retrieved from https://www.mckinsey.com/industries/automotive-and-assembly/our-insights/will-quantum-computing-drive-the-automotive-future#

Crosman, P. (2019, July 16). *Why banks like Barclays are testing quantum computing.* Retrieved from https://www.americanbanker.com/news/why-banks-like-barclays-are-testing-quantum-computing

Deutsch, D., & Jozsa, R. (1992). Rapid solutions of problems by quantum computation. *Proceedings of the Royal Society of London A: Mathematical, Physical and Engineering Sciences, 439*(1907), 553–558.

Dolev, S. (2018, January 6). *The quantum apocalypse is imminent.* Retrieved from https://techcrunch.com/2018/01/05/the-quantum-computing-apocalypse-is-imminent/

Egger, D. J., Gambella, C., Marecek, J., McFaddin, S., Mevissen, M., Raymond, R., ... Yndurain, Y. (2020). *Quantum computing for finance: State of the art and future prospects.* Retrieved from https://arxiv.org/abs/2006.14510

Entropica Labs. (2020). *A quantum perceptron.* Retrieved from https://qml.entropicalabs.io/

Farhi, E., Goldstone, J., Gutmann, S., & Sipser, M. (2000, January 28). *Quantum computation by adiabatic evolution.* Retrieved from https://arxiv.org/abs/quant-ph/0001106v1

Farhi, E., Goldstone, J., & Gutmann, S. (2014, November 14). *A quantum approximate optimization algorithm.* Retrieved from https://arxiv.org/abs/1411.4028

Feynman, R. P. (1982). Simulating physics with computers. *International Journal of Theoretical Physics, 21*, 467–488.

Foy, P. (2019, December 25). *Introduction to quantum programming with Qiskit.* Retrieved from https://www.mlq.ai/quantum-programming-with-qiskit/.

Furnham, A. (1986). Response bias, social desirability and dissimulation. *Personality and Individual Differences, 7*(3), 385–400.

Gambetta, J. (2020, September 15). *IBM's roadmap for scaling quantum technology.* Retrieved from https://www.ibm.com/blogs/research/2020/09/ibm-quantum-roadmap.

Gidney, C., & Eker, M. (2019, December 5). *How to factor 2048 bit RSA integers in 8 hours using 20 million noisy qubits.* Retrieved from https://arxiv.org/abs/1905.09749

Grover L. K. (1996). A fast quantum mechanical algorithm for database search. *Proceedings, 28th Annual ACM Symposium on the Theory of Computing,* 212–219.

Hackett, R. (2020, November 15). *IBM plans a huge leap in superfast quantum computing by 2023.* Retrieved from https://fortune.com/2020/09/15/ibm-quantum-computer-1-million-qubits-by-2030/

IBM.com (2020, August 20). *IBM delivers its highest quantum volume to date, expanding the computational power of its IBM cloud-accessible quantum computers.* Retrieved from https://newsroom.ibm.com/2020-08-20-IBM-Delivers-Its-Highest-Quantum-Volume-to-Date-Expanding-the-Computational-Power-of-its-IBM-Cloud-Accessible-Quantum-Computers.

Japan Gov News (2019, October 3). *Evolving innovation.* Retrieved from https://www.japan.go.jp/investment/evolving _innovation.html.

Jeste, D. V., Ardelt, M., Blazer, D., Kraemer, H. C., Vaillant, G., & Meeks, T. W. (2010). Expert consensus on characteristics of wisdom: A Delphi method study. *Gerontologist, 50*(5), 668–680.

Jupyter Book Community. (2020). *Hybrid quantum-classical neural networks with PyTorch and Qiskit*. Retrieved from https://qiskit.org/textbook/ch-machine-learning/machine-learning-qiskit-pytorch.html.

LaRose, R. (2019, March 22). *Overview and comparison of gate level quantum computing software platforms*. Retrieved from https://arxiv.org/abs/1807.02500

Lee, R. S. T. (2020). *Quantum finance: Intelligent forecast and trading systems*. Singapore: Springer.

Liao, Y., Ebler, D., Liu, F., & Dahlsten, O. (2019, November 20). *Quantum advantage in training binary neural networks*. Retrieved from https://arxiv.org/pdf/1810.12948.pdf

Loeffler, J. (2018, November 29). *No more transistors: The end of Moore's Law. InterestingEngineeering.com.* Retrieved from https://interestingengineering.com/no-more-transistors-the-end-of-moores-law.

Manin, Y. I. (1980). *Computable and Noncomputable (Vychisimloe y Nevchislimoe).* Moscow: Sov. Radio.

McKay, D. C., Alexander, T., Bello, L., Biercuk, M. J., Bishop, L., Chen, J., ... Gambetta, J. M. (2018, September 10). *Qiskit backend specifications for Open QASM and OpenPulse experiments*. Retrieved from https://arxiv.org/abs/1809.03452

Ménard, A., Ostolic, I., Patel, M., & Volz, D. (2020, February 6). *A game plan for quantum computing*. Retrieved from https://www.mckinsey.com/business-functions/mckinsey-digital/our-insights/a-game-plan-for-quantum-computing

Nielsen, M. A., & Chuang, I. L. (2010). *Quantum computation and quantum information* (10th anniversary edition). Cambridge: Cambridge University Press.

Nijs, D. E. L. W. (2015). Introduction: Coping with growing complexity in society. *World Futures: The Journal of New Paradigm Research, 71*(1-2), 1–7.

Orús, R., Enrique, M., & Lizaso, E. (2019). Quantum computing for finance: Overview and prospects. *Reviews in Physics, 4*, 100028.

Othmani, I., Ettl, M., & Guonaris, Y. (2020). *Exploring quantum computing use cases for airlines: A new technology is cleared for take-off*. Retrieved from https://www.ibm.com/thought-leadership/institute-business-value/report/quantum-airlines

Pednault, E. (2020, October 10). *On "quantum supremacy"*. Retrieved from https://www.ibm.com/blogs/research/2019/10/o"n-quantum-supremacy/

Peruzzo, A., McClean, J., Shadbolt, P., Yung, M. Y., Zhou, X. Q., Love, P. J., ... O'Brien, J. L. (2014). A variational eigenvalue solver on a photonic quantum processor. *Nature Communications, 5*(4213). DOI: 10.1038/ncomms5213

Roffe, J. (2019, July 25). *Quantum error correction: An introductory guide*. Retrieved from https://arxiv.org/abs/1907.11157

Rønnow, T. F., Wang, Z., Job, J., Boixo, S., Isakov, S. V., Wecker, D., ... Troyer, M., (2014). Defining and detecting quantum speedup. *Science, 345*(6195), 420–424.

Shor, P. W. (1997). Polynomial-time algorithms for prime factorization and discrete logarithms on a quantum computer. *SIAM: Journal on Scientific Computing, 26*(5), 1484–1509.

Simon, D. R. (1997). On the power of quantum computation. *SIAM Journal of Computing, 26*(5), 1474–1483.

Singh, S. (2019, December 28). *Quantum computing: Solving problems beyond the power of classical computing*. Retrieved from https://economictimes.indiatimes.com/tech/software/quantum-computing-solving-problems-beyond-the-power-of-classical-computers/articleshow/73011174.cms

Tacchino, F., Macchiavello, C., Gerace, D., & Bajoni, D. (2019). An artificial neuron implemented on an actual quantum processor. *npj Quantum Information, 5*(26). DOI: 10.1038/s41534-019-0140-4

Wittek, P. (2014). *Quantum machine learning: What quantum computing means to data mining*. New York: Academic Press.

Woerner, S., & Egger, D. J. (2019). Quantum risk analysis. *npj Quantum Information*, *5*(15). DOI: 10.1038/s41534-019-0130-6

Zhao, R., Tanttu, T., Tan, K. Y., Hensen, B., Chan, K. W., Hwang, J. C. C., … Dzurak, A. S. (2019). Single-spin qubits in isotopically enriched silicon at low magnetic field. *Nature Communication*, *10*, 5500.

APPENDIX A:GLOSSARY OF TERMS

TABLE A1

Quantum Computing Terms and Definitions

Term	Definition
Adiabatic quantum computing (AQC)	Emphasizes the resting state of a complex system by eigenvector minimization for Hamiltonian problems (Farhi et al., 2000).
Backend	Quantum hardware to run a quantum circuit and measure the final states of the qubits.
Coherence time	The amount of time a quantum state is stable form, such that it will be possible to conduct a solution procedure or optimization experiment before it is extinguished.
Fidelity	A metric for the similarity or closeness of two quantum states for a qubit.
Gate	A single or multi-qubit operation for the circuit model of quantum computing
Intelligence	Ability to obtain knowledge and skills and apply them to solve problems.
Noisy intermediate-scale quantum (NISQ)	An acronym for the current state of quantum computers that have 10-100 qubits (intermediate scale) without error correction (noisy).
NP Hard	Problems with a non polynomial time complexity
Qiskit	An open-source software developer kit (SDK) for writing and running gate-based quantum computing program on hardware such as IBM's Q Experience backends.
Qubit	The analog of a 0/1 bit in standard computing, but for which a quantum computer has different realizations using quantum particles.
Quantum approx. optim. algo. (QAOA)	An algorithm developed for quantum computing to estimate the solutions for hard combinatorial optimization problems using a gate-based approach.
Quantum error correction (QEC)	Approach used in quantum computing to correct data that has noise introduced from the quantum computer (Roffe, 2019).
Quantum machine learning (QML)	The analysis of data from complex settings for problems that can be simulated or optimized using the computation methods of a quantum computer.

(Continued)

TABLE A1 (Continued)

Quantum Computing Terms and Definitions

Term	Definition
Variational quantum eigensolver (VQE)	A quantum algorithm that enables the estimation of an eigenvalue for a Hamiltonian (Peruzzo et al., 2014).
Wicked problems	Problems with no stopping rule or a non-exhaustive set of potential solutions.
Wisdom	Exhibiting knowledge and experience to arrive at a good judgment.

2 Prediction Models for Accurate Data Analysis: Innovations in Data Science

Balwinder Kaur, Anu Gupta, and R. K. Singla
Department of Computer Science and Applications Panjab
University, Chandigarh, India

CONTENTS

DOI: 10.1201/9781003132080-2

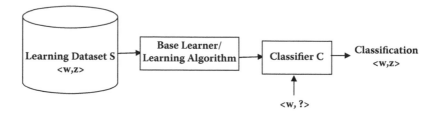

FIGURE 2.1   Generic classification process.

2.1   INTRODUCTION

The current era is the information era, with the need to find conferring benefits from the vast amount of information available which can be used for enhancing the process of decision making and tackle the issues related to Society 5.0 and humans. Data Mining, Machine Learning, and Statistical Techniques provide different methods to perform data analysis, for understanding and extracting new facts, patterns and knowledge which seems to be quite difficult manually when data is high dimensional and voluminous. Classification and prediction are two prominently used techniques for data analysis which help in extracting hidden information, patterns, predicting classes, and predicting future values or the trends (Han et al., 2011). Numerous prediction and classification techniques are suggested by researchers in the area like machine learning, data mining, statistics, and pattern recognition, etc.

Classification, a supervised learning technique that builds models that predicts the class or data labels of new instances based on available data sets. These models are known as Classifiers. There exist different classifiers like Decision tree, Rule-based, Association Rule Mining, k-Nearest-Neighbor, Bayesian classifiers, etc. There are also some other recent classifiers like Support Vector Machine (SVM), Backpropagation (a Neural Network technique) etc. (Han et al., 2011; Pujari, 2001). All these classification algorithms have their own pros and cons.

Classification is performed in two steps (Stefanowski, 2008), as shown in Figure 2.1: first, a learning algorithm is implemented to build a model based on predetermined classes or labels defined in the training set. This is known as the training or learning step. According to Han, "A tuple X, which belongs to a training dataset is represented by an attribute vector. Every tuple in the training set is presumed to belong to a predefined class as determined by another database attribute also known as class label attribute" (Han et al., 2011). This is the first step in the classification process. The current step is also seen as a mapping or function, $z = f(W)$, that can predict the class label z of a given tuple W.

$S = \{(w_1,z_1), (w_2,z_2), ..., (w_n,z_n)\}$, Set of learning examples/data set
f: $z = f(w)$, some unknown classification function
$wi = <w_{i1}, w_{i2}, ..., w_{im}>$ example described by m attributes
z – class label, value is drawn from a discrete set of predefined classes $\{z_1, ..., z_k\}$

Step two of the classification process is associated with the accuracy of the classifier. The predictive accuracy of the classifier is the percentage of correctly classified tuples in a test set. The test data set is selected randomly and independent of the training set. If classifier accuracy is acceptable, then the classifier can be utilized to classify future unseen data for which the class labels are unknown (Han et al., 2011).

Hence, the job of the learning model requires to estimate the function f(x) for constructing a classifier. The results obtained from the classifier can be of different forms depending upon the data used like continuous, nominal, etc. The classifier outputs are formally categorized as (Xu et al., 1992; Ranawana & Palade, 2006; Gargiulo et al., 2013; Kuncheva, 2014):

- *Abstract form*: The output of a classifier is a distinct single class.
- Rank level: The output classes are ranked by classifiers relative to its confidence of input data belonging to particular class. The most probable class is sent as preferred the final class according to classifier belief.
- *Measurement level*: These classifiers are probabilistic classifiers, which depict a value on the confidence there is an input value associated with each class in the classifier. The outcome of the classifier is in an array format containing the trust values.

There are many limitations like imbalanced data, noise, limitations of learning paradigms, and so on. Due to all these types of limitations, constructing a perfect prediction/classification model based on an individual classifier is difficult. But there exist several classifiers that can be implemented for a particular problem. These individual or traditional classifiers produce different data interpretations and results due to their learning behaviors. There exists no single method or technique to find out an individual classifier that provides results with optimal accuracy. The diversity in the learning paradigms and the variation in performance of classifiers on the same data sets are key factors that initiated the development of the Multi Classifier System (MCS). This MCS attempts to integrate the outcomes of separate individual classifiers to obtain results with optimal accuracy (Ranawana & Palade, 2006). Combining multiple classifiers for constructing a composite classifier is recommended as a new direction for providing computational excellence and improving prediction accuracy of prediction models as they provide an effective way to achieve near-to-optimal performance prediction.

2.1.1 OVERVIEW OF CHAPTER

Combining individual classifiers for creating ensembles for with optimal prediction accuracy is an accepted research area. It is known under names like ensemble system, committees of learners, mixtures of experts (Polikar, 2012), classifier ensembles, multiple classifier systems, etc. in literature (Polikar, 2006). In the chapter, the terms *multi classifier system* and *ensemble system* are used interchangeably. Section 2.2 formally introduces MCS, philosophy, history of MCS, justifies its need and its distinctive feature. Section 2.3 describe existing common ensemble techniques and fundamental differences among them. Section 2.4 gives a brief description of the Design/frameworks of MCS. Section 2.5 presents the common combination methods available. Section 2.6 gives summarised information on the different topological options available for connecting the classifiers in an MCS. Section 2.7 presents one of the key aspects of MCS i.e. MCS diversity and methods for the diversification of individual classifiers whereas Section 2.8 discusses the advantages and disadvantages associated with MCS. Section 2.9 summarizes various application areas. Section 2.10 point out emerging areas. Sections 2.11 and 2.12 various challenges associated with MCS and challenges w.r.t Society 5.0. Section 2.13 presents the conclusion and future work.

2.2   MULTI CLASSIFIER SYSTEM

The Multi Classifier System (MCS) or esemble approach, has evolved to address the issues related to single classifier-based prediction/classification models. One of the major advantages of MCS is that they are capable of enhancing the predictive/classification accuracy. As in the process of decision-making, where decision is taken based on consolidated opinions of all the board members or experts present, this is how exactly the multi classifier system works. The type of combination where diverse classifiers are utilized to classify the unseen data and the outcomes from individual classifiers are integrated using some combining approach to produce combined output is known as a Multi Classifier System. Intuitively, a mixture of experts or integration of classifiers provides superior results as compared to a singular decision-maker. The objective of MCS is to create a blended/compounded system that can surpass the accuracy of individual classifiers by merging the decisions of different classifiers leading to optimal prediction accuracy (Windeatt, 2005).

It seems quite natural and sensible that when different individual classifiers are integrated systematically, the combined outcome should be either better than or at least comparable to the outcomes of the finest performer within the ensemble of classifiers. Consequently, multiple classifiers are being integrated to attain the optimal prediction accuracy (Ranawanaet al., 2006).

2.2.1   PHILOSOPHY

There exist several practical and mathematical reasons for examining multi classifier systems as prediction models with computational excellence, but the inborn relation is to our day-to-day experiences. Whenever we as humans need to take some crucial decisions, we always seek a second opinion before making final decisions especially when the decisions are related to some medical, social, or financial matters. The reason behind taking multiple opinions is basically to enhance confidence regarding whether the decision made is correct or not. A similar approach can be followed in the case of data analysis and automated decision-making applications where the opinion of different experts i.e. classifiers can be considered which leads to optimal prediction accuracy and the approach is named a multi classifier system or ensemble system. The essence behind the multi classifier is to measure the performance of individual classifiers and merge them towards

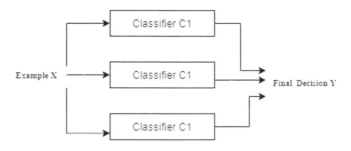

FIGURE 2.2 A multi classifier system.

obtaining a prediction model/classifier that surpasses individual classifiers (Polikar, 2006) in terms of performance and accuracy, as shown in Figure 2.2.

According to Thomas. G. Dietterich (2000), a collection of classifiers where the decision of individual classifiers is integrated by some method to classify new examples is known as ensemble of classifiers. An essential and acceptable precondition for an ensemble to be highly accurate as compared to its member classifiers is that if each member classifier is diverse and accurate.

According to Wolpert and Macready (1997) and as found from various research studies, there exists no single classifier approach which gives optimal results. MCS exploits the strengths of the individual classifier models for a classification task and produces a high-quality model. This high-quality ensemble model created by combining individual classifiers is at the cost of escalated complexity. So, the point arises,is the integration of classifiers is justified (Kuncheva, 2014). Intuitively, combining classifiers comes out as an elemental step, when a essential knowledge from classification models has been gathered. Although there are many questions related to equating classifiers to real-life problems, combining classifiers is expanding swiftly and getting recognition in the area of data sciences and machine learning (Kuncheva, 2014).

2.2.2 History

Taking into consideration the historical perspective, tracing the starting point of ensemble learning as one of the techniques of supervised learning is quite difficult. According to different studies, it started evolving in the late '70s when Dasarathy and Sheela in their research, discussed partitioning of the feature space using two classifiers; a k-NN and linear classifier (Dasarathy & Sheela, 1979; Polikar, 2006; Woźniak et al., 2014). From the literature, it is found that there are mainly three paths related to the early contributions in the field of ensemble techniques that have led to current methods; these are, the mixture of experts, an ensemble of weak learners, and combining classifiers.

The mixture of experts (Polikar, 2006) is prominently studied in the area of Neural Networks. In this, experiments and research studies performed are based on the divide-and-conquer strategy. The complex problem is divided into smaller parts and a mixture of parametric models try to learn jointly. The individual model outcomes are combined using some combination rules to get the final solution.

Ensemble of weak learners is studied by the machine learning community. Under this, the researchers try to boost the performance of weak learners. Algorithms like AdaBoost, Bagging, etc. are a few examples.

Combining classifiers is a technique widely used and studied for Pattern Recognition where researchers try to build *strong* classifiers by designing powerful *Combining rules.* As a result, a thorough understanding of the design and usage of diverse and strong combining rules has been collected (Zhou, 2012).

The field of ensemble learning or MCS made progress in the '90s with the work of Hansen and Salamon (Hansen & Salamon, 1990), where they considered a neural network ensemble (Polikar, 2006). Xu et al. (1992) considered joining multiple classifiers and their application to handwriting recognition. Almost in the same period, Schapire (1990) built the basis with the algorithm of AdaBoost. Later, Freund and Schapire in their algorithm showed that a powerful and computationally efficient classifier can be produced by *integrating* weak classifiers. Bagging is a very effective ensemble learning technique based on Bootstrap sampling. It is used for algorithms which are unstable e.g. Neural Networks, Decision Tree, etc. (Breiman, 1996). In 1996, Turner (Tumer et al., 1996) proved that taking the average of results of a large number of independent and unbiased classifiers can generate same outcomes as compared to optimal Bayes classifier. Whereas Ho et al. (1994) underlines the fact that each classifier's decision must receive useful representation in decision combination function. They examined many methods like Borda Count which are based on decision ranks. Ensemble techniques can also be implemented for enhancing the robustness and quality in the case of unsupervised techniques (Polikar, 2006; Rokach, 2010). The approaches used in these models are different from each other in terms of methods used for building the classifiers, and/ or the combination strategies applied (Polikar, 2006; Woods et al., 1997).

Further, a series of International annual Workshops on Multiple Classifier System (MCS), held annually since 2000, is *playing* a vital role in organizing and developingnew knowledge in the area of multi classifiers (Kuncheva, 2014).

2.2.3 NEED FOR MULTI CLASSIFIER SYSTEM

Multi classifier systems as prediction models are gaining popularity as they offer many options for handling real-world complex problems, which involve uncertainty, imprecision, noise in its high-dimensional data. MCS allows us to use raw data and prior knowledge to solve complex problems of Society 5.0 in advanced and encouraging ways. With advancement, this interdisciplinary research area is continuously expanding in the research community. There exists a large number of diverse classifiers, like k-Nearest-Neighbour, Decision Trees, etc. (Hanet al., 2011) All these traditional or individual classifiers have their own pros and cons as discussed in Table 2.1. It is hard to find a suitable classifier for a particular problem. If one finds a suitable classifier it won't be sufficient because as a lone classifier it may be unable to learn significant decision boundary for classification problems in-hand, or may perform poorly on unseen data (Freund & Schapire 1996) (Bellmann et al.,

TABLE 2.1

Comparison of Individual Classifiers

S. No.	Classifier	Features/Advantages	Limitation/Disadvantages
1.	Naïve Bayes	• Based on probability concept. • Implementation is simple. • Computationally efficient, and good for classification. • Assumes feature independence.	• Accuracy reduces on small data set.
2.	Decision Tree	• Models are easy to interpret. • Can be used with both continuous and discrete values. • Easily implemented. • Performance is not affected by non-linear relationship among the parameters. • Domain knowledge is not required.	• Models are less stable, small change in data changes the tree. • Pruning required. • Susceptible on outlier. • Requires to perform computation to find splitting criteria or branching attribute. • Can be duplication in the sub-trees
3.	k-Nearest Neighbor	• No requirement of classes to be linearly separate. • Instance-based learning. • Based on distance calculation according to pre-defined metrics. • Robust to noise. • Works on numeric value. • Tolerance to outliers.	• Take excessive time in training in voluminous and high dimensional data. • Performance is based on number of dimensions.
4.	Support Vector Machine	• It makes use of hyper-planes. • Very accurate. • Works well on linearly inseparable data.	• Works on large data set in both training and testing. • Complex in nature and large memory requirement.
5.	ANN	• Easily implemented with less features. • Ease to use. • Applicable to large no. of problems	• Processing time depends upon size of network. • Difficult to guess the size of neural network.
6.	Rule – Based	• Rules based on if-then are easy to interpret.	• Difficult to find single rule in case multiple rules are applicable.

2018). The multi classifier approach provides a solution by combining the base classifiers to exploit their local behavior and strengths to obtain a strong classifier with enhanced prediction accuracy and reliability. The acceptance of MCS is rising mainly because they provide an attractive solution along with better accuracy and computational excellence to various problems related to classification (Bellmann et al., 2018; Tsoumakas et al., 2008).

The foremost reason for using MCS is that these systems lead to an improvement in the accuracy in comparison to a single individual classifier. Achieving optimal prediction performance is the main motivation behind the development of MCS. The improvement in accuracy is achieved through the correction of uncorrelated errors. According to Hansen and Salamon (1990), for multi classifiers to be optimal inaccuracy, the individual classifiers are to be greater than 50% accurate, and another important aspect is that they are independent of each other. Secondly, ensembles allow scaling of algorithms to large high-dimensional databases as compared to other classification algorithms that are computationally complex, and when implemented on large databases they suffer memory problems. Ensemble provides a better approach to this by allowing the partition of the database and training the model on smaller manageable parts and combining the outcomes. Thirdly, there are situations when data collection at a single location is not possible due to issues related to the security or size of the data set. In such a situation, different base learners of an ensemble can learn from multiple physically distributed data sets and combine their results to get the outcome (Tsoumakas et al., 2008). The issues related to class imbalance have been producing hindrance in the performance of classifiers therefore, many methods like resampling, ensemble-based approaches like boosting, and bagging provide a solution for the same. Ensemble methods are also being proposed that are capable of transforming the imbalanced data into balanced by partitioning into multiple sets first and then build the classifier on these multiple sets and finally combine their outcomes (Tsoumakas et al., 2008; Sun et al., 2015).

Further, various experimental studies have been conducted by different research communities proving that integrating the outputs of multiple individual classifiers can reduce generalization error which is not possible with a single classifier (Zhou, 2012). Opitz and Maclin (1999) experimentally verified that ensembles can generalize well. Different research studies have proved that ensemble models are capable of reducing bias-error, as presented in the large margin classifiers theory (Rokach, 2010). As compared to a single classifier, the multi classifiers like Bagging and Boosting reduces bias-variance error and can also reduce the problem of overfitting (Baba et al., 2015). Bauer et al., (1999), in their experimental studies proved that different ensembles like AdaBoost, Bagging can reduce the error. All these issues justify the need for MCS as they can provide classification models with optimal prediction accuracies, reliability, and efficiency.

2.2.4 DISTINCTIVE FEATURES OF MULTI CLASSIFIER SYSTEM

Empirical observations and different research studies have confirmed that individual learning algorithms can outperform one and other when implemented for a particular scenario or on a particular data set, but it is difficult to find an individual classifier that achieves optimal results (Bauer & Kohavi, 1999; Valentini & Masulli, 2002). As a result, multiple learning classifiers or algorithms try to utilize the advantages of the base learners to achieve optimal accuracy and computational efficiency. There are various theoretical and experimental reasons given by the researchers to explain why multi classifier systems or ensemble classifiers are better

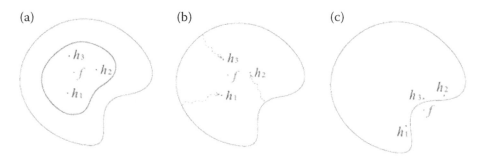

FIGURE 2.3   (a) Statistical reason; (b) computational reason; (c) representational reason.

than individual classifiers. Dietterich (2000), in his study, gave three important and fundamental reasons as depicted in Figure 2.3.

Statistical Reason: This reason appears when the training data set is small in size. It is often seen that different classifiers can be implemented on the available training data set and these classifiers may give similar performance as well as accuracy. Although, these classifiers may be indistinguishable in terms of an error on the training set, but may differ in generalization. Hence, there is a risk of choosing a classifier that may not be good in predicting the values of unseen data out of all the applicable classifiers. So, instead of picking one, it is good to create an MCS from all the accurate classifiers and take the average of their outputs leading to a reduction in the risk of selecting a wrong classifier.

Computational Reason: Even when there are large training data sets available, local search techniques are applied by some learning algorithms and get stuck in local optima. This leads to difficulty in finding the best classifier computationally. An MCS can be built by executing a local search from distinct starting points, which may give a superior estimation to an unknown hypothesis.

Representational Reason: In many ML tasks, the true classifier for unknown data cannot be found. So, by combining the classifiers it may be possible that the learning algorithm can give more exact or better predictions for unseen data.

A learning algorithm suffering from the statistical problem is considered to have a high "*variance,*" whereas an algorithm with representational issues is considered to have a high "*bias.*" Therefore, combining classifiers may reduce learning algorithms bias and variance (Zhou, 2012; Opitz & Maclin, 1999; Baba et al., 2015; Bauer & Kohavi, 1999). There are distinctive features of MCS related to their capabilities of handling huge amounts of high-dimensional heterogeneous data, learning behaviors, etc. which help them to achieve better accuracy and optimal prediction performance. These distinctive features are as follows.

Massive Amount of Data: In some specific applications, the data under analysis can be quite voluminous and may not be handled effectively by a single classifier. So, a better option is to fragment the data set into subsets of manageable size and train individual classifiers on distinct data subsets. Finally, the outputs from the individual classifiers of the ensemble are combined with help of some intelligent combination method. The prediction accuracy produced by MCS is better than the individual classifier.

Too Little Data: MCS can handle less amount of data available. The availability of illustrative and sufficient amounts of data is very much important for a classification algorithm to learn. If an adequate amount of training data is not available, then resampling techniques can be implemented to create a random subset of training data that may or may not be overlapping. Each of these subsets is used to train an individual classifier separatelyand the results of these individual classifiers can be integrated to get optimal accuracy with less amount of data using MCS.

Divide and Conquer: Due to complexity, some problems are tough for a single classifier to resolve irrespective of the data availability. Specifically, the reason may be inseparable class boundaries; i.e. the separating data into classes may be difficult. However, a suitable association of MCS with help of the divide and conquer strategy can divide the data space into a mini partition, where an individual classifier from the ensemble can learn on a partition. Hence, the difficult decision boundaries can be approximated more accurately with MCS.

Data Fusion: When data is collected from different sources, and it can contain different features that are heterogeneous. In such cases, it becomes difficult for a traditional classifier to learn from all the information contained in the data set. For example, the doctor can suggest different laboratory investigations, like ultrasound, blood investigations, ECG, etc. The investigations generated data holds distinctive features which can vary in numbers. With this type of data, it is difficult to train an individual classifier. In these types of cases, a data set from an individual investigation may be used to train a distinct classifier, whose results can be integrated later. Applications, where data from different individual sources are integrated to generate a better-informed decision, are named as data fusion applications. The MCS based approaches are being successfullyimplemented these type of applications (Polikar, 2012; Polikar, 2006).

Diversity: Diversity is a crucial factor for MCS to be successful. Its MCS diversity ensures that the integrated classifiers are independent of one and another. It implies that misclassifications do not occur simultaneously (Valentini & Masulli, 2002; Parvin et al., 2015). The disagreement between the individual classifiers or the independence between the individual classifiers is known as diversity.

Generalization: The underlying idea in ML is to specify up to what extent the outcomes learned from a training data set can be implemented on a unseen data set. The Ensemble results are more generalizable as compared to the individual classifier.

Bias–Variance Decomposition: In this decomposition, the classification error is decomposed into Bias and Variance (Kuncheva, 2014). The Bias-Variance decomposition helps in the reduction of the classification error leading to improvement in MCS accuracy as compared to the individual classifiers.

- A bias is an estimate of the averaged difference between the true and the predicted values.
- Variance measures the extent up to which the guesses of each of the learning

algorithms will vary concerning one and another i.e. the variance indicates the changeability of the classifier's guess regardless of the true state of nature.

There are different definitions and concepts given by various researches related to Bias-Variance decomposition depending upon the problem and ensemble systems. From different studies, it can be summarized that bias is linked with underfitting and variance is related to overfitting (Kuncheva, 2014). There exists a trade-off relationship between bias and variance; it is considered that classifiers with lower bias have higher variance and vice-versa (Polikar, 2012). Thus, MCS combines several classifiers with similar bias, and combining their outcomes leads to reduced bias.

This can be summarized from numerous experimental, and qualitative results obtained that the accuracies of ensemble systems are better in comparison to the individual classifier. These ensembles consist of a set of classifiers with complementary strengths and distinctive features which leads to better prediction performance and accuracy. There are different combination approaches available for integrating the decisions made by a set of accurate classifiers. Further, it justifies that MCS is not only giving better performance accuracy but is also better in terms of handling different sizes of data sets whether small or large. They are better at handling noisy data, imbalanced data, and prediction error. The MCS provides better generalization ability and diversity which makes them better than the individual classifier.

2.2.5   WORKING OF MULTI CLASSIFIER SYSTEM

A multi classifier system is proposed as a new direction towards achieving optimal prediction accuracy as compared to an individual classifier. Though these individual classifiers are based on different methodologies and can achieve different accuracies of classified samples. The main objective of integrating algorithms is to produce more accurate, precise, and confident prediction or classification results (Gargiulo et al., 2013). Dietterich (2000) gave informal and accessible reasoning explaining different viewpoints namely; computational, statistical, and representational, where the individual classifiers fail. MCS overcome these specified reasons to achieve near-to-optimal accuracy by exploiting the local behavior and strengths of the individual classifier models.

MCS emphasis integrating the classifiers from homogeneous or heterogeneous modeling backgrounds to give the final decision. Integrating the outputs of many diverse classifiers is beneficial only if there is a disagreement between them. A desirable and accepted criterion for MCS to be highly accurate as compared to its member classifiers is that the individual members are diverse and accurate (Hansen & Salamon, 1990), as integrating identical classifiers will not be effective and will produce no gain. Hansen and Salamon (1990), in their study, demonstrated that if in the ensemble the base classifiers are independent in the error production and the average error rate is less than 50%. Then, the expected error can be minimized to zero as the number of classifiers integrated increases to infinity; however, such assumptions are difficult to achieve and are rarely followed in practice. Thus, the enhancement in the final predictive accuracy of MCS occurs because of diversity among its components.

Further, the generalization ability i.e., to generalize the results on unseen data is one of the strongest properties of ensembles which makes them better than the individual classifier. Ensemble methods are being preferred mainly because they can boost the performance of weak learners. In previous years, various experiments and research studies were conducted by the machine learning community presenting that:

1. Generalization error is reduced by combining the outcomes from multiple classifiers.
2. The existence of different "inductive biases" in various types of classifiers makes ensemble methods very effective.
3. Ensemble methods can make effective utilization of diversity to minimize the variance-error without increasing the bias-error. In certain situations, bias-error is also minimized by ensemble.
4. Indeed, they can boost the performance of weak learners.

According to Thomas G. Dietterich (2000), classifiers are said to be diverse if error produced by them is different for unseen data instances. And for a classifier to be correct or precise the error rate has to be better than random guess for a new value X.

Assume a set of three classifiers, considered as hypothesis {H1, H2, H3} and a new object X:

- If all hypotheses are similar, then when H1(X) is incorrect, H2(X) and H3(X) will also be wrong i.e. they will be making the same decision.
- But, if the errors are not correlated, then if H1(X) is wrong, H2(X) and H3(X) may be correct → a majority vote will correctly classify X!

Many other researchers have suggested that the ensemble performance is based on two properties, which are the success rate of the individual base classifiers combined in the ensemble system and the independence between the results of the base classifiers (Woods et al., 1997). Some researchers even suggested that the diversity between individual models, the accuracy of an individual model, the size of an ensemble, and the combination strategy used for construction (Woźniak et al., 2014; Freund & Schapire 1996; Baba et al., 2015) are key factors for the successful working of an ensemble.

2.3   ENSEMBLE METHODS

To obtain a precise ensemble, the base classifiers should be possibly diverse and accurate in nature. This is being proved and emphasized by many researchers. There are different methods like the hold-out test, cross-validation, etc. for approximating the accuracy of base learners. But there is no precise definition of what is recognized as diversity. In practice, diversity is obtainable in different ways, such as by manipulating features, manipulating outputs, creating random samples of the training set, etc. The usage of the different base learners, combination methods, topologies, etc. helps in constructing different ensemble methods. Many effective ensemble methods are being used in various application areas. A few prominently used ensemble methods like Bagging, Boosting, Random Sub-space method, Stacking, etc. are briefly introduced here.

Boosting: Boosting is also called Adaptive Resampling and Combining (arcing). It is prominently used for boosting the performance of any weak classifier. In this method, the weak classifier like a decision tree is executed repeatedly on training data. The outcome produced from a weak classifier is integrated to produce a single ensemble model to achieve high accuracy in comparison to a weak classifier.

The Boosting algorithm was first proposed by Schapire in 1990 (Schapire, 1990). Later, the AdaBoost algorithm was introduced by Freund and Schapire in the year 1996 (Freund & Schapire 1996). The algorithm uses the idea of allocating weights to the instances. Initially, equal weights are assigned to all the instances, later with every round of execution, the reweighting is performed. While reweighting the instance, the weights of incorrectly classified instances is increased whereas the weight of correctly classified instances is reduced. Due to this, the algorithm is forced to concentrate on instances that are not correctly classified, so that they can be classified correctly during the next round. The process is repeated and it provides a series of classifiers complementing each other. Finally, the results are integrated for the final prediction (Schapire, 1990; Rokach, 2005; Pandey et al., 2014).

Boosting improves the performances mainly due to two reasons:

1. It creates a final classification model with a reduced error by integrating individual classifiers which may have more errors on the training data set.
2. The variance is also reduced significantly in a combined classifier as compared to variance produced by the weak classifier (Rokach, 2005).

There are also issues associated with boosting; occasionally boosting tends to produce deterioration in generalization performance (Rokach, 2005). According to Quinlan (1996), overfitting is one of the reasons for boosting failure. A large number of repetitions also lead to the creation of a complex classifier. One of the drawbacks associated with boosting is difficulty in understanding as several classifiers are required to be captured rather than a single classifier. Despite all issues mentioned, Breiman in 1996 refers to the concept of boosting as one of the most remarkable developments (Breiman, 1996).

Bagging: Also known as Bootstrap Aggregating (Breiman, 1996), is a common approach that can process the instances concurrently. Based on the Bootstrap (Efron and Tibshirani, 1994) sampling technique, the method focuses on creating an integrated classifier with improved accuracy. Every individual base learner is trained on a training data subset taken randomly with replacement from the training set. Bagging integrates the outputs of the classifiers with help of the voting technique to get the final prediction. As a result, bagging surpasses individual classifier model built from the original training data setin predictive performance (Rokach, 2005). According to Breiman L., bagging is a powerful technique for unstable classifiers as it is capable of eliminating instability. Classifiers, where a minor modification in the data set leads to major changes in the final predictions, are known as Unstable classifiers, for e.g., Neural Network, Decision Tree, etc. However, bagging methods are sometimes difficult to analyze.

Random Forests: Random forests (Breiman, 2001) is an effective classification technique. It is a blend of Random Sub space and Bagging method. Random forest is an assemblage of tree-structured classifiers/predictors, where every tree is expanded according to the bootstrap sample of the training set. Randomness is introduced in the splitting of each node where a feature subset is randomly selected and instead of considering all the splits, only the best split is selected from the subset. Final prediction results are obtained using the voting method. To introduce randomness, the features or parameters of the tree are changed. The Random forest produces outputs comparable with boosting and bagging, without changing the training set progressively. They reduce bias and are relatively robust to outliers and noise (Breiman, 2001). The generalization error of a forest tree classifier is based on the strong points of the individual trees and the correlation among them (Breiman, 1999).

Random Subspace Method: It is equivalent to Bagging. Rather than using the training data set, the Random Subspace Method (RSM) draws the samples from the feature space (Ho, 1998). In the Random Subspace Method, N-features are selected unsystematically in an arbitrary manner from the M-dimensional feature vector. Later, the base learners are trained on the training data set to create classifiers. The training subsets are built by replacing each instance in the training set into an n-dimensional vector. Finally, all the classifiers are integrated using some combination method. RSM is a good choice in the case when the training data set is small in size and even when data has many redundant features (Yang et al., 2004).

Stacking: A new concept was introduced to ensemble literature by stacking. Where it tries to improve the ensemble performance by "correcting" the errors. Stacking addresses the issue of classifier bias w.r.t training data set, with an objective of learning and using these biases for improving classification (Oza and Tumer, 2008). Wolpert, (1992) introduced the stacking framework. It creates a multi classifier by bringing into usage different base classifiers implemented on a lone training data set. In the first level, single classifiers or base learners are

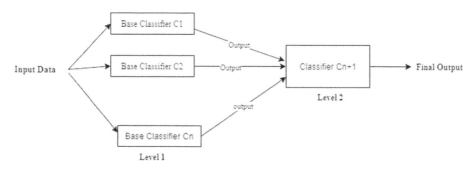

FIGURE 2.4   Generalized stacked approach.

produced by implementing individual learning algorithms on the same training data set. The output from these individual classifiers is then combined in the second level which is also known as meta-learner or meta-classifier (Džeroski and Žemko, 2004; Zhou, 2009). Figure 2.4 gives a diagrammatic representation of a stacked approach, where base classifiers C_1, ..., Cn are trained using input training parameters θ_1 through θ_n to produce hypotheses h_1 through hn. The outcomes of base classifiers and the relative true classes are later used as input for the second level classifier, C_{n+1} (Polikar, 2006).

2.3.1 COMPARISON AMONG ENSEMBLE METHODS

Taking into consideration the potential use of ensemble methods for prediction in different research areas, A large number of multi classifier methods are being developed. Researchers have compared various ensemble methods like Bagging, Boosting, Adaboost, Stacking, etc. Other than comparing the ensemble methods various basic combination methods like Voting, Stacking, Distribution summation, etc. are also compared which leads to the difference among the MCS. Theoretical analysis is performed for evaluating improvement in classification w.r.t accuracy and performance (Tumer et al., 1996). The studies have shown that there exist various factors differentiating one ensemble from another. Out of these the prime factors are summarized below (Rokach, 2005):

1. Combining method: It is one of the fundamental factors taken into consideration while designing an ensemble system. It is related to the strategy applied to combine the classifiers generated by learning algorithms. The basic combining approach integrates the output from the individual classifiers. Combining methods like Uniform voting and Majority voting, Stacking has been compared in different studies by researchers to see their effect on MCS.
2. Inter-classifiers relationship: It deals with how the classifiers interact with each other and their effect on one and another. It is associated with the type of topology used to create an ensemble system i.e. how one classifier is linked with another. Based on the inter-classifier relationship the MCS are divided into two categories: concurrent and sequential.
3. Ensemble size: ensemble size is defined as the number of classifiers combined to build an ensemble. There exists no generalization related to the size of the classifier. Different studies are performed to find an optimal size of the ensemble. In these studies, the size of the classifier is found by varying the number of classifiers from 10 to hundreds and even thousands to see the effect on the performance and accuracy. There is no fixed criteria or guideline for selecting the number of classifier models. The number of base classifiers selected to create an MCS, differentiate one MCS from another.
4. Diversity generator: Diversity is one of the crucial factors to make the multi classifier efficient. Diversity can be generated using different methods like varying input data, producing a variation in learner design, and so on. Diversity helps in creating complementary classifiers thus differentiating one ensemble system from another.

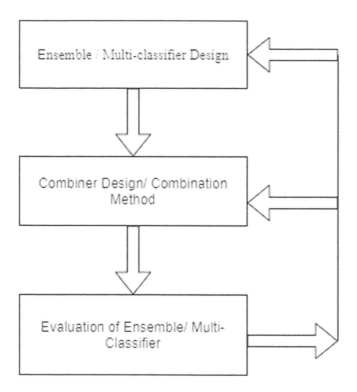

FIGURE 2.5   Generic design framework of multi classifier system.

2.4   DESIGN OF MULTI CLASSIFIER SYSTEM

In simple terms, the framework or the design of the MCS system is to integrate individual classifiers into one composite classifier using a suitable combination approach. The framework involves the basic stages or steps for building an MCS: the construction of base classifiers using a learning algorithm, combining individual base classifiers using a combination method (Bunke et al., 2002), and finally, the performance evaluation (Yang et al., 2004), as shown in Figure 2.5. The individual classifiers selected can be homogeneous or heterogeneous. A perfect ensemble can be devised from diverse and accurate classifiers using some suitable combination method.

The building blocks of an ensemble are explained below (Rokach, 2010):

1. Training set—It is a data set with known labels. The instances present in the data set are described by attributes denoted by A. The input attribute set consists of n attributes and is denoted as $A = \{a_1, ..., a_i, ..., a_n\}$ and y represents the class label or the target attribute.
2. Base Inducer—It is a learning algorithm implemented on a training data set and builds a classifier. The classifier presents a generic relationship among the input variables and the class label.

3. Diversity Generator—Diversity is very crucial for the success of MCS. It is generated between the classifiers with the help of a diversity generator.
4. Combiner—The combiner is the selected combination approach like Voting, responsible for integrating the classifiers to the generated final composite classification model.

The framework of the ensemble is for constructing a set of diverse and accurate classifiers. It aims to produce a set of collaboratively complementing, and generic classifiers that can be integrated to attain optimum accuracy as compared to individual classifiers (Yang et al., 2004).

2.5 COMBINATION TECHNIQUES

One of the important and fundamental steps in MCS is the approach applied to integrate the independent classifiers to achieve near-to-optimal prediction accuracy. Using various learning algorithms, a pool of classifiers C_1, C_2, ..., C_n, is created where, every individual classifier tries to estimate a function $f(x)$. An ensemble tries to integrate the results obtained in order to generate a better estimation of $f(x)$, i.e. it tries to integrate the outcomes $fC_1(x)$, $fC_2(x)$, ..., $fC_n(x)$ with the help of combinational function $fC(x)$ (Ranawanaet al., 2006). The approach followed at this stage depends on the classifier type used as a member for building ensembles. The classifiers used can be either homogeneous or heterogeneous based on the category they belong to is the same or different. Further, the type of combination approach may also depend on the type of outcomes i.e., whether the output is a label or continuous value (Valentini & Masulli, 2002). These combination techniques or combiners approaches can be used straight away once the training is done or additional training steps may be required in case of complex combinations like Stack generation (Polikar, 2012). These combination approaches are usually grouped into (i) non-trainable vs. trainable combiners, or (ii) based on output i.e., continuous outputs vs. class labels (Polikar, 2006).

In a trainable combiner approach, the weights (parameters) of the combiner are found with help of a separate training algorithm. The combination weights produced by trainable rules are generally instance-specific and are also known as Dynamic Combination Rules. Contrary, there is no separate training required in case of non-trainable rules like in weighted majority voting. Discussed below, are a few combination methods.

Uniform/Majority Voting: Voting is a prominently used combination method in MCS. This combiner approach is centered on voting, a concept used in democratic countries. Every classifier gives output and the final output is based on the number of votes received by a class. In this combination approach, the combiner counts the occurrences of each class and evaluates the vote count obtained by every class (Gargiulo et al., 2013). The classification of unknown instances is performed as per the maximum number of votes obtained by a class. Majority voting is categorized into three different versions, where the class is selected by the ensemble: unanimous voting, where all classifier agrees on a common class; simple majority, predicted by at least one more than half the number of classifiers votes; or plurality voting, where it receives the maximum number of votes, whether the total votes may or may not exceed 50% (Polikar, 2012; Polikar, 2006).

Distribution Summation: Clark and Boswell presented this combining method. The concept behind this is to add conditional probability vectors from individual classifiers (Rokach, 2005). The final class is selected on the basis of the maximum value in the sum vector (Rokach, 2010).

Bayesian Combination: It is based on a posteriori probability, where the outputs of base classifiers represent conditioning terms. This combination method is best suited to reduce the error probability (Gargiulo et al., 2013). But there is a problem associated as the knowledge of all the probabilities for classes available is frequently not known. To vanquish this issue, some decision rules can be used which are directly obtained from Bayesian formalism, the main combination rules are Sum rule, Max rule, Min rule, Product rule, Median rule, etc.

Borda Count: This approach uses ranking where a rank order corresponding to each class is created by the classifier. For creating a rank order, let the number of classes be c, the c-1 votes are given to first ranked, the second receives c-2 votes, similarly, the ith one gets c-i votes and the last ranked gets zero votes. For each class, the votes received by each classifier are summed and the class with maximum votes is the winner (Freund & Schapire 1996; Bellmann et al., 2018).

Behavior Knowledge Space (BKS): BKS obtains the information required to integrate the classifiers from a knowledge space. The knowledge space simultaneously records the decisions of all classifiers. A look-up table created on the basis of training data classification is used in this method. It keeps recording the frequency of labelling combinations generated by the classifiers in the training phase. While assigning a class to unseen data it traces which particular labelling combination occurs most during the training phase. During testing a class is assigned every time on the basis of the combination occurring most (Polikar, 2006).

Stacking: Stacking aims to attain the highest generalization accuracy (Rokach, 2005). It is a process where a classifier is trained to integrate different learners. The individual base learners/classifiers are known as the first-level learners, whereas the second-level learner or meta-learner are known as combiner. The concept behind this is to train the first-level learners using the training data whereas the second-level learners are trained using a new data set. The outputs of the first-level learners are treated as input features to second level learner (Zhou, 2012).

When there are more than one competing approaches to a problem as discussed and shown in Table 2.2, it becomes complex to select the most suitable one and it becomes difficult to find the best combination rule. Wolpert's *"no-free-lunch"* theorem has doubtlessly proven that there exists no single optimal classifier for various classification problems (Wolpert & Macready, 1997). The selection of the most suited algorithm is based on the structure of training data as well as prior knowledge. Studies have been conducted to compare various combination rules and ensemble generation techniques under different scenarios. The finest combination method is based on a particular problem (Polikar, 2006). Combination methods play a crucial role in achieving near-to-optimal prediction accuracy.

TABLE 2.2
Comparison of Combination Methods

Scheme	Architecture Used	Type (Trainable/ Non-trainable)	Adaptive	Level of Information	Remarks
Voting	Parallel	Non-trainable	No	Abstract	All classifiers are assumed to be independent
Sum, Median	Parallel	Non-trainable	No	Confidence	Assumes independent confidence estimator and is robust
Min, Max, Product	Parallel	Non-trainable	No	Confidence	Assumption of feature independence
Generalized ensemble	Parallel	Trainable	No	Confidence	Error correlation is considered
Stacking	Parallel	Trainable	No	Confidence	Training data is utilized properly
Borda Count	Parallel	Trainable	No	Rank	Rank to confidence conversion
Dempster-Shafer	Parallel	Trainable	No	Rank confidence	Non-probabilistic confidences are fused

2.6 THE TOPOLOGIES FOR MULTI CLASSIFIER SYSTEM

The integration of multiple classifiers is an interesting topic that can be looked into from different angles at both theoretical and implementation levels. At a theoretical level, there are two combination scenarios taken into consideration; in the first case, the same representation of the input pattern is used by all classifiers, and in the second case different representation is used by all. In the implementation, the classification of integration methods can be done on the basis of a combination of structures or topologies. The topologies are mainly categorized as Hierarchical, Multiple, Conditional, or Hybrid (Lam, 2000). Few researchers claim that the integration methods applied to combine the base learners is of lesser importance, as compared to the diversity which is a major factor in the success of multi classifier. While other researchers claim that the selection of a suitable fusion approach can improve ensemble performance. The selection of proper topology for constructing a multi classifier is generally based upon the type of problem (Ranawanaet al., 2006). According to Lu (1996), MCS topologies are broadly categorized into three categories, as shown in Figure 2.6: Parallel, Cascading, and Hierarchical.

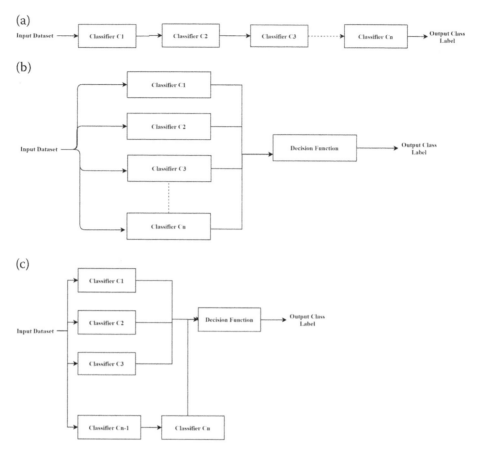

FIGURE 2.6 Visual representation of (a) cascading; (b) parallel; and (c) hierarchical topologies.

Parallel topology: The results of the individual classifier are integrated by a decision process in a single location. The main concern is the design of the decision process which represents the combination strategy applied. A well-structured and well-designed decision process can lead to attaining peak performance. Some of the prominently used combination approaches are Voting (majority), Stacking, etc. The parallel topology can be implemented in parallel processors to achieve real-time performance (Ranawana & Palade, 2006; Lu, 1996).

Cascading topology: The basic idea used in this topology is that the results produced by the base classifier at a lower level are used as input into the classifiers at the next higher level in succession until the final classifier is reached. The biggest issue related to this approach is that the errors made by classifiers at a lower level are not recovered by the classifiers at a higher level. Hence, the total error is the unrecovered error accumulated at each level (Lu, 1996).

Hierarchical topology: It combines both cascading and parallel approaches to achieve optimal performance. This type of approach complements the approaches used within and nullifies the disadvantages. Further, they can nullify the effect of classifiers performing poorly (Ranawana & Palade, 2006).

L. Lam (2000) gave a wider categorization of multi classifier topologies. He categorized the topologies as Conditional, Hierarchical, Hybrid, and Multiple-parallel topologies.

Conditional Topology: In this configuration, first a primary classifier is selected to perform the classification task. If the primary classifier fails to perform the classification correctly or classification is performed with low accuracy or confidence, a second classifier is implemented as presented in Figure 2.7. In most cases, the primary classifier is the first selected classifier, whereas the choice of the second classifier can be based on output or values obtained from the first classifier or it can be a static decision. The process continues till classifiers are available or instances are classified correctly. The queue is organized in a manner that computationally complex classifiers are at the end of the queue. The topology has the advantage of computational effectiveness when the foremost classifier is efficient and accurate. The second classifier, which is perhaps more detailed and time-consuming, is invoked only for more hard instances (Gargiulo et al., 2013; Lam, 2000). A complexity associated with such implementation is in the choice of process to judge the success or failure of a classifier. The complexity of the problem increases with an

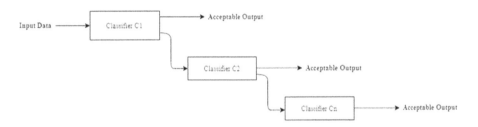

FIGURE 2.7   Visual representation of conditional topology.

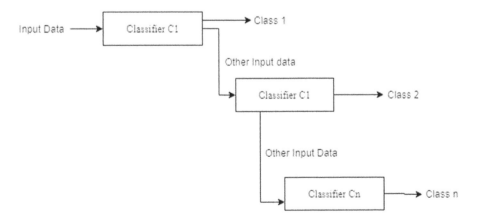

FIGURE 2.8   Visual representation of hierarchical (serial) topology.

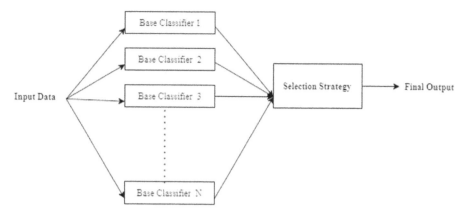

FIGURE 2.9   Visual representation of hybrid topology.

increase in the number of classifiers used in MCS (Ranawana & Palade, 2006; Gargiulo et al., 2013).

Hierarchical (Serial) Topology: In Serial/Hierarchical topology, the classifiers are implemented in a sequence as shown in Figure 2.8. Each classifier implemented reduces the possible set of classes to which the data set belongs. This results in the individual expert or classifiers to be more focused on decision making (Lam, 2000). The usual way to design a queue of classifiers is to place the classifier with respect to decreasing error values, that is, the classifier with large error will appear first in the queue, whereas the classifier with low error will appear later. This process progressively reduces a complicated problem into a simple problem (Gargiulo et al., 2013).

Hybrid topology: Hybrid topology is considered to be a trade-off between serial and parallel topologies (Lam, 2000). Systems based on hybrid topology integrate mechanisms for selecting the most suitable classifier for a given input data set. The selected classifier is later used to perform the final classification as presented in Figure 2.9. As the performance of classifiers varies according to the given data set. Consequently, the selection of a suitable classifier will help to streamline the complete classification process. The biggest disadvantage of this configuration is its complexity irrespective of better performance as compared with other topologies (Gargiulo et al., 2013).

Multiple (Parallel) topology: It is also known as selection-based or fusion-based topologies and is a prominently used topology (Gargiulo et al., 2013). In this approach, different classifiers operate in parallel to generate a classification for a given data set. In next step, the decisions of individual classifiers are integrated to produce the final classification. Advantages of this topology is that it can be implemented under parallel hardware architectures. The major disadvantage is that it sustains additional computational cost, as each classifier is executed before final fusion is performed. Parallel topologies can be implemented using various combiner methods based on the type of information generated by the classifiers (Lam, 2000). Multiple (parallel) topology is shown in Figure 2.10.

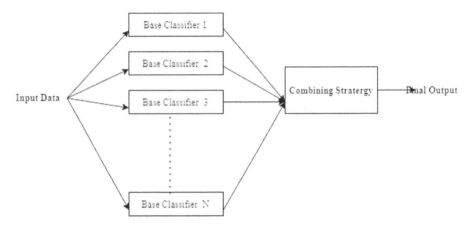

FIGURE 2.10   Visual representation of multiple (parallel) topology.

2.7   ENSEMBLE DIVERSITY

Diversity acts as a directional measure in the process of designing MCS and achieving optimal accuracy. The success of MCS is based on the accuracy and diversity of its individual base learners (Woźniak et al., 2014). Integrating the results of various classifiers is useful if they disagree in terms of errors produced corresponding to the data set available. Hu (2001) showed that classification accuracy is improved due to uncorrelated errors generated in diversified classifiers. The diversity in the classifier results is an essential prerequisite for the success of the ensemble. There are various methods to introduce diversity in the ensemble method as given by different researchers.

Brown et al. (2005) recommended two approaches that can be used to generate diversity; the explicit or implicit approach. The explicit approach focuses on the diversity metric optimization whereas implicit approach consists of techniques of the independent generation of individual classifiers. In explicit approach, classifier training is performed conditional to the previous classifier to exploit the advantages of other classifiers (Woźniak et al., 2014). Bagging is an example of an implicit technique where diversity is generated with help of random sampling of the training data set. In Boosting, which is an example of explicit technique diversity is produced by focusing on the errors made by the current classifier whereas, in the case of random forest, diversity is achieved by selecting different branches of the Decision tree (Opitz & Maclin, 1999; Baba et al., 2015; Brown et al., 2005). Diversity can be generated in different ways. The following are techniques mentioned in various research studies used in generating ensemble diversity:

1. Manipulating the Training Sample: It is the most prominently used method for generating diversity (Zhou, 2012). Ensemble methods generate a pool of classifiers by manipulating the training instances. Every classifier in the pool is trained on a different training subset produced by resampling based on the assumption of creating complementary classifiers. The training samples can

be manipulated either by manipulating the instances, by manipulating the features, or by taking into attention the local behavior or specialization of individual classifiers. Bagging and Boosting are two examples.

2. Manipulating Classifier Outputs: Diversity can be imposed by manipulating results produced by the individual classifiers. So, every classifier is trained to acknowledge only a subset of the classes. The combination method used should be able to retrieve the complete set of class labels. For example, in the case of a multi-class problem, it can be divided into two-class classification problems.

3. Manipulating the learning parameters: Simplest method for achieving diversity is to manipulate the parameter settings of the individual base learner. For example, different split conditions can be implemented to individual Decision trees, or varying initial weights can be allocated to Neural networks individually.

4. Manipulating target attribute: In target attribute manipulation, in place of inducing an individual complex classifier, many classifiers with simple distinct representations of the target attribute are induced. These manipulations are on the basis of aggregation of the original target's values (called Concept Aggregation) or more complex functions (called Function Decomposition) (Rokach, 2010).

5. Search Space Partition: The basic idea behind this approach is that every member in the ensemble explores an instance sub space produced by dividing the original instance space. The subspaces are considered to be independent and the final model is integrated from simpler models. There are two common approaches under search space manipulations: feature subset-based and divide and conquer approaches.

6. Hybridization: It is also known as the Multi Inducer strategy because diversity is induced with help of distinct inducers. Each of these inducers contain implicit or explicit bias (Rokach, 2010).

There is no confirmed formal formulation and measures for ensemble diversity. There are examples where different diversity generation approaches are used together like in Random forest which embraces both input data manipulation and input feature manipulation to generate diversity. All diversity generation methods are based on an effective heuristic approach, to introduce randomness into the training process.

2.8   ADVANTAGES AND DISADVANTAGES OF MCS

As in the case of the individual classifiers, there are advantages and disadvantages related to the MCS also. Based on the various studies and researches being conducted by the researchers and the widespread acceptance of MCS in different research and application fields, the advantages of MCS are summarized:

1. Improvement in predictive accuracy, as the classifiers combined, are diverse and complementing each other which leads to improved accuracy.

2. A stable and more robust model is created using a multi classifier approach as the results from individual classifiers used in MCS are aggregated. The aggregation reduces the effect of noise in the data set as compared to the individual models. This leads to better performance and model stability.

3. Both non-linear and linear data relationships can be captured using ensemble model.

4. MCS can work extremely well under two extreme cases of data availability. One, when the training data set is extremely small and secondly, when a huge data set is available. In the case of less data, MCS uses methods like Bootstrapping, Bagging, or Boosting. In the case of voluminous data, MCS trains the classifiers on partitioned data and then combines the decisions.

5. Many popular machine learning algorithms such as Decision tree are actually heuristic search algorithms. These search approaches sometimes do not assure optimal solutions. Hence, in ensemble methods whereclassifiers are integrated, each classifier can begin its working from different starting points of the search space. It is similar to a multi-start local random search. This type of multi-start search increases the chances of finding an optimal model leading to better performance accuracy.

6. MCS are easily implemented in efficient computing environments like multithread computer architecture, parallel, and distributed computing systems. Especially, when databases are partitioned for privacy and security reasons. Individual classifiers can be implemented on these subsets of databases to obtain a partial solution. These partial solutions computed on each partition can be integrated to obtain a final decision, as the combination of the networked decision available.

7. As per "no free lunch," which states that there is no single classifier that gives the best solution to various problems. Every classifier has its own competence domain; thus, it is difficult to design an individual classifier that outperforms others for each classification task. MCS tries to always select the local optimal model from the pool of trained classifiers.

As compared to advantages, there exist a few disadvantages associated with MCS. These are:

1. It is difficult to understand an ensemble of classifiers. As multiple individual classifiers having their specific learning strategy are combined using various combination approaches, which leads to an increase in complexity and difficulty in understanding.

2. Using ensemble methods sometimes reduces the model interpretability due to an increase in the complexity of the model. It leads to difficulty in drawing any important business insight.

3. Computation and design time is high in MCS as individual classifiers are integrated to get a single composite model. These individual base classifiers are trained individually on the training data set leading to an increase in computation time.

4. Due to increased complexity of ensemble method there is increase in processing time since they are more complex than single classifiers.
5. Thereexists no single model in MCS that applies to different problems. Every application has its domain, associated requirements, and issues, MCS cannot be generalized to a global level that provides an optimal solution to all problems related to various application areas.

2.9   APPLICATIONS OF ENSEMBLE

Ensemble methods have appeared to be effective and successful in many applicative domains related to Data sciences and Society 5.0. It is considered to be the current directions in data sciences, machine learning and other related research areas. With the latest trends and application of Data sciences and predictive analysis is leading to digital transformation of the society as a whole. The application of MCS in this area has increased incredibly in recent years due to their increased computational excellence. The increase in the processing power has allowed usage of high-dimensional and large training data sets as well as training of different classifiers combinations (Woźniak et al., 2014). Oza & Tumer (2008), Gargiulo et al. (2013), Polikar (2006) in their studies presented frequently used ensemble techniques as well as broad categories of application areas. Many works exist that combine different types of classifiers belonging to different groups; heterogeneous classifiers or belonging to the same group, that is, homogeneous classifiers. Ensemble learning is being used in diverse applications to solve real world problems like classification of music, person recognition, credit scoring and banking, recommender system, class-imbalance, security treats and intrusion detection (Cruz et al., 2018), a few of them are discussed briefly where ensemble-based prediction models are being used efficiently in Society 5.0.

1. Remote Sensing: In remote sensing, a large amount of high-dimensional data is collected with various features. The data suffers from issues related to multiple outputs and classes. Because of all these issues, individual classifiers are not able to perform well. The MCS uses the divide-and-conquer approach, where individual classifiers with different properties handle data generated by remote sensing systems and extract beneficial knowledge from the input data which can be combined by these ensemble systems. The aspects covered by MCS related to remote sensing mainly belong to land cover mapping and change detection domains. Land cover mapping is used to identify different materials existing on the surface area under study. Whereas change detection is used to identify the surface areas where the land cover has changed with time. MCS can handle the multi-class issue of remote sensing (Gargiulo et al., 2013; Oza & Tumer, 2008; Woźniak et al., 2014). Land classification is an example of the applicability of classifier ensembles. Ensemble techniques like Random forest are being used in remote sensing mountainous terrain.
2. Computer security: Computer security is a vital service these days. Almost every sector like banking, universities, health care, industries, communication, etc. are collecting and storing large volumes of data which need to be

secured. To get a secure environment is a prime concern, and the ensemble approaches are trying to provide a secure environment and avoid negative impact (Gargiulo et al., 2013; Oza & Tumer, 2008).

Distributed Denial of Service: one of the most threatening attacks isDDoS that an ISP faces. Many services are sensitive to these types of attacks like military applications, e-governance, etc., which can lead to revenue loss as well as performance degradation. A generic framework is required for the auto-detection of DDoS, where MCS is used for attack detection.

Malware: Malicious code, such as viruses, trojan, spyware, etc. ensemble-based predictive approaches can be used here where the classifiers can learn patterns in the known malicious codes extrapolating to yet unseen codes better than an individual algorithm which leads to lesser loss.

There are many other services like Intrusion Detection and Intrusion Prevention etc. where MCS can be applied to get better results.

3. Fraud detection and banking: In the present financial conditions, the analysis and information processing, for problems like credit risks, frauds and other associated issues have become the main concerns of government and society. The development of tools that can predict credit risk and other related problems is the need of society. Ensemble-based systems are very crucial in developing these types tools (Yu et al., 2010).

Fraud detection deals with identification of fraud at earliest after its perpetration. Application of ML in fraud detection provide methods to counteract fraudsters. MCS has been applied successfully in this domain (Kim & Sohn, 2012). It uses real-life transaction data and with help of resampling techniques which are used to deal with class imbalance data.

Similarly, prediction models for credit risk helps to predict whether a loan applicant will repay on a loan or not. The ML application to this problem uses approaches like Stacking, Bagging, Boosting, as well as other conventional classifiers (Woźniak et al., 2014).

4. Recommender system: Recommender systems have become a prime area of research. These systems provide assistance to consumers in selecting the products according to their interests and choices. Many interesting works apply the multi classifier approach to recommender systems showing that the ensemble-based approaches improve the recommendation (Woźniak et al., 2014).

5. Medicine: A vast area for application for different innovative techniques. There are several medical applications where Machine learning is implemented. These examples are related to various types of problems, like analyzing the human genome, X-ray image analysis, etc. The field of medicine has massive amounts of data which is sometimes ambiguous and imprecise in certain situations. These types of issues can create difficulties especially for classification algorithms of machine learning. So, MCS provides a solution to these types of problems. Clinical Decision Support Systems (CDSS) is one of the examples where the ensemble system is used successfully (Gargiulo et al., 2013; Oza et al., 2008).

6. Person recognition: Person recognition is prominently used applications of ensemble learning. It deals with the problem of identity verification of a

person using personal characteristics, it is being typically used in security applications. The typical applications include fingerprint recognition, face recognition, iris recognition, and behavior recognition which includes hand-writing and speech. Combining these types of diverse features into a single recognizer is hard as they differ in scales. Ensembles consist of individual recognizers for each modality which work better as they integrate atdecision level where the scales are similar (Oza & Tumer, 2008).

Other than a few application areas of real world which affect the human society discussed above there are many more areas like Computer-aided medical diagnosis (Belciug, 2016), Gene expression analysis, document analysis (Gargiulo et al., 2013), text categorization, etc. are quite useful and related to the Society 5.0. Ensemble learning can be used to provide better analysis and computational ex-cellence in the field of Data sciences wherever machine learning is implemented.

2.10 EMERGING AREAS

The emergence of new trends in data science and data analytics is leading to di-gitalization of various application areas as well as taking the society through rapid digital transformation. There are new developments in almost every field. All these new developments contribute in innovative applications leading to digital revolution of society. In this direction, in order to build better predictive models in the area of data sciences and machine learning, MCS are developed. The main force behind using the MCS is to curtail the risk of selecting a poor performing individual classifier, or it can be said that to improve upon the performance of an individual classifier with the help of an intelligently combined MCS. The intense research in this field has led to increase in understanding, maturity, and increased confidence level. Research in this field is still growing with new emerging areas of data sci-ences are being benefited by MCS and there is an increase in the trust related to the field in handling different issues. The emerging areas of applications are as follows (Polikar, 2006):

1. Incremental learning: In many applications, the entire data set is not available completely, it is available in small batches with time. In this type of condition, it is required by the classifier to learn new information from new instances without losing or forgetting previous knowledge. The ability of the classifier to learn in these types of situations is known as incremental learning. MCS is successfully being used to address such issues.
2. Data fusion: Data collected from different sources may have different feature sets with heterogeneous composition. MCS fits in this situation, as individual classifiers can be produced by training base algorithms on these hetero-geneous data sets collected from different sources. Finally, they are integrated to get necessary data fusion. Many fusion approaches have been proposed for the same.
3. Feature selection: It is a way to enhanceensemble diversity. In this process the classifier is trained with data that consists of distinct feature subsets.

Ensembles for feature selection are being proposed as a new direction (Bolón-Canedo & Alonso-Betanzos, 2019; Kashef & Nezamabadi-pour, 2019).

4. Data filtering: Usage of ensemble in filtering the training data set for enhancing the accuracy is coming up as a new line of research (Zhou, 2009).

There are many other areas such as Educational Data Mining, Ensemble for Clustering, Ensembles for Feature selection, Dynamic Classifier Selection (Cruz et al., 2018), Dynamic Ensemble Selection (Lustosa Filho et al., 2018), etc. where ensemble methods are being brought into use and/or proposed, and the list is increasing continuously.

2.11 CHALLENGES IN BUILDING MULTI CLASSIFIER SYSTEM

The main challenges in building an MCS lie in the correct selection of the combination method and diversification of the classifier. Further, there are other critical challenges like introducing a suitable base classifier and to provide generalized approach to ensure diversity (Woźniak et al., 2014). Some of these challenges' issues are discussed:

- System topology: It is associated with the interconnection between the individual classifier, i.e. how to connect the individual classifiers to produce near-to-optimal accuracy.
- Ensemble design: It is associated with the creation of an ensemble. The issue is related to strategies that must be selected for choosing the base classifiers.
- Fusion design: It deals with the building of decision combination functions (fuser) that can utilize the strengths of individual classifier and integrate them perfectly.
- Number and size of training subsets: It is an issue concerned with the selection of the optimal number of instances and the size of the training data set.
- Combination strategies: There are indeed many combination strategies for arbitration and combining the classifiers to create a composite classifier. Selecting the most suitable technique based on the problem at hand, size of the data set available, type of data set, etc. to construct an MCS is an important issue. The accuracy, reliability, and performance of the MCS are dependent on the combination strategy selected.
- Size of the ensemble: Several empirical studies have been conducted by various researchers, but no fixed criteria or guidelines exist for finding the optimum number of classifiers required. Therefore, more in-depth analysis is required to explore how and under what conditions the ensemble size will affect the accuracy or what will be the perfect size for getting optimal accuracy from ensembles.

2.12   ISSUES AND CHALLENGES RELATED TO SOCIETY 5.0

There exist various challenges and issues related to the building and implementation of MCS. But other than technical issues discussed, there are many other ethical issues which need to be addressed.

- Privacy: It is a fundamental issue related to data sciences which need to be addressed. In this digital age, preservation of privacy is a challenge. As sensitive and crucial data is analyzed, it should be taken care that it is not being misused by a person or organization which can pose a threat to society.
- Security: It is a big concern of data sciences as data to be analyzed by the prediction model is either collected at a single location before processing or is used as such being available at distributed locations. In both the cases the security checks need to be imposed. Lack of security effects the integrity, availability, and confidentiality, etc. of the data.
- Data quality: It is very important to check upon the quality of data available for the analysis in the field of data sciences. Required are certain checks which can see into the quality of data available for analysis using predictive models build using MCS.
- Cost: With the exponential growth of data in various different forms, which may be at distributed locations, the analysis of data becomes costly. The cost is involved in the collection, pre-processing, and processing of the data before the application of predictive model.

2.13   CONCLUSION AND FUTURE WORK

The digitalization of society has led to exponential growth of data. This data needs to be analyzed for the purpose of extracting beneficial knowledge, facts, pattern, and predicting future trends. Various data analysis tools and techniques are used for this purpose. A multi classifier system is a leading approach for generating prediction models/classifiers with optimal prediction accuracy and computational excellence. Also known under different names like ensemble system, a mixture of experts, classifier ensemble, etc. in literature. It follows the human nature of seeking multiple opinions before reaching a final optimal decision. The basic principle is to obtain classification from many individual classifiers and to integrate the classification outcomes to get better classification results as compared to the individual classifier. Ensemble learning or multi classifier is a strong ML paradigm that shows evident advantages and effectiveness in various application domains of Society 5.0. The paper brings in a summarized way various aspect related to MCS. It explains the philosophy and needs behind the MCS. It explains the design as well as how MCS works and also the reasons behind achieving optimal prediction accuracy. The paper presents in brief existing MCS like Bagging, Boosting, etc., and also gives an insight into various combination approaches, topologies, diversity generation methods which are key factors in achieving optimal accuracy. It also discusses the advantages-disadvantages, applications, emerging areas, and challenges associated with MCS. It brings forth the issues related to Society 5.0. It gives a future direction for further research in various emerging areas. The current study will be applied to build an optimal prediction modelbased on MCS/ensemble approach. The optimal prediction model build will be implemented for predicting the academic performance of students based on the data collected by the academic institutions.

To summarize, ensemble systems give simple, intelligent, powerful, and optimal solutions to a variety of problems related to data sciences and machine learning. Ensemble systems are proving to be successful in number of problem areas that were difficult to deal with using a single classifier-based models. The success of MCS in achieving computational excellence and optimal predication accuracy is largely characterized by the proper selection of classifiers, topology adapted, combinatorial methodology, and diversity generation. Altogether, the paper provides information required by a user to understand MCS, its components, application areas, issues and gives a future direction. The combination of classifiers is a vast area of research with many different perspectives. The MCS is an approach that helps to achieve optimal prediction accuracy, with enhanced performance and robustness.

REFERENCES

Baba, N. M., Makhtar, M., Fadzli, S. A., & Awang, M. K. (2015). Current issues in ensemble methods and its applications. *Journal of Theoretical and Applied Information Technology, 81*(2), 266.

Bauer, E., & Kohavi, R. (1999). An empirical comparison of voting classification algorithms: Bagging, boosting, and variants. *Machine Learning, 36*(1–2), 105–139.

Belciug, S. (2016). Machine learning solutions in computer-aided medical diagnosis. In Andreas Holzinger (Ed.), *Machine learning for health informatics* (pp. 289–302). Cham, Switzerland: Springer.

Bellmann, P., Thiam, P., & Schwenker, F. (2018). Multi-classifier-systems: Architectures, algorithms and applications. In *Computational intelligence for pattern recognition* (pp. 83–113). Cham, Switzerland: Springer.

Bolón-Canedo, V., & Alonso-Betanzos, A. (2019). Ensembles for feature selection: A review and future trends. *Information Fusion, 52*, 1–12.

Breiman, L. (1996). Bagging predictors. *Machine Learning, 24*(2), 123–140.

Breiman, L. (1999). Random forests—random features. Technical Report 567, Statistics Department, University of California, Berkeley.

Breiman, L. (2001). Random forests. *Machine Learning, 45*(1), 5–32.

Brown, G., Wyatt, J., Harris, R., & Yao, X. (2005). Diversity creation methods: A survey and categorisation. *Information Fusion, 6*(1), 5–20.

Bunke, H., & Kandel, A. (2002). *Hybrid methods in pattern recognition (Vol. 47)*. London: World Scientific.

Cruz, R. M., Sabourin, R., & Cavalcanti, G. D. (2018). Dynamic classifier selection: Recent advances and perspectives. *Information Fusion, 41*, 195–216.

Dasarathy, B. V., & Sheela, B. V. (1979). A composite classifier system design: Concepts and methodology. *Proceedings of the IEEE, 67*(5), 708–713.

Dietterich, T. G. (2000, June). *Ensemble methods in machine learning*. International Workshop on Multiple Classifier Systems (pp. 1–15). Springer, Heidelberg, Berlin.

Džeroski, S., & Ženko, B. (2004). Is combining classifiers with stacking better than selecting the best one?. *Machine Learning, 54*(3), 255–273.

Efron, B., & Tibshirani, R. J. (1994). An introduction to the bootstrap. New York: CRC Press.

Freund, Y., & Schapire, R. E. (1996, July). Experiments with a new boosting algorithm. *Machine Learning: Proceedings of the Thirteenth International Conference*, 148–156. DOI: https://dl.acm.org/doi/10.5555/3091696.3091715weblink: https://cseweb.ucsd.edu/~yfreund/papers/boostingexperiments.pdf

Gargiulo, F., Mazzariello, C., & Sansone, C. (2013). Multiple classifier systems: theory, applications and tools. In *Handbook on Neural Information Processing* (pp. 335–378). Heidelberg, Germany: Springer.

Han, J., Pei, J., & Kamber, M. (2011). *Data Mining: Concepts and techniques*. Waltham, MA: Elsevier.

Hansen, L. K., & Salamon, P. (1990). Neural network ensembles. *IEEE Transactions on Pattern Analysis and Machine Intelligence, 12*(10), 993–1001.

Ho, T. K. (1998). The random subspace method for constructing decision forests. *IEEE Transactions on Pattern Analysis and Machine Intelligence, 20*(8), 832–844.

Ho, T. K., Hull, J. J., & Srihari, S. N. (1994). Decision combination in multiple classifier systems. *IEEE Transactions on Pattern Analysis and Machine Intelligence, 16*(1), 66–75.

Hu, X. (2001, November). Using rough sets theory and database operations to construct a good ensemble of classifiers for data mining applications. *Proceedings 2001 IEEE International Conference on Data Mining* (pp. 233–240). IEEE.

Jain, A. K., Duin, R. P. W., & Mao, J. (2000). Statistical pattern recognition: A review. *IEEE Transactions on Pattern Analysis and Machine Intelligence, 22*(1), 4–37.

Kashef, S., & Nezamabadi-pour, H. (2019). MLIFT: Enhancing multi-label classifier with ensemble feature selection. *Journal of AI and Data Mining, 7*(3), 355–365.

Kim, Y., & Sohn, S. Y. (2012). Stock fraud detection using peer group analysis. *Expert Systems with Applications, 39*(10), 8986–8992.

Kuncheva, L. I. (2014). *Combining pattern classifiers: Methods and algorithms*. New York: John Wiley & Sons.

Lam, L. (2000, June). *Classifier combinations: implementations and theoretical issues*. International Workshop on Multiple Classifier Systems (pp. 77–86). Springer, Berlin.

Lu, Y. (1996). Knowledge integration in a multiple classifier system. *Applied Intelligence, 6*(2), 75–86.

Lustosa Filho, J. A. S., Canuto, A. M., & Santiago, R. H. N. (2018). Investigating the impact of selection criteria in dynamic ensemble selection methods. *Expert Systems with Applications, 106*, 141–153.

Opitz, D., & Maclin, R. (1999). Popular ensemble methods: An empirical study. *Journal of Artificial Intelligence Research, 11*, 169–198.

Oza, N. C., & Tumer, K. (2008). Classifier ensembles: Select real-world applications. *Information Fusion, 9*(1), 4–20.

Pandey, M., & Taruna, S. (2014). A comparative study of ensemble methods for students' performance modeling. *International Journal of Computer Applications, 103*(8). DOI: 10.5120/18095-9151

Parvin, H., MirnabiBaboli, M., & Alinejad-Rokny, H. (2015). Proposing a classifier ensemble framework based on classifier selection and decision tree. *Engineering Applications of Artificial Intelligence, 37*, 34–42.

Polikar, R. (2006). Ensemble based systems in decision making. *IEEE Circuits and Systems Magazine, 6*(3), 21–45.

Polikar, R. (2012). Ensemble learning. In *Ensemble machine learning* (pp. 1–34). Boston, MA: Springer.

Pujari, A. K. (2001). *Data mining techniques*. Universities Press.

Quinlan, J. R. (1996, August). Bagging, boosting, and C4. 5. In *Aaai/iaai, 1*, 725–730.

Ranawana, R., & Palade, V. (2006). Multi-classifier systems: Review and a roadmap for developers. *International Journal of Hybrid Intelligent Systems, 3*(1), 35–61.

Rokach, L. (2005). Ensemble methods for classifiers. In *Data mining and knowledge discovery handbook* (pp. 957–980). Boston, MA: Springer.

Rokach, L. (2010). Ensemble-based classifiers. *Artificial Intelligence Review, 33*(1–2), 1–39.

Schapire, R. E. (1990). The strength of weak learnability. *Machine Learning, 5*(2), 197–227.

Stefanowski, J. E. R. Z. Y. (2008). *Multiple classifiers.* PowerPoint slides.

Sun, Z., Song, Q., Zhu, X., Sun, H., Xu, B., & Zhou, Y. (2015). A novel ensemble method for classifying imbalanced data. *Pattern Recognition, 48*(5), 1623–1637.

Tsoumakas, G., Partalas, I., & Vlahavas, I. (2008, July). *A taxonomy and short review of ensemble selection.* Workshop on Supervised and Unsupervised Ensemble Methods and Their Applications, Patras, Greece, (pp. 1–6).

Tumer, K., & Ghosh, J. (1996). Analysis of decision boundaries in linearly combined neural classifiers. *Pattern Recognition, 29*(2), 341–348.

Valentini, G., & Masulli, F. (2002, May). Ensembles of learning machines. Italian Workshop on Neural Nets (pp. 3–20). Springer, Heidelberg, Berlin.

Windeatt, T. (2005). Diversity measures for multiple classifier system analysis and design. *Information Fusion, 6*(1), 21–36.

Wolpert, D. H. (1992). Stacked generalization. *Neural Networks, 5*(2), 241–259.

Wolpert, D. H., & Macready, W. G. (1997). No free lunch theorems for optimization. *IEEE Transactions on Evolutionary Computation, 1*(1), 67–82.

Woods, K., Kegelmeyer, W. P., & Bowyer, K. (1997). Combination of multiple classifiers using local accuracy estimates. *IEEE Transactions on Pattern Analysis and Machine Intelligence, 19*(4), 405–410.

Woźniak, M., Graña, M., & Corchado, E. (2014). A survey of multiple classifier systems as hybrid systems. *Information Fusion, 16*, 3–17.

Xu, L., Krzyzak, A., & Suen, C. Y. (1992). Methods of combining multiple classifiers and their applications to handwriting recognition. *IEEE Transactions on Systems, Man, and Cybernetics, 22*(3), 418–435.

Yang, L. Y., Qin, Z., & Huang, R. (2004, August). Design of a multiple classifier system. In Proceedings of 2004 International Conference on Machine Learning and Cybernetics (IEEE Cat. No. 04EX826) (Vol. 5, pp. 3272–3276). IEEE.

Yu, L., Yue, W., Wang, S., & Lai, K. K. (2010). Support vector machine based multiagent ensemble learning for credit risk evaluation. *Expert Systems with Applications, 37*(2), 1351–1360.

Zhou, Z. H. (2009). Ensemble learning. *Encyclopedia of Biometrics, 1*, 270–273. Springer, Berlin.

Zhou, Z. H. (2012). *Ensemble methods: Foundations and algorithms.* CRC Press.

3 Software Engineering Paradigm for Real-Time Accurate Decision Making for Code Smell Prioritization

Randeep Singh, Amit Bindal, and Ashok Kumar
Department of Computer Science & Engineering Maharishi
Markandeshwar (Deemed to be University), Mullana-Ambala,
India

CONTENTS

DOI: 10.1201/9781003132080-3

3.1 INTRODUCTION

Software maintenance is a necessary and inevitable activity in software engineering. It helps in keeping a software system healthy with the continuous change mandated by the changing technical and business environments. Different kinds of change include bug fixing, capacity enhancement, removal of outdated functions, and performance improvement. Software maintenance is an ongoing activity and it constitutes about 75% of the total involved cost especially for large and complex software systems (April & Abran, 2012; Seacord, Plakosh, & Lewis, 2003). Different developer's activities like reading, navigating, searching, and editing that is involved during the process of software maintenance possess a direct link along with the structural attributes of the underlying source code of the system (Soh, Yamashita, Khomh, & Guéhéneuc, 2016). Moreover, usually, strict deadlines along with various external constraints usually guide these activities. The external constraints includes: bad design decisions, inexperienced developers' team, limited/ no business knowledge, high workload and pressure on developers, little importance given to the quality aspect of the software system (Tufano et al., 2015; Vale, Souza, & Sant'Anna, 2014). Therefore, such activities/constraints result in the introduction of technical debts (Shull, Falessi, Seaman, Diep, & Layman, 2013) that ultimately negatively affects software maintenance. Data Science is a modern interdisciplinary computer science field that aims at extracting useful knowledge out of typically large-sized data (Hayashi, 1998). Software system plays important role in this data analysis phase of data science that handles complex mathematical and statistical data analysis, information visualization, etc. (Akerkar & Sajja, 2016). It is the opinion of the authors that such software systems are complex in design and maintaining them is a challenging task due to the frequently changing requirements of users. Moreover, code smell prioritization can further reduce involved maintenance efforts.

According to Fowler et al., presence of bad smells (generally termed code smells) are the main factors behind technical debts of the software system (Fowler et al., 1999). These different underlying structural shortcomings in the source code are termed as code smells and they affect a software system negatively by enhancing its involved maintenance efforts. The part of the software system that is affected with code smells is generally recommended to tackle in order to improve the quality of these code elements and hence, reduce the future maintenance efforts. Different authors conclude that different code smells directly affects its maintainability (Yamashita & Moonen, 2012; Hermans & Aivaloglou, 2016). Moreover, Hermans et al. based upon carried out controlled experimentation conclude that novice programmers can significantly improve the quality of the underlying source code by removing different code smell (Hermans & Aivaloglou, 2016). Further, the list of code smells belonging to a system can be numerous due to the large volume of the source code especially for large and complex software systems (Fontana & Zanoni, 2017; Marinescu, 2012). Tackling such numerous code smells imposes feasibility and other business-specific constraints that hinder the developers from solving all code smells. Therefore, the maintenance team is always interested in screening the list of code smells and focused on prioritizing them in order to filter a set of code smells that are important and must be addressed at the given instance of time. Moreover, an empirical investigation needs to be carried out in order to determine the effect of tackling the prioritized list as compared to mitigating all the available code smells in the software system. This screening and filtering of code smells is termed prioritization of code smells and it helps in optimizing maintenance efforts devoted by the maintenance team.

Even though several approaches exist in the literature for identifying a list of possible code smells belonging to a system (Arcoverde, Guimaraes, Macía, Garcia, & Cai, 2013; Azeem, Palomba, Shi, & Wang, 2019; Fernandes, Oliveira, Vale, Paiva, & Figueiredo, 2016; Oliveira et al., 2017; Fontana & Zanoni, 2017; Marinescu, 2012; Vidal, Marcos, & Díaz-Pace, 2016; Palomba, Panichella, Zaidman, Oliveto, & De Lucia, 2017; Fontana, Ferme, Zanoni, & Roveda, 2015). However, there exist a large number of areas in which such approaches can be improved and the code smell detection approach can be made better. The existing approaches do not fully reflect the actual need of the developers' team. There is no clear evidence regarding the underlying factor based on which the filtering and screening of code smell are done. Moreover, most such approaches do not gather the practitioner's view about the particular code smell, instead, they silently ignore this important parameter. Nevertheless, most of the existing approaches do not even prioritize the obtained code smells and the developers are left to performing this tedious task. Only a few prioritization and code smell screening approaches are proposed by (Fontana & Zanoni, 2017; Arcoverde et al., 2013; Vidal et al., 2016). This activity is time-consuming and does not always fruitful because not all developers are fully aware of the context and business domain of the software. Further, there exist various standard automatic/semi-automatic tools (Moha, Gueheneuc, Duchien, & Le Meur, 2009; Lanza & Marinescu, 2007), however, their use is limited by the fact that they usually return numerous code smells, and handling all of them is a challenge for the maintenance team. Moreover, these different tools use different screening concepts and they also do not directly correlate and reflect the developer's context about different code smells. Most of the existing approaches rely on static analysis of the source code and do not include dynamic and/or semantic aspects of the software. Therefore, this chapter carries out the task of efficient identification of code smells in the software and prioritizes them in order to reduce the maintenance team's efforts, time, and invested maintenance cost and efforts devoted to handling different code smell using various software artifacts.

The rest of this book chapter is organized into the following different sections. A detailed literature survey is presented in Section 3.2. Section 3.3 provides details about the proposed code smell detection and prioritization approach. Section 3.4 gives details about the experimental setup and discusses various obtained results. Section 3.5 gives concluding remarks and possible future works.

3.2 LITERATURE SURVEY

Code smell detection and refactoring had remained a topic of interest for the IT industry and researchers had presented several techniques. This section presented a detailed discussion regarding various recent key research linked with code smell detection, refactoring, and/or prioritization. This discussion would help any researcher to set future work directions while performing prioritization of different code smells.

Several approaches had been proposed since the 80s by various researchers for handling different dynamics/aspects of code smell detection and refactoring. Broadly, such existing approaches can be categorized into different approaches such as graph-based, metaheuristic based, metric-based, and/or metric-based. Authors in (Counsell et al., 2010; Lacerda, Petrillo, Pimenta, & Guéhéneuc, 2020; Singh & Kumar, 2018) performed detailed surveys about various existing code smells and also mentioned different refactoring opportunities to handle these code smells. Khomh et al. carried out a study to find a relationship between the existence of code smells and underlying change proneness of classes in software systems (Khomh, Di Penta, & Gueheneuc, 2009). They determined that smelly classes were mostly involved with the change related to a software system. The dependency relations existing among different nodes of a graph were studied by Abdellatief et al. and a set of dependency metrics were presented in (Abdellatief & Md Sultan, 2011). These metrics are helpful in identifying various code smells. The authors in (Al Dallal, 2012) utilized different object-oriented quality metrics and proposed various models to determine opportunities to extract subclass refactoring. Gamma et al. carried out a study to determine the role and feasibility of different design patterns in identifying different code smells. Several design patterns were proposed that were commonly adopted by different developers in order to improve the quality of the software system. The prioritization of code smells is considered important in (Marinescu, 2012). The authors coined the term *severity* and computed it based on various static code metrics. Based on this severity score, they ranked different code smells belonging to a system. A systematic literature survey is carried out in (Singh & Kaur, 2018). The authors in this article studied 238 research articles and revealed information about code smell identification along with anti-pattern approaches. A similar recent literature survey is also carried out in (Agnihotri & Chug, 2020) aiming at identifying actively used code smells and refactoring approaches in the past decade.

The authors in (Sae-Lim, Hayashi, & Saeki, 2018) find 5 Ws that are related to code smells. The first W (which) focused on that only some of the smell are studied and only a few of the smell is interrelated independently. The second W (when) is related to how the research is done over time, the third W (what) is related to the findings, aims, and experiments. In the fourth W (who), the authors involved and continue in finding the code smells, and finally, the fifth W (where) is related to the circulation of the research on duplicate code while others had disseminated their knowledge on other types of code smells. Fontana et al. utilized the concept of intensity index computed from the distribution of code metrics of software (Fontana et al., 2015). The intensity value assigned to a software element ranged from 1 (Very Low) to 10 (Very High). Bibiano et al. carried out an extensive study on 57 software systems and analyzed 19 different types of code smells along with 13 different transformation types present and belonging to the software (Bibiano et al., 2019). Freire et al. investigated the effect of carrying out refactoring at the UML class diagram level and its influence on involved technical debt (Freire et al., 2020). Ibrahim et al. carried out an investigation to determine the influence of traces of code smell and refactoring on redundancy in the test case generation process (Ibrahim et al., 2020). They carried out their study of android applications and concluded that code smell detection and refactoring can significantly reduce the redundancy in generated test cases.

Yamashita et al. further extended the approach of Fontana et al. and proposed a Context Relevance Index (CRI) metric by considering a factor that represents the developer's maintenance effort (Yamashita & Moonen, 2013). The idea of prioritization of code smells at the component level was carried out by Vidal et al. in (Vidal et al., 2016). Besides these, many automated/semi-automatic tools were proposed in the literature by different researchers (Marinescu et al., 2005; Paiva, Damasceno, Figueiredo, & Sant'Anna, 2015; Tsantalis et al., 2008). (Liu & Zhang, 2018) proposed a detection tool known as DT which is able to detect the duplicated code smell by using a dynamic programming algorithm. Dynamic programming uses the resemblance among relative lines to find traces ofduplicated code smell in the source code. Furthermore, during experimentation, the detection tool DT is further compared with four well-known detection tools CheckStyle, JDeodorant, iPlasma, and PMD, and the experimentation reveals that the detection precision is more than the above tools. Authors in (Palomba, et al., 2013; Palomba et al., 2014) proposed an approach by utilizing change history information of a software and termed it as HIST (Historical Information and smell detection). They further applied HIST on eight software projects with source code in java language and finally compared the result with existing smell detectors. The result obtained from the HIST ranges from 61% to 80% for the precision value and for recall the values go up to 100%. (Li, 2019) used DLFinder, the automated static analysis tool, which is used to detect the problem of duplicated logging code smells. The authors performed an extensive study by taking the result of over 3k logging statements and feedback from the developers and finds that DLFinder is able to find 85% of the occurrence. The authors in (Guggulothu & Moiz, 2019, February) proposed a dependency finder tool to analyze and detect the relationships between four code smells. This prioritizing order is recommended to developers by the author it is based on the quality of the design. The evaluation is done on java source code and the result shows improvement in the design quality. (Kamaraj & Ramani, 2019) combines different methods for the detection of code smells by finding a number of parameters in the source code. This model is also used to generate test cases on the web as well as GUI applications. Martini et al. performed analysis of various code smells on four industrial projects and evaluated the technical difficulty that arises due to various code smells, difficulty in refactoring, and difficulty faced while applying code smell detection tool. They prioritize refactoring based on the negative impact (Martini, Fontana, Biaggi, & Roveda, 2018). Yoshida et al. carried out a study to check whether there exists any relation between code smells and refactoring patterns present in a software system (Yoshida, Saika, Choi, Ouni, & Inoue, 2016). The impact of refactoring on fixing bugs present in a software system is carried out by (Ma, Chen, Zhou, & Xu, 2016). The authors in (Techapalokul & Tilevich, 2019) carried out a study on block-based programming methodology and proposed four types of Scratch refactoring approaches that ultimately help in improving the underlying code quality. These approaches are Extract Constant, Extract Custom Block, Reduce Variable Scope, and Extract Parent Sprite.

The authors in (Vidal, Oizumi, Garcia, Pace, & Marcos, 2019) rank different code smells based on five criteria and an evaluation on 23 versions of 4 software systems using the tool *JSpIRIT*. Then a comparison between prioritization is done using two factors, namely based on developers and another based on chosen criteria. Obtained results were compared and prioritized at 80 code locations. Guggulothu and Moiz (2019, January) proposes an approach to order code smell to save developers efforts which are based on code smells relevant metrics with the selection technique and finding internal relations between these code smells in java source code. The authors in (Oliveira et al., 2019) done two studies to proposed heuristics first based on criteria which are chosen by developers to detect and prioritize bad smells and second by the evolution of these heuristics and find results helped the developers to prioritize the design level code smells. Garg et al. analyzed the design metrics for clone detection by finding the structural properties related within the classes which are useful for the quality and maintenance of the software (Garg & Singh, 2020). The authors in (Guimaraes, Vidal, Garcia, Diaz Pace, & Marcos, 2018) proposed three prioritization criteria based on different elements with code smells to improve the accuracy, automation, and effectiveness of the software and tested these criteria on the JSpIRIT tool by evaluating two applications. (Husien, Harun, & Lichter, 2017) proposed a harmfulness model for the detection of bad smells and prioritization of refactoring to remove code smells. The models combine, change history and severity for the detection of GOD class smell by using the open-source project JHotDraw. The authors in (Katbi, Hammad, & Elmedany, 2020) proposed a framework for visualization and prioritization of three types of code smells by testing different versions of the open-source system. Different key research findings available in the literature are summarized in Table 3.1.

3.3 PROPOSED METHODOLOGY

Code smells negatively affect quality and maintainability of a system. Their detection and mitigation are necessary in order to systematically maintain software over a period. These manageable software systems are vital for multidisciplinary branches of computer science such as data science, which requires the intelligent computation of underlying data. Moreover, these code smells can be prioritized in order to further reduce the devoted maintenance efforts, time, and cost. Figure 3.1 of this chapter shows the proposed methodology for efficiently detecting various code smells that are direct factors behind the quality as well as maintenance efforts and prioritize these various key code smells that ultimately helps in improving quality in an optimum manner. The proposed quality improvement and maintenance reduction approach consists of three main steps, namely code smell detection, code smells prioritization, and improving quality by performing refactoring operations.

TABLE 3.1

Summarization of Different Key Approaches Proposed in Literature

Reference	Category	Research Contribution	Research Gap
(Singh & Kaur, 2018)	Code Smells, Refactoring	Carries out a systematic survey on 238 research articles. Authors focus on revealing in-depth knowledge about code smell identification approaches and the presence of anti-patterns.	-----
(Agnihotri & Chug, 2020)	Code Smells, Refactoring, Software Metrics	Carried out a systematic literature survey aiming at determining various key code smells identified by different researchers, corresponding refactoring approach used, and their relationships with different software metrics. Feature Envy, Long Methods, and Data Class code smells are found to be actively targeted by different researchers in past decades. Extract Class refactoring is majorly applied to mitigate code smells followed by the move and/or extract method refactoring approaches.	-----
(Fontana et al., 2015)	Code Smells Prioritization	Considers prioritizing code smells as important to determine key potential maintenance problems. Proposed intensity index based metric approach to prioritize code smells.	Utilize a very basic and trivial set of software metrics.
(Vidal et al., 2016)	Code Smells Prioritization	Proposed a semi-automated approach to rank different code smells by utilizing three factors, namely past modification history, code relevance, and modifiability scenario of the system.	Use of trivial software metrics combination, ignorance of expert knowledge for code smell prioritization
(Guggulothu & Moiz, 2019, January)	Code Smells Prioritization	Proposed a code smell prioritization approach based on the design quality	Ignores code smell relevance, and expert opinion.

(Continued)

TABLE 3.1 (Continued)
Summarization of Different Key Approaches Proposed in Literature

Reference	Category	Research Contribution	Research Gap
		metrics for the software system.	
(Oliveira et al., 2019)	Code Smells Prioritization	Proposed prioritization heuristics for shortlisting design relevant smells based on developers' expertise.	Ignores the importance of software metrics and relies only on expert opinion.
(Guimaraes et al., 2018)	Code Smells Prioritization	Proposed code smells prioritization approach based on the available architectural blueprints of a software system.	Rely on the expertise of involved participants for identifying underlying architectural blueprints.

These different phases of the proposed approach in Figure 3.1 are further discussed in detail in the subsequent subsections.

3.3.1 CODE SMELLS DETECTION

This step of the approach shown in Figure 3.1 in this chapter aims at efficiently identifying various code smells belonging to an object-oriented software system. These code smells are the key factors degrading underlying quality of the software system as well as involved maintenance efforts of the maintenance team. The proposed approach is robust and is not dependent only on the underlying source code. The proposed approach makes

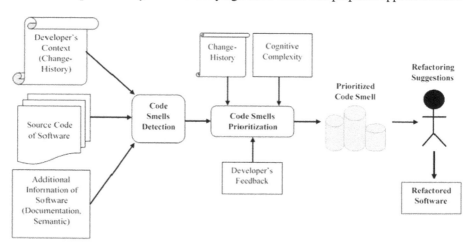

FIGURE 3.1 Proposed methodology for efficient code smell detection and maintenance reduction by prioritizing code smells.

use of various other dynamics of the software system in order to further enhance the code smells detection accuracy. These other information sources include documentation, semantic information of the source code, and the developer's perception about code smell reflected in the form of available change-history of system. Different code smells targeted in this chapter are Long Parameter List, God Class, Long Method, Feature Envy, Duplicated Code, Shotgun Surgery, and Refused Bequest. These considered seven code smells are mostly targeted by different researchers in the recent past (Arcoverde et al., 2013; Azeem et al., 2019; Fernandes et al., 2016; Oliveira et al., 2017; Fontana & Zanoni, 2017; Marinescu, 2012; Vidal et al., 2016; Palomba et al., 2017; Fontana et al., 2015) and these code smells also lie in the top-ranked/prioritized list of code smells (Vidal et al., 2016; Sae-Lim et al., 2018; Pecorelli et al., 2020). Different approaches proposed by authors in this chapter for these different code smells are discussed.

3.1.1.1   Feature Envy Code Smell

The key characteristics of this code smell is that different class members are wrong placed in a software system and are identified by imbalanced coupling and cohesion relations among different classes. Feature envy arises due to the lack of responsibilities of different classes and it sometimes results in a long sequence of unanticipated changes in many classes. Inappropriate coupling arising among different class pairs is the main criteria used to detect feature envy code smell. The main idea is to keep the class interdependence at a minimum wherever possible. The authors in this book chapter proposes the concept of usage patterns and applied it at the method granularity level. Usage patterns represent the set of different properties (member variables of different classes in a system) used at different levels (methods/class/ statements) at any instant in time. The usage patterns concept is also used for software remodularization by Rathee et al. (Rathee & Chhabra, 2017). The proposed feature envy code smell detection approach consists of four main steps. In the first step, the system is viewed as a collection of methods (rather than classes) and usage patterns are identified at the method level by parsing the static source-code of the software along with identifying the usage of different variables (present in different classes) available in the system. The set of member variables names that are used by a method will become the output of this step and these names may belong to one or several classes belonging to the original architecture of the software system. In the second step, the interdependence among different methods used in the first step and various classes belonging to the software is computed by modeling software as a dependency graph. In this graph, different methods and classes become nodes and edges represent belongingness (coupling) among different pairs of methods and classes present in a software system. The belongingness is computed based onidentified usage patterns. The mathematical expression used for computing the belongingness score for a method M_i with a class C_j, as shown below:

$$Belongingness\,(M_i\,C_j) = \frac{Ref\,(UP_j)}{Ref\,(UP_j) + \Sigma_{k=1;k\neq j}^{t}\,Ref\,(UP_k)} \qquad (3.1)$$

Here, $Ref\,(UP_j)$ denotes different member variables belonging to the jth class present in the usage pattern set of method M_i. Similarly, $Ref\,(UP_k)$ is the total number of member variables present in the usage pattern set of method M_i of the rest of the other classes present in the software system. Moreover, t is the total number of classes or software elements belonging to the software system.

In the third step, the modeled system is divided into different groups based on the criteria of computed belongingness score among different pairs of methods and classes. Hierarchical Agglomerative Clustering algorithm with complete linkage criteria is used to perform clustering of different methods. Finally, in the fourth step, the obtained clusters (set of methods) are analyzed to determine opportunities for moee method refactoring (traces of feature envy code smell).

3.1.1.2   GOD Class Code Smell

GOD classes are generally a set of classes that tend to centralize the functionality of the software and generally very big in size. Authors in this chapter utilize multiple criteria to detect traces of GOD class code smell. The proposed GOD class detection approach in this paper makes use of three criteria, namely 1) usage pattern based coupling of a class, 2) size of the class measured using well-known KLOC metric, and 3) the extent of foreign member variables used by a class. Usage pattern-based coupling value for a class is measured using a metric that is mathematical expressed as follows:

$$Coupling\,(i) = \frac{Ref\_Inter\,(i)}{Ref\_Total\,(i)} \qquad (3.2)$$

Here, $Ref\_Inter\,(i)$ denotes all member variables' name of classes other than i^{th} class that belongs to the usage pattern set of ith class. Similarly, $Ref\_Total\,(i)$ is the total number of references made by the ith class in the usage pattern set i.e. it represents the cardinality of the usage pattern set of ith class. The extent of foreign member variables used by i^{th} class in a software system is measured using the following mathematical formula:

$$Foreign\_Extent\,(i) = \sum_{j=1;j\neq i}^{N} \frac{Usage_i(j)}{N}; \; Where \qquad (3.3)$$

$$Usage_i(j) = \frac{Number\ of\ member\ variables\ from\ Class\ j\ used\ by\ Class\ i}{Total\ number\ of\ member\ variables\ belonging\ to\ Class\ j}$$
$$(3.4)$$

Here, the proposed $Foreign\_Extent\,(i)$ metric is based on the measurement of $Usage_i(j)$ factor for the rest of the other classes present in the software system. This usage factor is basically defined as the ratio of member variables of jth class used by the considered i^{th} class to the total number of member variables present in the jth class. Finally, the proposed god class detection approach stresses utilizing expert opinion for determining different threshold values for considered three GOD class identification criteria.

3.1.1.3 Long Method Code Smell

The long method code smell detection approach in this book chapter is also based on multiple factors in order to enhance the detection accuracy as compared to other existing approaches in the literature. The four criteria used include 1) cyclomatic complexity of the method, 2) functional relatedness among different statements of the method, 3) semantic relatedness of the method, and 4) size of the method. Here, the LOC metric is used to measure the size of a method. Functional relatedness among different statements of a method is used to measure diversity among statements. In this chapter, measuring functional relatedness is based on the concept of usage patterns. However, now statements are considered as new granularity level for computing usage pattern and while computing usage pattern at the statement level, the usage pattern set contains member variables name that is directly or indirectly (through method calls) used by the considered statement. Once the usage patterns are identified at the statement level, the method is modeled asa graph where nodes in the graph will represent different statements of the method, and edges present in the graph represents the usage pattern-based coupling among different statements. Based on this graph representation and the coupling relations among different statements, the functional relatedness is measured using the following metric:

$$FunctionalRelatedness = Cohesion_{FR} = \left(1 - \frac{T_{IPC}}{N}\right) \qquad (3.5)$$

Here, T_{IPC} is the total number of independent paths/cycles existing in the graph and N is the total number of statements in the graph.

The McCabe Cyclomatic Complexity (CC) (McCabe, 1976) metric is further used to measure the understandability and maintainability complexity of the method in this chapter. The semantic relatedness of the method is further computed by parsing different statements of the method and extracting independent tokens. These extracted tokens are further analyzed using the well-known multilingual database called WordNet (Fellbaum, 1998). WordNet is a state of art corpus that can easily handle different semantic relations existing among different words (tokens) in the form of noun, adjectives, verbs, adverbs, etc. The actual semantic relatedness for a method M_k is computed using the following equation proposed in (Zhang, Sun, & Zhang, 2017):

$$Semantic\ \ Relatedness\,(M_k) = SR\,(M_k)$$
$$= Average\left(\Sigma_{i=1}^t \Sigma_{j=1}^t \frac{2 * depth\,(C_1, C_2)}{len\,(C_1, C_2) + 2 * depth\,(C_1, C_2)}\right) \qquad (3.6)$$

Here, $C_1 \& C_2$ are the different tokens belonging to the methods and $depth\,(C_1, C_2)$ represents the depth among different concepts in the synset tree representation of $C_1 \& C_2$. Similarly, $len\,(C_1, C_2)$ is the shortest path distance between $C_1 \& C_2$.

Finally, the proposed long method code smell detection approach stresses utilizing expert opinion for determining different threshold values for considered four long method code smell identification criteria.

TABLE 3.2
Complexity Weight Assignment Based on Parameter's Data Type Categorization

	Data Type Categories			
	Simple	Medium	Complex	Highly Complex
Assigned Weight	0.10	0.40	0.70	1.00

3.1.1.4 Long Parameter List Code Smell

In literature, Long Parameter List (LPL) code smell is most commonly identified by parameter counting approach. However, the authors in this chapter stress that the complexity of those parameters is equally responsible for enhancing the understandability and complexity of the method. Therefore, the proposed long parameter list code smell detection approach in this chapter is based on two key factors, viz 1) parameters count, and 2) the complexity of the considered parameter list of a method. The authors propose a complexity metric for computing the complexity of the considered parameter list of a method. This metric is mathematically represented as follows:

$$Complexity\,(M_i) = \frac{\sum_{k=1}^{N} DataTypeComplexity\,(k)}{N} \tag{3.7}$$

Here, $DataTypeComplexity\,(k)$ is the complexity weight assigned to the kth parameter of a method based on the underlying complexity of its corresponding data type. Table 3.2 shows the categorization of different data types assigned to different parameters of a method. This categorization is considered to classify understandability and maintainability efforts devoted by developers. Moreover, it shows the considered weight value. This value is randomly selected by authors in order to represent the underlying understandability and maintainability efforts required to modify the method.

Finally, the proposed LPL detection approach stresses utilizing expert opinion for determining different threshold values for considered two LPL identification criteria in this chapter.

3.1.1.5 Refused Bequest Code Smell

Inheritance design flaws in the software system are designated as refused bequest code smells. In this chapter, the authors consider two critical key criteria, namely 1) inheritance adoption, and 2) domain closeness among different classes involved in the inheritance. The first criteria, namely inheritance adoption, is used to measure the degree to which inherited properties are used in the derived class. The authors propose a metric named PIUR (Penalized Inheritance Usage Ratio) in this chapter for this purpose. This metric is defined as the ratio of the total number of properties inherited by the derived class to the total properties provided by the parent class to the outside world and is inversely weighted by the Depth of Inheritance Tree (DIT) at a particular level of inheritance tree conceptualized in the software system. The mathematical formulation of this metric is presented below:

$$IUR = \frac{1}{DIT} * \frac{\# Parent\ Class\ Properties\ Inherited}{Total\ Properties\ Present\ in\ Parent\ Class} \qquad (3.8)$$

The second criteria, namely domain closeness, is used to measure the domain-based closeness among different methods of a class. In this paper, the domain-based closeness is measured by modeling the interactions among methods as a graph where nodes represent different methods and edges represent the interdependence/functional similarity measured in terms of member variable usage and/or method call. Using this graph, authors also propose a metric called DCR (Domain Closeness Ratio) that measures the degree of functional closeness among methods of a class is based on graph theory and is defined using the following equation:

$$DCR = \frac{1}{TFG} \qquad (3.9)$$

Where *TFG* denotes the total number of connected components (independent subgraphs) present in the above-modeled graph (aka forest in graph theory). Here, the ideal value of TFG is 1 that represents that all the methods are functionally related to each other. Any higher value represents the case of the presence of functionally unrelated methods. Finally, the proposed refused bequest code smell detection approach stresses utilizing expert opinion for determining different threshold values for considered two refused bequest code smell identification criteria in this chapter.

3.1.1.6 Shotgun Surgery Code Smell

Shotgun surgery code smell represents the case of linked modifications that need to be carried out by software developers/maintainers in a software system due to change performed in one of the software elements (class). The proposed shotgun surgery code smell detection approach of this paper makes use of two interlinked criteria for accurately detecting shotgun surgery code smell. The first criteria considered is known as a functional association and it measures relatedness arising due to the usage pattern based similarity between different classes. Determining this kind of functional association is important for detecting shotgun surgery code smell because such associated classes together provide specific functionality within a system. Changes done to any one of such associated classes can produce a ripple effect within the software system and hence a strong predictor of shotgun code smell. The Degree of Functional Association (DFA) between any two classes, say, C_i and C_j, whose usage pattern sets are given by $UsagePattern(C_i) = \{V_k, V_l, ..., V_m\}$ and $UsagePattern(C_j) = \{V_p, V_q, ..., V_r\}$ is given by the following equation:

$$DFA(C_i,\ C_j) = \frac{|UsagePattern(C_i) \cap UsagePattern(C_j)|}{|UsagePattern(C_i) \cup UsagePattern(C_j)|} \qquad (3.10)$$

The second criteria considered is known as *co-change based association* and it measures relatedness due to co-evolution (modified together) of code belonging to different classes. Palomba et al. stressed that shotgun surgery is inherently a change history based (Palomba et al., 2013). The proposed approach considers this criterion in combination with the functional association because it is our belief that considering only co-change-based association can sometimes give misleading results if underlying change history is not available or properly maintained. In the proposed shotgun surgery code smell identification approach of this paper, the Degree of Co-change based Association (DCA) is measured using the *Support & Confidence* metric commonly used in association rule mining.

Now, based on the above-mentioned two criteria, the probabilistic score for two classes undergoing modification together is given by the following equation:

$$Probability_{SS}(C_i,\ C_j)$$

$$= \begin{cases} DFA(C_i,\ C_j)_*Confidence(C_i,\ C_j); & DFA(C_i,\ C_j) \\ & > \alpha AND\ Confidence \qquad (3.11) \\ & (C_i,\ C_j) > \beta \\ 0; & Othewise \end{cases}$$

The above-mentioned probability score for shotgun surgery code smell is used to build a directed graph in which the nodes represent different classes and the edges represent the probability score for shotgun surgery assigned to any two classes in the system. From this directed graph, traces of shotgun surgery code smell is determined by using graph theory and extracting independent path chains. Now, the shotgun surgery for a class C_i based on a threshold γ is given by the following equation:

$$Shotgun\ Surgery(C_i) = \begin{cases} 1; & if\ Chain_{Length(C_i)} > \gamma \\ 0; & Otherwise \end{cases} \qquad (3.12)$$

Here, the expression notes the length of the independent chain starting at node C_i.

3.1.1.7 Duplicated Code Smell

The duplicated code smell is the result of copy-pasting of code fragment at different portions in a software system. Traces of duplicated code in a software system denotes the presence of increased understandability and maintainability. The proposed approach here aims at identifying duplicated code smell using two criteria. Firstly, the underlying code is denoted in an abstract form that is independent of underlying language syntactic rules. It helps to detect modified copy-pasted code. Secondly, the semantic aspects of the underlying code are used for detecting exact copy-pasted code. In this paper, the abstract form consists of two steps, namely 1) lexical analysis, and 2) syntactical analysis (parsing). The lexical analysis works at the class level and it converts the underlying source-code statements in the form of elementary tokens. These tokens have the characteristic that they help in representing the underlying code in a language-independent manner and thus help in detecting modified copy-pasted code. The syntactical analysis aims at representing the language-independent tokens in the form of a tree. Once, different classes of the software system are represented in the form of a tree, these trees are compared with each other. If any two trees are found identical or there is a sub-tree similarity, then there is an indication of duplicated code. In graph theory, this can be easily accomplished by determining the inorder/postorder tree traversals of both trees and determining if one is a subset of another. In our proposed approach, the role of semantic aspects comes when the user tries to change logic by either replacing loops/operators in the duplicated code. This results in an unmatched abstract syntactic tree. The semantic similarity between any two classes is obtained by using a well-known TF-IDF based cosine similarity metric. The algorithmic steps to detect duplicated code is shown.

Algorithm_for_DUPLICATED_CODE (Class C1, Class C2)

1. Perform lexical analysis of the source code of class C1 and C2 separately
2. Perform syntactic analysis (parsing) based on the lexical information obtained in step-1 in order to have abstract parse tree.
3. Perform inorder/postorder traversals of abstract parse trees.
4. Let S1 and S2 are the corresponding tree traversal sets.
5. Length = Longest_Common_Substring (S1, S2)
6. if (Length > 0)
7. {
8. Display the code identified as Common Substring to be "Duplicated Code"
9. }

3.3.2 CODE SMELLS PRIORITIZATION

This second phase of the proposed approach of this chapter aims to prioritizethe obtained list of code smells identified in the first phase of the approach shown in Figure 3.1. The proposed code smells prioritization mechanism considers multiple criteria in order to predict ranking with higher accuracy. These three considered criteria are 1) severity of the code smell; 2) involved maintenance efforts; and

3) user perception for the code smell. The authors in (Olbrich et al., 2010) determined that not all code smells equally affects the quality of a system. Some of them require urgent attention and remedy while others can be simply ignored, as they are not directly associated with degrading quality aspects of the system. Therefore, it is necessary to identify which of the obtained code smells are require immediate attention as compared to others. The considered severity criteria of the proposed approach handle this issue for a code smell. In the proposed approach of this chapter, the change-history of software is used to measure the maintenance efforts devoted by the maintenance team in the past. These devoted maintenance efforts reflect the importance of the code smells in order to improve the quality. Depending on the change-history, the severity of a code smell S_i is computed using a metric that is mathematically represented as follows:

$$Severity\,(S_i) = \frac{T_C}{T_N} \tag{3.13}$$

Here, T_C is the total number of independent transactions that involve classes engaged in ith code smell S_i. Similarly, T_N denotes independent transactions available in the underlying change-history of a software system.

Lack of understandability and complexity factors of a software system can create problems during software maintenance. These factors are directly linked to the quality. Therefore, any class that contains complexity issues and is involved with any kind of code smell must be dealt with higher importance. Hence, the chosen second criterionthat measures involved maintenance efforts of the maintenance team handles this issue. This paper considers the CC (Cognitive Complexity) metric available in literature (Mishra et al., 2018). This metric helps in measuring the current understandability and complexity at the class level. Mathematically, this metric is represented as follows:

$$CC = \sum_{p=1}^{s} \left(\sum_{j=1}^{q} \left[\prod_{k=1}^{m} W_c(j,\,k,\,l) \right] \right) \tag{3.14}$$

Here, the cognitive complexity of a class is measured based on the presence of three kinds of control statements in its source code, namely branch, sequence, and iterative. This weight in the proposed equation is represented with W_c.

Further, the domain knowledge and intelligence of developers cannot be completely ignored while prioritizing code smells. Therefore, the user knowledge is taken as another key criterionin the proposed code smell prioritization technique shown in Figure 3.1 of this chapter. The user feedback is taken based on multiple factors namely, smell relevance, refactoring cost, quality, and localization of multiple code smells. A team of experts is chosen and their opinion is received via a questionnaire shared with them. Their remarks are recorded and finally aggregated in order to get a final opinion. Finally, based on these considered three criteria, the ranking of a code smell is obtained using the following equation:

$$RankCS = Severity * CC * UserFeedback \qquad (3.15)$$

3.3.3 REFACTORING AND QUALITY IMPROVEMENT

The aim of this phase of the proposed approach in Figure 3.1 is to perform subsequent appropriate refactoring for mitigating the prioritized code smells by suggesting various refactoring operations to the maintenance team members. The feature envy and long method code smell are handled using the proposed move method/field refactoring suggestions. The LPL code smell is handled by suggesting replace parameters with a method call. Extract class and/or extract interface refactoring is suggested for god class code smell. The refactoring done for the refused bequest code smells is termed an Extract Class that aims at putting all unimplemented methods and member variables of the base class into another separate class. Similarly, the shotgun surgery and duplicated code smell are handled using the Extract Class refactoring applied at the class level. This is done in order to keep the common code that is distributed at different places at a single place (newly extracted class).

3.4 EXPERIMENTATION AND RESULTS

Experimentation conducted and the obtained result interpretation step provides sufficient strength to the proposedapproach. Therefore, this part gives experimentation design setup details such as different software considered during experimentation, experimental planning, interpretation is done using various formulated research questions. These different points are detailed in the following subsections.

3.4.1 EXPERIMENTAL PLANNING AND SETUP

This chapter considers using a proven systematic Goal-Question-Matrix (GQM) approach (Caldiera & Rombach, 1994). In this approach, first, the purposeful goals are specified and then based on the goals appropriate data is selected (metric values based on the information present in the considered software systems). Finally, the goals are operationally evaluated from the actual obtained experimental results. Further, the approach in this chapter is extensively evaluated with different open-source Java software systems that are commonly being used in literature (Harman et al., 2012; Moha et al., 2009; Paiva et al., 2017; Rathee & Chhabra, 2019). Table 3.3 gives statistics about various studied targeted systems for experimentation purposes. The considered targeted systems are from wide domains and are of varied sizes. Different considered dependencies information and considered metric values are identified using a self-designed program by the authors.

3.4.2 RESEARCH QUESTIONS AND EVALUATION METHOD

The present research work in this chapter focuses on correctly identifying and mitigating various code smells and performs their prioritization for reducing the involved maintenance efforts. Further, this chapter formulates the following research questions in order to test and evaluate the proposed approach:

TABLE 3.3

Statistics About Various Considered Experimental Studied Target Systems

S.No.	Studied Software	Total No. of Classes	Description
1.	ArgoUML *version 0.26*	1358	A UML diagram Java application
2.	Gantt *version 1.10.2*	245	A Java project scheduling and management application
3.	Xerces *version 2.7*	991	An Apache library for handling XML data in an application
4.	Health Watcher *version 0.9*	115	A web-based application to be used in health departments
5.	JUnit *version 4.2*	24	A Java-based unit testing framework

RQ1 *What is the accuracy of the proposed approach in detecting various code smells?*

The aim behind considering this Research Question (RQ) is to test the performance of the proposed approach of this chapter. The evaluation is done using state-of-art precision, recall, and F-measure information retrieval metrics (Baeza-Yates & Ribeiro-Neto, 1999). These considered metrics evaluate by doing a comparison between the obtained code smells and the standard/actual code smells present in the considered studied systems. As standard code smell list is not available in the literature. Therefore, such a list is obtained by aggregating the results of different state-of-art tools widely used by IT industries.

RQ2 *What is the effect of prioritization on devoted maintenance efforts?*

The aim of this formulated RQ is to test the usefulness of doing prioritization on identified code smells for a software system as depicted in the proposed technique shown in Figure 3.1. For this purpose, it is necessary to measure the maintenance efforts devoted by the maintenance team. Since maintenance efforts is a qualitative term and there is no direct measure for it. However, the change-history of a software system can be easily used to indirectly measure the devoted maintenance efforts by the maintenance team. The change-history of a software system contains detailed information about different software elements that have undergone modification (insertion/deletion/mutation) at a different instance of time by the maintenance team. These modified software elements also indicate changes in response to the quality improvement of a system by mitigating code smells belonging to a system.

RQ3 *Does refactoring help in improving the quality of the software system?*

The reason behind considering this RQ is to determine the performance of refactoring proposed in this chapter. Here, the evaluation is done using a well-known modularization metric known as TurboMQ that helps in computing software system's quality in two situations viz original and refactored system. The original system is the system that contains considered code smells. The refactored system is the software that is obtained after applying the proposed code smell detection and mitigation approach of this chapter.

3.4.3 RESULTS ANALYSIS AND INTERPRETATIONS

This subsection presentsa detailed discussion and interpretation of experimental results obtained in this chapter. The obtained results and their interpretations are divided in the form of various RQs formulated in this chapter in Section 3.4.2.

RQ1 *What is the accuracy of the proposed approach in detecting various code smells?*

For analyzing the obtained results and to determine the feasibility of the proposed approach of this chapter, the obtained results are compared with the gold-standard list of various code smell belonging to a softwarethat is considered to be the actual code smells belonging to a system. Because there is no available standard gold standard available in the literature, therefore, this chapter prepares this gold standard using various state-of-art code smell detection tools, namely InFusion, iPlasma, DÉCOR (Moha et al., 2009), CheckStyle, and PMD. The obtained code smells by these tools are analyzed and a common list is prepared to obtain the gold standard used in this chapter. For combining individual code smell lists of different tools and to create a single code smell list (to be used as the gold standard), this chapter considers expert opinion. We constituted two teams for this purpose. The first team consisting of five undergraduate and postgraduate students runs different considered tools on different datasets considered in this chapter and collects individual code smell lists. In the second step, these individual lists are further tailored and filtered by presenting them to the second team of experts consisting of experts from the IT industry. These experts analyzed individual code smell lists based on different critical factors such as code smell relevance, quality improvement factor, and maintenance cost involved. Table 3.4 shows various values of the obtained results as part of the comparison between the obtained experimental code smells and the gold standard code smells. This table shows precision, recall, and f-measure values.The obtained value of precision metric varies from 69% to 81% and is sufficiently high to support the accuracy of the proposed code smell identification and prioritization approach of this chapter. Moreover, the value of the recall metric

TABLE 3.4
Obtained Experimental Results

S.No.	Studied Software	Precision	Recall	F-Measure
1.	ArgoUML	75%	95%	84%
2.	Gantt	72%	92%	81%
3.	Xerces	81%	94%	87%
4.	Health Watcher	69%	96%	80%
5.	JUnit	74%	91%	82%

varies between 91% to 96% and the obtained values strongly reveal that the pro-posed approach of this chapter is capable of sustaining high sensitivity. In other words, the proposed approach is capable of identifying nearly every code smell present in the considered code smell's gold standard list. Further, depending on our experimentation and obtained results, we observed that there exists some code smells that are present in our obtained experimental code smells list but missing from the considered gold standard code smell list. The answer to this situation where our proposed approach detects more smelly instances from the underlying software system is that the proposed different code smell approaches in this chapter are based on modern, relevant, and dynamic software characteristics rather than mere primitive metrics. Hence, it can be clearly concluded that the obtained results confirm that the proposed approach has sufficiently and significantly higher recall and f-measure values for different considered software systems.

RQ2 *What is the effect of prioritization on devoted maintenance efforts?*

Another aspect considered in this chapter is to evaluate the effect of different code smell prioritization and tackling only a subset of code smells. The aim is to determine the effect of prioritization on involved maintenance efforts. To answer this RQ, the underlying change-history of a software system is used to measure the devoted maintenance efforts by the maintenance team. To do this, the change-history of software is extracted from different platforms namely SVN, Subversion, GitHub, and the following equation is used to measure the maintenance efforts devoted for the ith code smell present in the system:

$$Maintenance \ \ Effort = ME\,(i) = \frac{T_C}{T_N} \qquad (3.16)$$

Here, T_C denotes different transactions present in the change-history of a software system that involves changes to different software elements involved in the ith code smell. T_N. denotes transactions present in the change-history of a software system. Three scenarios are considered to evaluate this RQ, namely, (1) code smells are prioritized and involved maintenance efforts are measured (SCENARIO-1), and (2) all code smells are tackled and involved maintenance efforts are measured (SCENARIO-2). Table 3.5 depicts the obtained results measured in considered two scenarios using the proposed maintenance effort measurement metric in this chapter. The average value for change in maintenance efforts is only 4.34%. It indicates that when only prioritized code smells are tackled then the involved maintenance efforts measured using the proposed equation of this chapter are nearly equal to the maintenance efforts involved in tackling all *available code smells*. The change-history of different considered soft-ware systems contains changelogs mostly involved with software elements involved with prioritized code smell lists. Hence, it is necessary to prioritized code smells and they actually represent how developers tackle code smells belonging to a software system while improving its underlying architecture/quality.

Figure 3.2 shows the comparison of the involved maintenance efforts considered in SCENARIO-1 & SCENARIO-2 using a bar plot chart. From the plot, it is clearly observed that the involved maintenance efforts (computed using corresponding

TABLE 3.5
Comparing the Effect of Prioritization of Different Identified Code Smells

| S.No. | Studied Software | Change History based Effort Measurement | | |
		Scenario –1	Scenario –2	% Difference
1.	ArgoUML	0.83	0.85	2.35%
2.	Gantt	0.75	0.76	1.32%
3.	Xerces	0.65	0.72	9.72%
4.	Health Watcher	0.87	0.87	0.00%
5.	JUnit	0.77	0.84	8.33%
	Average % Difference			**4.34%**

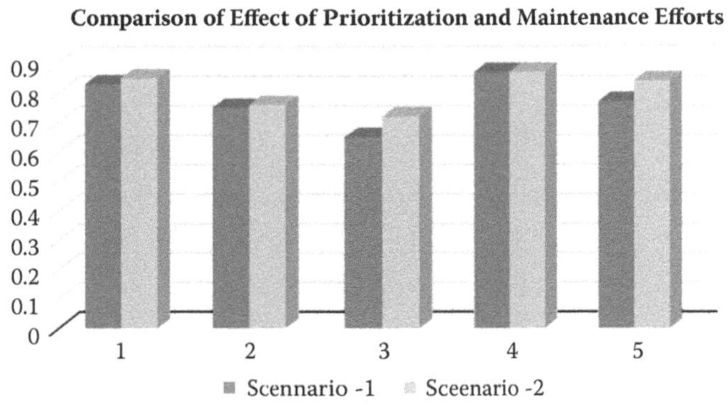

FIGURE 3.2 Comparison of the effect of prioritization and involved maintenance efforts.

change-history of system) are nearly comparable in both cases. It means developers generally never target removing each and every code smell belonging to a software system. However, at the different instant of time, the only subset of possible code smells are necessary from optimal quality improvement aspects and thus are only targeted by developers.

RQ3 *Does refactoring help in improving the quality of the software system?*

Another aspect considered in this chapter is to evaluate the effect of prioritizing different code smells and their underlying effect on software quality. The aim is to determine the effect of refactoring on software quality and its involved maintenance efforts. Table 3.5 shows the obtained experimental results that compare two scenarios. The considered scenario-1 considers tackling all the obtained code smells and scenario-2 considers prioritizing code smells and tackling the only subset of identified code smells. In both scenarios, the TurboMQ modularization metric is used to compare quality of the system. The involved maintenance efforts are also measured in two scenarios and the obtained results are presented in the last three columns of Table 3.6. From the presented results, it is clear that although tackling

TABLE 3.6

Effect of Prioritizing Code Smells on Software Quality and Maintenance Efforts

S.No.	Studied Software	TurboMQ Metric-Based Software's Quality Measurement			Change History-BasedEffort Measurement		
		Scenario –1	Scenario –2	% Difference	Scenario –1	Scenario –2	% Difference
1.	ArgoUML	1.63	1.61	–1.30%	0.83	0.65	–27.69%
2.	Gantt	0.87	0.82	–6.10%	0.76	0.54	–40.74%
3.	Xerces	3.21	3.09	–3.88%	0.65	0.52	–25.00%
4.	Health Watcher	2.62	2.54	–3.15%	0.82	0.71	–15.49%
5.	JUnit	1.96	1.87	–4.81%	0.91	0.77	–18.18%
	Average			**–3.85%**		Average	**–25.42%**

only prioritized code smells results in decreased quality of the software system (with very less factor). The average value for the decrease in the overall quality of a software system is only −3.85%. However, at the same time, the reduction in maintenance efforts is remarkable (with a sufficiently higher factor) with a reduction in overall maintenance efforts by 25.42%. In other words, the obtained prioritized code smells list contains key code smells that are the main contributing factors in the quality of the software. Simultaneously, involved refactoring/maintenance cost and time are strongly reduced. This shows the usefulness and feasibility of the proposed code smell prioritization approach of this chapter.

3.5 CONCLUSION AND FUTURE WORK

Different types of code smells may affect a software system and degrade its quality.Degraded quality of a software system results in increased maintenance efforts and ultimately hinders their applications in the field of data science. Manageable software systems are helpful in accelerating involved computation intelligence and can significantly reduce burdens at the human end. The approach for identifying these code smells with higher accuracy is proposed in this chapter. Moreover, prioritization is considered important in this chapter and its effect on involved maintenance efforts is analyzed. The obtained experimental results conclude that prioritizing is important from a maintenance point of view. It results in significantly reducing maintenance efforts. The future works related to this chapter can go in multiple directions. Firstly, the proposed approach to detect and mitigate different code smells can be represented in the form of fully/semi-automated tool support. Secondly, the proposed approach can be tested or evaluated for the suitability of other combinations of metrics in order to determine any improvements in the results. Thirdly, a feasibility test for the proposed metrics of this chapter can be carried out in identifying other kinds of code smells. Finally, but not least, the effectiveness and hence usability of the proposed approach can be tested in the field of bug prediction and pattern discovery in source code.

REFERENCES

Abdellatief, M., & Md Sultan, A. (2011, October). Component-based software system dependency metrics based on component information flow measurements. *The Sixth International Conference on Software Engineering Advances ICSEA2011*, Barcelona, Spain (pp. 76–83).

Agnihotri, M., & Chug, A. (2020). A systematic literature survey of software metrics, code smells and refactoring techniques. *Journal of Information Processing Systems, 16*(4), 915–934.

Akerkar, R., & Sajja, P. S. (2016). *Intelligent techniques for data science*. Cham, Switzerland: Springer International Publishing.

Al Dallal, J. (2012). Constructing models for predicting extract subclass refactoring opportunities using object-oriented quality metrics. *Information and Software Technology, 54*(10), 1125–1141.

April, A., & Abran, A. (2012). *Software maintenance management: Evaluation and continuous improvement* (Vol. 67). New York: John Wiley & Sons.

Arcoverde, R., Guimaraes, E., Macía, I., Garcia, A., & Cai, Y. (2013, October). Prioritization of code anomalies based on architecture sensitiveness. *2013 27th Brazilian Symposium on Software Engineering* (pp. 69–78). IEEE.

Azeem, M. I., Palomba, F., Shi, L., & Wang, Q. (2019). Machine learning techniques for code smell detection: A systematic literature review and meta-analysis. *Information and Software Technology, 108*, 115–138.

Baeza-Yates, R., & Ribeiro-Neto, B. (1999). *Modern information retrieval* (Vol. 463). New York: ACM Press.

Bibiano, A. C., Fernandes, E., Oliveira, D., Garcia, A., Kalinowski, M., Fonseca, B.,...& Cedrim, D. (2019, September). A quantitative study on characteristics and effect of batch refactoring on code smells. *2019 ACM/IEEE International Symposium on Empirical Software Engineering and Measurement (ESEM)* (pp. 1–11), Porto de Galinhas, Brazil. IEEE.

Caldiera, V. R. B. G., & Rombach, H. D. (1994). The goal question metric approach. In *Encyclopedia of software engineering*, 528–532.

Counsell, S., Hamza, H., & Hierons, R. M. (2010). An empirical investigation of code smell 'deception'and research contextualisation through paul's criteria. *Journal of Computing and Information Technology, 18*(4), 333–340.

De Paulo Sobrinho, E. V., De Lucia, A., & de Almeida Maia, M. (2018). A systematic literature review on bad smells—5 W's: which, when, what, who, where. *IEEE Transactions on Software Engineering, 47*(1), 17–66.

Fellbaum, C. (1998). WordNet: An electronic lexical database and some of its applications. Cambridge: MIT Press.

Fernandes, E., Oliveira, J., Vale, G., Paiva, T., & Figueiredo, E. (2016, June). A review-based comparative study of bad smell detection tools. *Proceedings of the 20th International Conference on Evaluation and Assessment in Software Engineering* (pp. 1–12). Association for Computing Machinery, New York, NY, USA. DOI:https://doi.org/10.1145/2915970.2915984

Fontana, F. A., Ferme, V., Zanoni, M., & Roveda, R. (2015, October). Towards a prior-itization of code debt: A code smell intensity index. *2015 IEEE 7th International Workshop on Managing Technical Debt (MTD)* (pp. 16–24). IEEE.

Fontana, F. A., & Zanoni, M. (2017). Code smell severity classification using machine learning techniques. *Knowledge-Based Systems, 128*, 43–58.

Fowler, M., Beck, K., Brant, J., & Opdyke, W. (1999). Refactoring: improving the design of existing code, Google Scholar Digital Library.

Freire, S., Passos, A., Mendonça, M., Sant'Anna, C., & Spínola, R. O. (2020, August). On the influence of UML Class Diagrams Refactoring on Code Debt: A family of re-plicated empirical studies. *2020 46th Euromicro Conference on Software Engineering and Advanced Applications (SEAA)* (pp. 346–353), Portoroz, Slovenia. IEEE. DOI: 10.1109/SEAA51224.2020.00064.

Gamma, E. (1995). *Design patterns: elements of reusable object-oriented software*. Noida: Pearson Education India.

Garg, R., & Singh, R. K. (2020). Analysis and Prioritization of Design Metrics. *Procedia Computer Science, 167*, 1495–1504.

Guggulothu, T., & Moiz, S. A. (2019, January). Prioritize the Code Smells Based on Design Quality Impact. *International Conference on Intelligent Computing and Communication Technologies* (pp. 406–415). Springer, Singapore.

Guggulothu, T., & Moiz, S. A. (2019, February). An approach to suggest code smell order for refactoring. *International Conference on Emerging Technologies in Computer Engineering* (pp. 250–260). Springer, Singapore.

Guimaraes, E., Vidal, S., Garcia, A., Diaz Pace, J. A., & Marcos, C. (2018). Exploring architecture blueprints for prioritizing critical code anomalies: Experiences and tool support. *Software: Practice and Experience, 48*(5), 1077–1106.

Harman, M., Mansouri, S. A., & Zhang, Y. (2012). Search-based software engineering: Trends, techniques and applications. *ACM Computing Surveys (CSUR), 45*(1), 1–61.

Hermans, F., & Aivaloglou, E. (2016, May). Do code smells hamper novice programming? A controlled experiment on Scratch programs. *2016 IEEE 24th International Conference on Program Comprehension (ICPC)* (pp. 1–10). IEEE.

Husien, H. K., Harun, M. F., & Lichter, H. (2017). Towards a severity and activity based assessment of code smells. *Procedia Computer Science, 116*, 460–467.

Hayashi, C. (1998). What is data science? Fundamental concepts and a heuristic example. In C. Hayashi, K. Yajima, H. Bock, N. Ohsumi, Y. Tanaka, & Y. Baba (Eds.), *Data science, classification, and related methods* (pp. 40–51). Tokyo: Springer.

Ibrahim, R., Ahmed, M., Nayak, R., & Jamel, S. (2020). Reducing redundancy of test cases generation using code smell detection and refactoring. *Journal of King Saud University-Computer and Information Sciences, 32*(3), 367–374.

Jiang, L., Misherghi, G., Su, Z., & Glondu, S. (2007, May). Deckard: Scalable and accurate tree-based detection of code clones. *29th International Conference on Software Engineering (ICSE'07)* (pp. 96–105). IEEE.

Kamaraj, N., & Ramani, A. V. (2019). Search-based software engineering approach for detecting code-smells with development of unified model for test prioritization strategies. *International Journal of Applied Engineering Research, 14*(7), 1599–1603.

Katbi, A., Hammad, M., & Elmedany, W. (2020). Multi-view city-based approach for code-smell evolution visualisation. *IET Software, 14*(5), 506–516.

Khomh, F., Di Penta, M., & Gueheneuc, Y. G. (2009, October). An exploratory study of the impact of code smells on software change-proneness. *2009 16th Working Conference on Reverse Engineering* (pp. 75–84). IEEE.

Lacerda, G., Petrillo, F., Pimenta, M., & Guéhéneuc, Y. G. (2020). Code smells and re-factoring: a tertiary systematic review of challenges and observations. *Journal of Systems and Software*, 110610. DOI: 10.1016/j.jss.2020.110610

Lanza, M., & Marinescu, R. (2007). *Object-oriented metrics in practice: Using software metrics to characterize, evaluate, and improve the design of object-oriented systems*. Physica-Verlag, Springer Science & Business Media. DOI: 10.1007/3-540-39538-5.

Li, W., & Shatnawi, R. (2007). An empirical study of the bad smells and class error probability in the post-release object-oriented system evolution. *Journal of Systems and Software, 80*(7), 1120–1128.

Li, Z. (2019, May). Characterizing and detecting duplicate logging code smells. *2019 IEEE/ACM 41st International Conference on Software Engineering: Companion Proceedings (ICSE-Companion)* (pp. 147–149). IEEE.

Liu, X., & Zhang, C. (2018, April). DT: an upgraded detection tool to automatically detect two kinds of code smell: duplicated code and feature envy. *Proceedings of the International Conference on Geoinformatics and Data Analysis* (pp. 6–12).

Ma, W., Chen, L., Zhou, Y., & Xu, B. (2016, November). Do we have a chance to fix bugs when refactoring code smells?. *2016 International Conference on Software Analysis, Testing and Evolution (SATE)* (pp. 24–29). IEEE.

Marinescu, C., Marinescu, R., Mihancea, P., Ratiu, D., & Wettel, R.: iplasma: An integrated platform for quality assessment of object-oriented design. *Proceedings of 21st International Conference on Software Maintenance (ICSM)* (2005).

Marinescu, R. (2012). Assessing technical debt by identifying design flaws in software systems. *IBM Journal of Research and Development, 56*(5), 9–1.

Martini, A., Fontana, F. A., Biaggi, A., & Roveda, R. (2018, September). Identifying and prioritizing architectural debt through architectural smells: A case study in a large software company. *European Conference on Software Architecture* (pp. 320–335). Cham, Switzerland: Springer.

McCabe, T. J. (1976). A complexity measure. *IEEE Transactions on software Engineering*, (4), 308–320.

Misra, S., Adewumi, A., Fernandez-Sanz, L., & Damasevicius, R. (2018). A suite of object oriented cognitive complexity metrics. *IEEE Access*, *6*, 8782–8796.

Moha, N., Gueheneuc, Y. G., Duchien, L., & Le Meur, A. F. (2009). Decor: A method for the specification and detection of code and design smells. *IEEE Transactions on Software Engineering*, *36*(1), 20–36.

Olbrich, S. M., Cruzes, D. S., & Sjøberg, D. I. (2010, September). Are all code smells harmful? A study of God Classes and Brain Classes in the evolution of three open source systems. *2010 IEEE International Conference on Software Maintenance* (pp. 1–10). IEEE.

Oliveira, R., Sousa, L., de Mello, R., Valentim, N., Lopes, A., Conte, T., & Lucena, C. (2017, May). Collaborative identification of code smells: A multi-case study. *2017 IEEE/ACM 39th International Conference on Software Engineering: Software Engineering in Practice Track (ICSE-SEIP)* (pp. 33–42). IEEE.

Oliveira, A., Sousa, L., Oizumi, W., & Garcia, A. (2019, September). On the Prioritization of Design-Relevant Smelly Elements: A Mixed-Method, Multi-Project Study. *Proceedings of the XIII Brazilian Symposium on Software Components, Architectures, and Reuse* (pp. 83–92), Salvador, Brazil.

Paiva, T., Damasceno, A., Padilha, J., Figueiredo, E., & Sant'Anna, C. (2015). Experimental evaluation of code smell detection tools. In: 3rd workshop on software Visualization, Evolution, and Maintenance (VEM), pp 17–24

Paiva, T., Damasceno, A., Figueiredo, E., & Sant'Anna, C. (2017). On the evaluation of code smells and detection tools. *Journal of Software Engineering Research and Development*, *5*(1), 7.

Palomba, F., Bavota, G., Di Penta, M., Oliveto, R., De Lucia, A., & Poshyvanyk, D. (2013, November). Detecting bad smells in source code using change history information. *2013 28th IEEE/ACM International Conference on Automated Software Engineering (ASE)* (pp. 268–278). IEEE.

Palomba, F., Bavota, G., Di Penta, M., Oliveto, R., Poshyvanyk, D., & De Lucia, A. (2014). Mining version histories for detecting code smells. *IEEE Transactions on Software Engineering*, *41*(5), 462–489.

Palomba, F., Panichella, A., Zaidman, A., Oliveto, R., & De Lucia, A. (2017). The scent of a smell: An extensive comparison between textual and structural smells. *IEEE Transactions on Software Engineering*, *44*(10), 977–1000.

Pecorelli, F., Palomba, F., Khomh, F., & De Lucia, A. (2020, May). Developer-driven code smell prioritization. *International Conference on Mining Software Repositories*. In Proceedings of the 17th International Conference on Mining Software Repositories (pp. 220–231).

Rathee, A., & Chhabra, J. K. (2017). Restructuring of object-oriented software through cohesion improvement using frequent usage patterns. *ACM SIGSOFT Software Engineering Notes*, *42*(3), 1–8.

Rathee, A., & Chhabra, J. K. (2019). A multi-objective search based approach to identify reusable software components. *Journal of Computer Languages*, *52*, 26–43.

Sae-Lim, N., Hayashi, S., & Saeki, M. (2018). Context-based approach to prioritize code smells for prefactoring. *Journal of Software: Evolution and Process*, *30*(6), e1886.

Seacord, R. C., Plakosh, D., & Lewis, G. A. (2003). Modernizing legacy systems: Software technologies, engineering processes, and business practices. Switzerland: Addison-Wesley.

Shull, F., Falessi, D., Seaman, C., Diep, M., & Layman, L. (2013). Technical debt: Showing the way for better transfer of empirical results. *Perspectives on the Future of Software Engineering* (pp. 179–190). Springer, Berlin.

Singh, R., & Kumar, A. (2018). Identifying Various Code-Smells and Refactoring Opportunities in Object-Oriented Software System: A systematic Literature Review. *International Journal on Future Revolution in Computer Science & Communication Engineering*, 8(March), 62–74.

Singh, S., & Kaur, S. (2018). A systematic literature review: Refactoring for disclosing code smells in object oriented software. *Ain Shams Engineering Journal*, 9(4), 2129–2151.

Soh, Z., Yamashita, A., Khomh, F., & Guéhéneuc, Y. G. (2016, March). Do code smells impact the effort of different maintenance programming activities? *2016 IEEE 23rd International Conference on Software Analysis, Evolution, and Reengineering (SANER)* (Vol. 1, pp. 393–402). IEEE.

Techapalokul, P., & Tilevich, E. (2019, October). Code quality improvement for all: Automated refactoring for Scratch. *2019 IEEE Symposium on Visual Languages and Human-Centric Computing (VL/HCC)* (pp. 117–125). IEEE.

Tsantalis, N., Chaikalis, T., & Chatzigeorgiou, A. (2008, April). JDeodorant: Identification and removal of type-checking bad smells. *2008 12th European Conference on Software Maintenance and Reengineering* (pp. 329–331). IEEE.

Tufano, M., Palomba, F., Bavota, G., Oliveto, R., Di Penta, M., De Lucia, A., & Poshyvanyk, D. (2015, May). When and why your code starts to smell bad. *2015 IEEE/ACM 37th IEEE International Conference on Software Engineering* (Vol. 1, pp. 403–414). IEEE.

Vale, T., Souza, I. S., & Sant'Anna, C. (2014). Influencing factors on code smells and software maintainability: A cross-case study. *2nd Workshop on Software Visualization, Evolution and Maintenance (VEM'14)*, Maceio, AL, Brazil; 2014:86–93.

Vidal, S. A., Marcos, C., & Díaz-Pace, J. A. (2016). An approach to prioritize code smells for refactoring. *Automated Software Engineering*, 23(3), 501–532.

Vidal, S., Oizumi, W., Garcia, A., Pace, A. D., & Marcos, C. (2019). Ranking architecturally critical agglomerations of code smells. *Science of Computer Programming*, 182, 64–85.

Yamashita, A., & Moonen, L. (2012, September). Do code smells reflect important maintainability aspects? *2012 28th IEEE international conference on software maintenance (ICSM)* (pp. 306–315). IEEE.

Yamashita, A., & Moonen, L. (2013, May). Exploring the impact of inter-smell relations on software maintainability: An empirical study. *2013 35th International Conference on Software Engineering (ICSE)* (pp. 682–691). IEEE.

Yoshida, N., Saika, T., Choi, E., Ouni, A., & Inoue, K. (2016, May). Revisiting the relationship between code smells and refactoring. *2016 IEEE 24th International Conference on Program Comprehension (ICPC)* (pp. 1–4). IEEE.

Zhang, X. G., Sun, S. Q., & Zhang, K. J. (2017). A novel comprehensive approach for estimating concept semantic similarity in wordnet. *arXiv preprint arXiv:1703.01726.*

4 Evaluating Machine Learning Capabilities for Predicting Joining Behavior of Freshmen Students Enrolled at Institutes of Higher Education: Case Study from a Novel Problem Domain

Pawan Kumar and Manmohan Sharma
Lovely Professional University, Phagwara, Punjab, India

CONTENTS

DOI: 10.1201/9781003132080-4

4.1   INTRODUCTION

Education becomes necessary for each individual to live and contribute to society. The vision of the Ministry of Education, Government of India, is to reduce the illiteracy rate in India. For that, they have started many new educational institutions (government as well as private) across the country. As per the University Grant Commission (UGC) report, there are almost 900 universities, including government and private, across India (UGC, 2020). Most of these private universities spend hugely in competing to reach out to students who are exploring different options to take admission into. Due to an increase in competition, institutions start their admission process for the next session quite early, usually at the start of the year. Every admission year, from the students, enrolled in an educational institution, some students do not want to join the same institution. The reasons include getting a higher scholarship in some other institution, preferred college, or discipline of interest in some other institution. Each student who takes admission but does not join is a loss to the concerned educational institute in terms of resources invested. Ability to foresee such students is important for an educational institute. Institutes can use this information to plan activities towards improving retention of enrolled students.

During recent years, machine learning (ML) has been a consistent buzzword in the field of Information technology. A subfield of artificial intelligence, ML is a combination of computer science, mathematics, statistics and domain knowledge. In ML, the learning refers to inference rules from the training data set for predicting class labels of the targeted data (Mitchell, 2006). In recent years, ML-based models in terms of prediction accuracy have proven their worth in solving complex problems. A majority of these advanced ML models are opaque in the sense that their prediction decisions are not easy to interpret (Wagstaff, 2012). The ability of an ML-based model to explain its behavior to its human users is termed human interpretability and is crucial for the growth of machine learning (Ribeiro, Singh, & Guestrin, 2016). Interpretability offers multiple advantages including facilitating trust-building, finding new insights towards the underlying process, debugging of the model, and ensuring fairness in the model.

This study aims to explore the applicability of ML in helping educational institutions predict the joining behavior of their freshmen students. The idea is to formulate this research question as a binary classification ML problem. The target variable is the actual joining status of a student with 'Joined' or 'Lost' as possible outcomes. The first objective is to learn an ML-based model using admission details and actual joining status of freshmen students of previous batches for learning. Such a model can be used for predicting the joining behavior of freshmen students of the new batch. The second objective is to analyze important factors that contribute towards the joining behavior the most. This problem domain is dynamic as student joining behavior is affected by multiple factors that include changes in admission policies and trends in the education sector every year. So, an additional objective has been to study variations in importance of different factors over the admission years. The outcomes of this study are helpful to an educational institution to improve its admission-related processes.

Several research studies aiming at building ML models for predicting student academic performance or employability exist in literature. Predictive models have been developed to predict dropouts or academic performance in case of distance learning. This research study is different from the existing work in the sense that it is aiming to analyze joining behavior of freshmen students enrolled in regular mode of formal education. As per our knowledge, this is the first such kind of analysis for educational institutions. Table 4.1 gives a summary of papers related to building ML-based models for students.

TABLE 4.1
Review of the Related Work

Author	Aim	Key Methods	Dataset(s)	Pros	Cons
(Simpson, 2006)	Predictive academic performance in distance learning	LR, statistical methods	Napier University, Edinburgh, UK	Use of instructor feedback. Side effects of predictive models	Need to set benchmarks in student support
(Kaur, Singh, & Josan, 2015)	Predicting slow and fast learners	Multilayer Perceptron, Naïve Bayes, SMO, J48 and REPTree	152 high school students	Internal grades and attendance found as top rank variables. Classification accuracy of 75% using Multilayer Perception.	Need to explore new factors
(Ahmad, Ismail, & Aziz, 2015)	Predicting academic performance	NB, Decision trees and association-based rules	397 first-year computer science students	Classification performance of 71% using association-based rules	Small data set size
(Hughes & Dobbins, 2015)	Identification of probable MOOC drop-outs	Using the level of engagement, interaction, and attendance as attributes	Liverpool John Moores University, UK	Case study of eRegister, an attendance monitoring system	Need to make eRegister adapt to MOOC environment
(He, Baileyt, Rubinstein, & Zhang, 2015)	Identifying At-Risk Students in MOOCs	Logistic regression	Case study of Discrete Optimization on Coursera.	Suggested suitable interventions for marginal students. Weekly predictive models.	Collaboration with course instructors for planning timely interventions

(Continued)

TABLE 4.1 (Continued)
Review of the Related Work

Author	Aim	Key Methods	Dataset(s)	Pros	Cons
(Marbouti, Diefes-Dux, & Madhavan, 2016)	Early detection of at-risk students	LR, SVM, Decision tree, Multilayer perceptron, NB, kNN, model ensemble	The first-year engineering students at a midwestern US university	Uses only in-semester performance data	Identifying research questions for effective use of predictive models
(Pandey & Taruna, 2016)	Integration of multiple classifiers for predicting student performance	DT, kNN, Aggregating One-Dependence Estimators	3 student performance data sets of an engineering college in India	Integrates three complementary algorithms.	Need to adapt to become a decision support system
(Altujjar, Altamimi, Al-Turaiki, & Al-Razgan, 2016)	Predicting critical courses affecting student performance	ID3 Decision tree	King Saud University, Riyadh, Saudi Arabia of the 2013–14 year	Year-wise model	Small data set size
(Hoffait & Schyns, 2017)	Identification of freshmen likely to face major difficulties	LR, RF, ANN	Undergraduate students at University of Liege, Belgium	A decision support system to identify at-risk freshmen students	Exploring applicability in other domains
(Krasilnikov & Smirnova, 2017)	Investigating factors related to social-adaptation of first-year students	Logit regression, Social network indicators	68 students at a Russian University	Correlation between social-adaptation during the term and academic performance	Small data set size
(Abu Zohair, 2019)	feasibility of creating a prediction model with small data set	SVM, LDA	50 graduated students of a master's program	SVM and LDA found efficient in training small data set size.	Utilized student's administration records only

TABLE 4.1 (Continued)
Review of the Related Work

Author	Aim	Key Methods	Dataset(s)	Pros	Cons
(Lau, Sun, & Yang, 2019)	Predicting academic performance using conventional statistical analysis and NN	Statistical Analysis; ANNs	1,000 undergraduate university students of China	Uses a combination of statistical analysis and ANNs. Prediction accuracy of 84.8%.	Additional predictors need to be explored
(Kumar, Pawan & Sharma, 2020)	Predicting academic performance of international students	Logistic Regression, NB, CART, RF, LIME	921 international students studying at Lovely Professional University, India	Evaluates multiple ML algorithms for classification, uses LIME for interpreting ML models	Additional predictors needed, the extension to regression

This research work has the following contributions:

i. Illustrate, predict, and highlight the freshmen student joining behavior.
ii. Present a ML-model-based freshmen student joining behavior prediction framework to overcome institutional losses.
iii. Investigate the ML-based model behavior to identify parameters that affect freshmen student joining behavior.
iv. Uncovering variations in important factors affecting joining behavior of freshmen students over successive admission years.

The organization of the rest of the chapter is as follows: Section 4.2 details about the demography of the subjects, experiments designed and machine learning techniques used. Section 4.3 explains the proposed mathematical framework including equations used, framework diagram, and pseudo-code representation. Section 4.4 compiles the outcomes of the experimental work and its discussion. Section 4.5 concludes the chapter and suggests possible future lines of work.

4.2 METHODS AND MATERIALS

This section talks about the structure of the data set, the novelty in preparation of the data set, machine learning techniques used for learning models, and understanding their behavior.

4.2.1 DEMOGRAPHY OF THE SUBJECTS USED

For training and evaluation of ML modes, the data set consisted of freshmen students of a university in North India. The data of 2017 and 2018 admission batches were considered and total number of records was 28,114 (Table 4.2).

4.2.2 NOVELTY AND MODIFICATIONS IN PREPARING THE DATA SET

The target variable is JoiningStatus with two possible classes as Joined or Lost. Using feature selection techniques, the following features were shortlisted for the learning ML model.

    i. AdmissionMonth
   ii. MarksCategory

TABLE 4.2
Structure of the Data Set

Attribute	Datatype Type	Description
RegistrationNumber	Int	Unique Identification Number for each student
AdmissionMonth	Int	Month in which student took admission
Gender	Factor	{Female, Male}
State	Factor	State which a student hail from
HomeTownType	Factor	Rural, Urban
BatchYear	Factor	Year of taking admission e.g. 2017
ProgramName	Factor	Examples: BCA, BTech., BSc
Discipline	Factor	Computer Applications, Chemistry, etc.
QualifyingExam	Factor	Examples: 10th, 10+2
MarksPercent	Numeric	Percentage of marks in qualifying program
CategoryCode	Factor	Category of the student e.g. General
TransportAvailed	Factor	Did the student avail transport?
LoanLetter	Factor	Did the student applied for loan letter?
PreviouslyStudied	Factor	Studied earlier at same Institute? [Yes/No]
HostelAvailed	Factor	Did the student avail hostel?
MessAvailed	Factor	Availed Mess facility or not?
ScholarshipPercentage	Numeric	Percentage of Scholarship offered to the student
ScholarshipBracket	Factor	{H: High, L: Low, M: Medium}
EconomicCondition	Factor	How is the economic condition of student
FeePaidPercentage	Numeric	What percentage of the tuition fee has been paid?
MediumOfStudy	Factor	Medium of study in qualifying exam
FeePaidCategorized	Factor	{H: High, L: Low, M: Medium}
MarksCategory	Factor	Categorization of student based on qualifying exam marks
HostelOrTransport	Factor	Did the student avail any of the two facilities?
StudentStatus	Factor	{Joined, Lost} [Target variable]

    iii. LoanLetter
    iv. HomeTownType
    v. QualifyingExam
    vi. PreviouslyStudied
    vii. ScholarshipBracket
    viii. FeePaidCategorized
    ix. HostelOrTransport

The eliminated features are either redundant or strongly correlated with any of the shortlisted features. As a student opts for either hostel or transport facility, a new variable "HostelOrTransport" is derived with values as logical OR operation. "ScholarshipBracket," "FeePaidCategorized," and "MarksCategory" are categorical variables derived from their continuous counterparts "ScholarshipPercentage," "FeePaidPercentage," and "MarksPercent," respectively.

4.2.3 EXPERIMENTS DESIGNED

To explore the applicability of ML in this problem domain, identifying factors that affect freshmen students' joining behavior, and variations across different admission years, the following experiments were designed:

    i. Learning an ML-based model from 2017
    ii. Learning an ML-based model from 2018 admissions
    iii. Learning an ML-based model from 2017+2018 admissions together
    iv. An evaluating model trained using 2017 admissions for predicting joining behavior of freshmen students of 2018 batch
    v. Understanding the behavior of our model black-box model
    vi. Identifying variations in factors affecting freshmen student joining behavior across different admission years.

Table 4.3 mentions the count of admissions and the percentage of "Joined" and "Lost" cases for each year. It also mentions the baseline accuracy which is the classification accuracy if the majority class is always predicted. It is used as a benchmark for evaluating the worthiness of employing an ML solution.

4.2.4 CLASSIFICATION ALGORITHMS AND PERFORMANCE EVALUATION

As listed in Table 4.4, four different ML algorithms for classification were explored to develop a model to predict the joining behavior. Logistic Regression (LR) is based on the use of the logistic function and Naïve Bayes (NB) assigns probability of each class using Bayes theorem. CART algorithm has advantage of being easily interpretable (Breiman et al., 1984). CART tree is used to identify year-wise significant features. Random Forest (RF)makes uses of a collection of typically hundreds of trees and helps improve the prediction accuracy of CART (Breiman, 2001). Classification accuracy (Acc), sensitivity (Se), specificity (Sp), and area under the

TABLE 4.3
Batch-wise Details

Batch Year	Freshmen Students	Students Joined	Students Lost	Students Joined (%)	Students Lost (%)	Baseline Accuracy
2017	14989	8731	6258	58	42	0.58
2018	13125	8374	4751	64	36	0.64
2017+2018	28114	17105	11009	61	39	0.61

TABLE 4.4
Merits and Demerits of Methods Used

Method	Application Area	Merits	Demerits
LR	Binary classification problems. A special case of linear regression.	Gives classification as well as probability. Can be extended from binary to multi-class.	Can be outperformed in predictive performance. Restrictive expressiveness.
NB	Mainly used in text classification and multi-class problems	Training and prediction are quick.	Sensitive to skewed data
CART	Useful in classification and regression tasks	Rapid classification of new examples. Easily Interpretable. Implicit feature selection.	Overfitting. High Variance. Low bias.
RF	Can be used for classification as well as regression problems	High predictive performance. Estimate feature importance.	Less interpretable. Computationally expensive.
Variable Importance	Identification of features affecting the model outcome most	Nice interpretation. Global insight.	Need labeled data.

ROC curve (AUC) were used as performance evaluation metrics of ML models developed using the previous four algorithms.

4.2.5 INTERPRETABILITY TECHNIQUES

Advanced ML models like RF lack in human interpretability despite being very good in prediction accuracy. Lack of human interpretability means looking at the outcomes of these models it is not easy to interpret why a particular target class is labelled for a given instance (Lipton, 2018). Variable importance measures and variable importance plots have been used for extracting global insights into the behavior of a black-box model. Table 4.4 describes the merits and demerits of the methods used.

4.3 PROPOSED MATHEMATICAL FRAMEWORK

For evaluating the performance of ML-based models, the following set of equations are used to compute performance metrics:

$$\text{Acc} = \frac{\text{TP} + \text{TN}}{\text{N}} \tag{4.1}$$

$$\text{Se} = \frac{\text{TP}}{\text{TP} + \text{FN}} \tag{4.2}$$

$$\text{Sp} = \frac{\text{TN}}{\text{TN} + \text{FP}} \tag{4.3}$$

$$\text{P} = \frac{\text{TP}}{\text{TP} + \text{FP}} \tag{4.4}$$

Where TP = True Positive, TN = True Negative, FP = False Positive, FN = False Negative, and N = total predictions. TP and TN together comprise correct classifications. FP and FN together comprise incorrect classifications.

Variable importance measure is a measure of the importance of a variable as rendered by an ML model in its decision making. It helps understand what the variables are that affect the decision outcomes the most. These measures were computed for RF model using the following equations:

Mean minimum depth (mmd) for each node =

$$\text{mmd} = \frac{\sum_{t=1}^{k} (\text{md})^t}{k} \tag{4.5}$$

where (md)t represents minimal depth for a feature in tth tree of the forest and k represents the number of trees in the forest.

gini_d for a feature is computed as below:

$$1 - \sum (p(x) * p(x)) \tag{4.6}$$

where p(x) is the probability of a class.

Increase in error rate i.e. acc_d is computed as below:

$$= 1 - \text{AUC} \tag{4.7}$$

where AUC lies between 0 to 1 and is area under the ROC curve.

times_a_root = Number of times afeature is picked as a

root node while constructing trees of the forest   (4.8)

The pseudo code of the proposed work is given as Pseudo code 1.

To identify year-wise variations in significant-factors, the data set was separated into two subsets, one each for admission batch 2017 and 2018. For each algorithm and data set, a classification model was developed and evaluated using performance evaluation metrics. For understanding the model behavior, mmd, acc_d, gini_d, and times_a_root were computed for each feature used in learning an ML-based model. For mmd, the factor with minimum value is considered most important. For acc_d, gini_d, and times_a_root, the factor with maximum value is considered most important. Plot a CART tree for data of 2017 and 2018 to understand the year-wise difference in important factors affecting joining behavior.

PSEUDO CODE 1   LEARNING AND INTERPRETING ML MODEL

ALGO = {LR, NB, CART, RF}
    NODE = {Set of all predictors}
    VIM = {mmd, acc_d, gini_d, times_root}
    data = data set of freshmen students
    Input: data
    Output: MLmodel, Model behavior
    procedure Learn_Interpret_Model
    Step 1. Split data year wise into D= {d_{2017}, d_{2018} and $d_{2017+2018}$}
    Step 2. **for** each d in D
    for each algo in ALGO
    Learn a $ML_{d,algo}$ model
    Compute Acc, Se, Sp, P, and AUC for $ML_{d,algo}$
    end for
    end for
    Step 3. **for** each node in NODE
    Compute mmd, acc_d, gini_d, times_root
    end for
    Step 4. Identify top nodes with:
    min(mmd);**max**(acc_d); **max**(gini_d); **max**(times_root)
    Step 5. **for** each d in {d_{2017}, d_{2018}}
    plot CART tree $CART_d$
    end for
    Step 6. Identify differences in $CART_{2017}$ and $CART_{2018}$
    end procedure

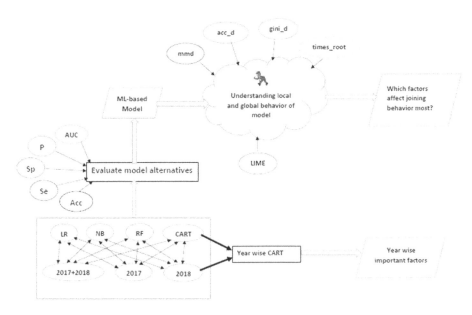

FIGURE 4.1   Proposed framework diagram.

Figure 4.1 provides a graphical representation of the proposed framework. The input to the framework is the data set and a set of classification algorithms. The proposed framework builds and evaluates model alternatives. The outcomes of the proposed work are a predictive model along with an understanding of its global behavior. Additionally, it identifies admission-year-wise variations in factors that affect the joining behavior the most.

4.4   RESULTS AND DISCUSSION

This section presents the outcomes from the experiments conducted in the context of the research questions formulated in the introduction section.

4.4.1   COMPARISON OF MODELS USING PERFORMANCE METRICS

Tables 4.5 and 4.6 compile the results for performance evaluation metrics for ML models using data of the admission years 2017 and 2018, respectively. Table 4.7 presents performance metrics for ML models developed using 2017 and 2018 admissions put together.

It was observed that in terms of classification accuracy, RF outperformed all models learned irrespective of training data (or admission batch) used. All the models generalized well indicating no major overfitting issues. Performance of models for 2018 was better than those for 2017 by around +6%. CART and NB were found to be most sensitive to "Lost" class for admission batch 2017

TABLE 4.5

Comparison of ML Models Using Data of 2017 Admission Batch

Model	Accuracy	Sensitivity (Se)	Specificity	Area Under ROC
LR	.76	.56	.90	.82
NB	.74	.66	.80	.80
CART	.73	.72	.73	.76
RF	.76	.66	.82	.81

TABLE 4.6

Comparison of ML Models Using Data of 2018 Admission Batch and Comparison of ML Models Using Data of 2017 Admission Batch

Model	Accuracy	Sensitivity (Se)	Sensitivity (Se)	Sensitivity (Se)
LR	.82	.66	.91	.86
NB	.80	.67	.88	.85
CART	.81	.62	.91	.79
RF	.82	.63	.93	.84

TABLE 4.7

Comparison of ML Models Using Data of 2017 + 2018 Admission Batches and Comparison of ML Models Using Data of 2017 Admission batch

Model	Accuracy	Sensitivity (Se)	Sensitivity (Se)	Sensitivity (Se)
LR	.78	.63	.88	.83
NB	.75	.67	.81	.82
CART	.76	.60	.86	.75
RF	.78	.61	.89	.82

and 2018, respectively. All models were found good in terms of Specificity and AUC values.

Table 4.8 compiles the performance metrics for predicting joining behavior of 2018 admissions using the model learned using admissions of 2017 batch. All models except CART achieved an accuracy of around 77%. Moreover, the accuracy of predicting the joining behavior of 2018 admissions is more using 2018-based model as compared to that using 2017-based model. It is an indicator of the dynamic nature of this problem domain as joining behavior of students is affected by the changes in admission policies over the different admission years.

4.4.2 INTERPRETING MODEL BEHAVIOR

The RF model had given the best performance in terms of accuracy. To explain the behavior of this black box model, variable importance measures were used. Table 4.9 mentions variable importance measures for RF algorithm-based ML model. These include mean minimum depth(mmd), decrease in accuracy(acc_d), decrease in gini (gini_d), and number of times a node is picked as a root node (times_root) while constructing a tree of RF. For each measure, the most significant four nodes are shown in bold. It is observed that the top four variables with minimum mean minimum depth are ScholarshipBracket, MarksCategory, HostelorTransport, and FeePaidCategorized. Moreover, the variables with max-imum acc_d,gini_d, and times_a_root are ScholarshipBracket, MarksCategory, HostelorTransport, and FeePaidCategorized.

TABLE 4.8
Predicting Joining Behavior for 2018 Using the 2017 Model

Model	Accuracy	Sensitivity	Specificity
LR	.77	.60	.86
NB	.77	.69	.81
CART	.72	.73	.71
RF	.77	.69	.81

TABLE 4.9
Variable Importance Measures

Variable	mmd	acc_d	gini_d	times_a_root
AdmissionMonth	2.596	0.014	247.4	0
FeePaidCategorized	**1.786**	**0.03**	322.5	70
HomeTownType	2.634	0.007	130.2	14
HostelorTransport	**1.758**	**0.028**	**356.6**	**98**
LoanLetter	2.528	0.008	106.6	38
MarksCategory	**1.348**	**0.043**	**467.8**	**105**
PreviouslyStudiedatLPU	3.25	0.004	42.4	6
QualifyingExam	2.598	0.006	109.5	0
ScholarshipBracket	**1.028**	**0.125**	**1,147.8**	**169**

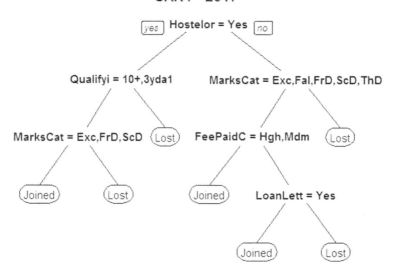

FIGURE 4.2   CART tree for 2017.

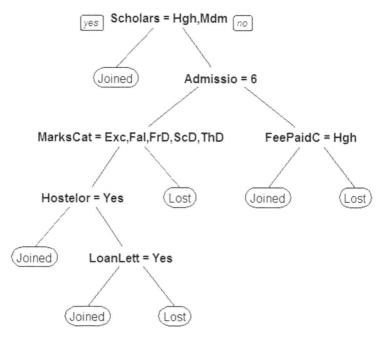

FIGURE 4.3   CART tree for 2018.

4.4.3 Year-wise Significant Features

Figures 4.2 and 4.3 represent tree representations of CART models for admission batch 2017 and 2018, respectively. Many of the features(nodes) comprising these trees are common for batch 2017 and 2018. However, "AdmissionMonth" and "ScholarshipBracket" are new entries in 2018 and the feature "QualifyingExam" is no longer found contributing. Also, the order of nodes while traversing these trees is not the same for common features, which are again an indicator of variations in feature importance over the different admission batches.

4.5 CONCLUSION

Classification accuracy achieved by our ML-based models is around 80%. It is a considerable improvement when benchmarked against the 60% baseline accuracy. The RF algorithm outperformed others by giving the best classification accuracy. "ScholarshipPercentage," "MarksCategory," "HostelorTransport," and "FeePaid Categorized" are identified as the most significant features affecting joining behavior of freshmen students. Variable importance measures are useful in understanding the behavior of a black-box model. These important factors vary from one year to next year owing to the dynamic nature of this problem domain. The outcomes of such research studies can be placed as a feedback mechanism for the management of an educational institution. These outcomes help in planning customized student engagement activities. These activities can in turn help an educational institute improve retention ratio of enrolled freshmen students. The variation in factors affecting the joining behavior over successive admission years is helpful in assessing the impact of changes or innovations in their admission-related processes. Such studies can also help in assessing the impact of external trends prevailing in the education sector.

There are several lines for future work. First, there is a need of additional innovative and relevant features to be incorporated for learning, based on discussion with different stakeholders. Also, there is a need to capture the response of enrolled students towards student engagement activities planned by an educational institute prior to the start of actual classes. Feedback from prospective students regarding factors affecting their choice of an educational institution can be collected and incorporated into learning of the model.

REFERENCES

Abu Zohair, L. M. (2019). Prediction of student's performance by modelling small data set size. *International Journal of Educational Technology in Higher Education, 16*(1). https://doi.org/10.1186/s41239-019-0160-3

Ahmad, F., Ismail, N. H., & Aziz, A. A. (2015). The prediction of students' academic performance using classification data mining techniques. *Applied Mathematical Sciences, 9*(129), 6415–6426. https://doi.org/10.12988/ams.2015.53289

Altujjar, Y., Altamimi, W., Al-Turaiki, I., & Al-Razgan, M. (2016). Predicting critical courses affecting students Performance: A case study. *Procedia Computer Science, 82*(March), 65–71. https://doi.org/10.1016/j.procs.2016.04.010

Breiman, L., Friedman, J., Stone, C. J., & Olshen, R. A. (1984). *Classification and regression trees.* CRC press.

Breiman, L. (2001). Random forests. *Machine Learning, 45*(1), 5–32. https://doi.org/10.1 023/A:1010933404324

He, J., Baileyt, J., Rubinstein, B. I. P., & Zhang, R. (2015). Identifying at-risk students in massive open online courses. *Proceedings of the National Conference on Artificial Intelligence, 3,* 1749–1755.

Hoffait, Anne-Sophie, and Schyns, M. (2017). Early detection of university students with potential difficulties. *Decision Support Systems, 101,* 1–11.

Hughes, G., & Dobbins, C. (2015). The utilization of data analysis techniques in predicting student performance in massive open online courses (MOOCs). *Research and Practice in Technology Enhanced Learning, 10*(1). https://doi.org/10.1186/s41039-015-0007-z

Kaur, P., Singh, M., & Josan, G. S. (2015). Classification and prediction based data mining algorithms to predict slow learners in education sector. *Procedia Computer Science, 57,* 500–508. https://doi.org/10.1016/j.procs.2015.07.372

Krasilnikov, A., & Smirnova, A. (2017). Online social adaptation of first-year students and their academic performance. *Computers and Education, 113,* 327–338. https://doi.org/10.1016/j.compedu.2017.05.012

Kumar, P. & Sharma, M. (2020). Predicting academic performance of international students using machine learning techniques and human interpretable explanations using LIME—Case study of an Indian university. In *International Conference on Innovative Computing and Communications,* (pp. 289–303). Springer, Singapore.

Lau, E. T., Sun, L., & Yang, Q. (2019). Modelling, prediction and classification of student academic performance using artificial neural networks. *SN Applied Sciences, 1*(9), 1–10. https://doi.org/10.1007/s42452-019-0884-7

Lipton, Z. C. (2018). The mythos of model interpretability. *Communications of the ACM, 61*(10), 35–43. https://doi.org/10.1145/3233231

Marbouti, F., Diefes-Dux, H. A., & Madhavan, K. (2016). Models for early prediction of at-risk students in a course using standards-based grading. *Computers and Education, 103,* 1–15. https://doi.org/10.1016/j.compedu.2016.09.005

Mitchell, T. M. (2006). The Discipline of Machine Learning. *Machine Learning, 17*(July), 1–7. Retrieved from http://www-cgi.cs.cmu.edu/~tom/pubs/MachineLearningTR.pdf

Pandey, M., & Taruna, S. (2016). Towards the integration of multiple classifier pertaining to the Student's performance prediction. *Perspectives in Science, 8,* 364–366. https://doi.org/10.1016/j.pisc.2016.04.076

Ribeiro, M. T., Singh, S., & Guestrin, C. (2016). "Why should i trust you?" Explaining the predictions of any classifier. *Proceedings of the ACM SIGKDD International Conference on Knowledge Discovery and Data Mining, 13-17 August,* 1135–1144. https://doi.org/10.1145/2939672.2939778

Simpson, O. (2006). Predicting student success in open and distance learning. *Open Learning, 21*(2), 125–138. https://doi.org/10.1080/02680510600713110

UGC. (2020). No Title. Retrieved from University Grants Commission website: https://www.ugc.ac.in/oldpdf/PrivateUniversity/Consolidated_List_Private_Universities.pdf

Wagstaff, K. L. (2012). Machine learning that matters. *Proceedings of the 29th International Conference on Machine Learning, ICML* 2012, *1,* 529–534.

5 Image Processing for Knowledge Management and Effective Information Extraction for Improved Cervical Cancer Diagnosis

S. Jaya[1] and M. Latha[2]
[1]Research Scholar, Sri Sarada College for Women (Autonomous), Salem-16, India
[2]Associate Professor, Sri Sarada College for Women (Autonomous), Salem-16, India

CONTENTS

DOI: 10.1201/9781003132080-5

5.1   INTRODUCTION

5.1.1   DIGITAL IMAGE PROCESSING

Digital image processing plays an important role in various fields. Imaging is everything, which can be applicable to improve and develop the any technologies as
well as algorithms. Many researchers has been developing different concepts related
to image processing. Image is one of the essential evidence to perform some operations for the purpose of applying mathematical equations, statistical methods,
and algorithm models to improve the visualize of the image. Several image processing tools help to extract the information from the image which are used to move
further for image analysis. Generally image processing has some basic steps, they
are image attainment, image preprocessing, segmentation, feature extraction, feature selection, and recognition. These are the fundamental steps that are followed
for each and every process. Image processing has no end it is real core and limitless.
Basically an image has a two-dimensional matrix that represents rows and columns.
The spatial coordinates are horizontal, vertical f(x,y), and the image is represented
with the combination of intensity and pixel value. The operations can be done over
the image by adjusting the intensity and pixel value. The range between 0 to 255 is
the 2D array of the image. Based on the pixel range, the image will be displayed.
Basically image processing has three steps:

1. Image acquisition tool for capture the object (input)
2. Accessing and manipulating the image (adjust, update)
3. Resulted image after made some alteration (output)

(Figure 5.1)
Basically the image has three color models: binary image, gray-scale image, and
color image/RGB color models. Binary image is the combination of black and
white, gray image is the combination of spatial gray colors, and color image is the
combination of red, green, and blue. Making some modifications on a particular
image will make a good appearance as well as reach good quality. Digital image
processing is used by several applications that help to improve their working goal.
Agriculture, medicine, remote sensing, pattern recognition, biometric, and forensic
departments are the different fields that utilize image processing technology. At
present machine learning and artificial intelligence are the most welcoming technology. Virtual reality and augmented reality are the trending fields that make
modern society. These modern technologies are developing only because of digital
image processing. The necessary image processing is very essential to solve the
problem in various applications.

5.1.2   FUNDAMENTAL KEY STAGES IN DIGITAL IMAGE PROCESSING

There are more important steps to be used for the purpose of enhancing the image in
order to improve the pixel information (Figure 5.2).

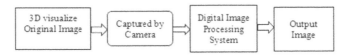

FIGURE 5.1   Representation of digital image.

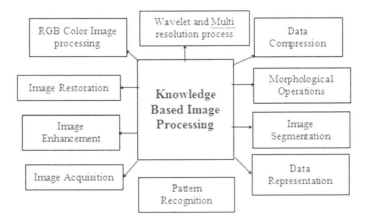

FIGURE 5.2   Life cycle process in digital image processing.

a. **Image Acquisition**

Image capturing is a transformation of optical image into numerical data in form of matrix which can be used by the system in future to make some process. The captured image can be taken from any camera based on the different application. It is the very first step in image processing, where manipulation of the image applicable in this application. Energy is important source for capturing the image and it is described by E(Energy). The frequency can be called as f that is H_2. Wave length is measured to determine the photo λ(m).

$$E = (hc)/\lambda$$

$$E = hf$$

Quantum detector is a very essential mechanism for image sensing and image acquisition. These are the elements playing a vital role during image acquisition.

b. **Image Enhancement**

Image enhancement is mostly used by several applications. Contrast enhancement is generally used to enhance the visualization of the image by focusing the object in a image. It deals with the difference appearance between the object and the background. Image enhancement is adjusting the spatial colors and brightness of the entire image. This image enhancement considers two areas:

1. Kind of spatial domain
2. Kind of frequency domain

Spatial domain focuses on the intensity value to manipulate the image to improve the brightness and increase the contrast level. It is to be represented by:

$$g(x, y) = T[f(x, y)]$$

Frequency domain can be manipulated by the Fourier transform of a image. Histogram equalization, image smoothing, and sharpening techniques are widely used for image enhancement.

c. **Image Restoration**

Image degradation may happen due to some illuminations and camera defects. Various kinds of noise are involved in an image like, blurring, sunlight illuminations, misfocus, and many more aspects. Degradation model considers x as original image and y denotes data (Figure 5.3).

Spatial method: $g(x,y) = h(x,y) * f(x,y) + \eta(x,y)$
Frequency method: $G(u,v) = H(u,v) F(u,v) + \eta(u,v)$
Image representation: $G = HF + \eta$

These are all degradation functions that are very essential and approximation of H is also necessary in image restoration model. The value of H can be determined by three ways:

1. By observation
2. By experimentation
3. By mathematical modeling

Completion of the degradation function will move towards to Blind Convolution to regenerate the original image. Restoration techniques are inverse filtering, wiener filtering, constrained least square filtering, and non-linear filtering.

d. **Color Image Processing**

The color image processing consists of brightness, hue, and saturation. It simplifies feature extraction and feature identification. Color is a powerful tool for

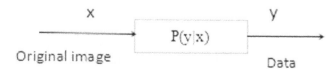

FIGURE 5.3   Image restoration model.

segmentation and object recognition. This color image processing is derived into two steps: full color processing and pseudo color processing. RGB color image processing is applied in full color space like color television, color scanner, and pseudo color processing is grays cale color converted into full color images for visualization purposes. RGB colors are red, green, and blue and essential colors are cyan, magenta, and yellow. The combination of primary colors produces the black color. Each RGB colors are represented in terms of 0 to 255. The conversion of RGB to CMY color performs the equation:

$$\begin{bmatrix} R \\ G \\ B \end{bmatrix} = \begin{bmatrix} 1 \\ 1 \\ 1 \end{bmatrix} - \begin{bmatrix} C \\ M \\ Y \end{bmatrix}$$

Generally the colors will be normalized in the range of 0 to 1. RGB color space turns over into HSI (Hue, Saturation, Intensity) model:

$$H\,(Hue) = \begin{Bmatrix} \theta & if\ B \le G \\ 360 - \theta & if\ B > G \end{Bmatrix} \tag{5.1}$$

$$S\,(Saturation) = 1 - \frac{3}{(R + G + B)}[\,min(R,\ G,\ B)] \tag{5.2}$$

$$I\,(Intensity) = \frac{1}{3}(R,\ G,\ B) \tag{5.3}$$

Color transformation performs by this equation:

g(x, y) = T[f(x, y)]
By given f(x,y) - color input image
g(x,y) - output image
T - color transform.

e. **Wavelet and Multi-Resolution Process**

Wavelet transform used to provide time frequency information also it is based on the small waves. Fourier transforms are used to provide only frequency information. The application of wavelet and multi-resolution is image compression, image analysis, and image transmission. Small objects in the image can be displayed at high resolution whereas large objects in the image can be displayed at coarse resolution. Wavelet transform is a combination of features are pyramids, subband coding, and Haar transform. Image pyramid at the level of P + 1 for P > 0:

$$N^2(1) + \frac{1}{(4)^1}f + \frac{1}{(4)^2} + \cdots + \frac{1}{4^p}\Big) \le \frac{4}{3}N^2 \tag{5.4}$$

The original image can be decomposed into a group of band limited components called as Subband. Down and up sampling are important factor to be applied in subband equation. After applying down and up sampling substitution on the subband equation:

$$X'(z) = \frac{1}{2}G_0(z)[H_0(z)X(z) + H_0(-z)]X(-z)] + \frac{1}{2}G_1(z)[H_1(z)X(z)$$

$$+ H_1(-z)X(-z)]$$

$$= \frac{1}{2}[G_0(z)H_0(z) + G_1(z)H_1(z)]X(z) + \frac{1}{2}[G_0(z)H_0(-z) + G_1(z)H_1(-z)]X(-z)$$

$$(5.5)$$

Haar wavelet transform is the simplest and oldest method used for signal and image compression. The basic form of Haar wavelet is:

$y_n = H_n x_n$ Whereas n- input function
The inverse of Haar transform is,

$$x_n = H^T y_n$$

f. Image Compression

Image compression refers to reducing the size of the image for storing purposes in the huge database. It is considered when maintaining large database for storing memory. Lossless and lossy image compression techniques are widely used by various applications. Especially in medical image diagnosis these techniques will be used. Lossless compression is to reduce the file size without loss of information and image quality will be remain same. Lossy compression is permanently removing the information and not storing the original image. PNG, JPEG, and GIF are the types of compressed image files (Figure 5.4).

g. Morphological Processing

Morphological operations are working based on the set theory. The main objectives of the morphological operations are removing the imperfection structure

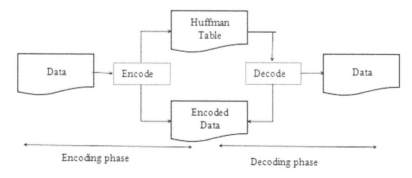

FIGURE 5.4 Huffman table work process during image compression.

and shapes in the image. Opening, Closing, Erosion, and Dilation operations play the essential roles to solve the structuring problem. Structural element is the small set of matrix that is considered a tiny part from the image which considers the value of 0 or 1.

Erosion simply defines the erode, shrink, and reduce. It deals with removing pixels on the object boundary. Minimum neighborhood pixel set is 0 in the case of binary image.

The erosion can be done by,

$$A \ominus B = \{z | (B)_z \subseteq A \tag{5.6}$$

A derives set,

B derives structure element, diminution object.

Dilation is increasing the size of the object whereas filling holes in shape and size based on the structuring element. The objects will be more visible. Maximum neighborhood pixel indicates 1 in the case of a binary image.

Opening and Closing Operations

These operations are generally used to restore the shrunk images. Opening operation performs to bring the original image to the extreme possible extend. It works erosion, and then dilation can be done. Closing operation can be done by eliminating the holes as well as used for smoothing the contour.

Opening operation denoted by,

$$A \circ B = (A \ominus B) \oplus B \tag{5.7}$$

Closing denoted by

$$A \cdot B = (A \oplus B) \ominus B \tag{5.8}$$

h. Segmentation

Image segmentation is dividing the image into a multiple parts or required objects. Segmentation can be done for further image analysis and object detection. This techniques applied in several applications like image retrieval, medical image analysis, tumor segmentation, object detection, pedestrian detection, face, iris and biometric detection, and traffic surveillance video. Threshold is one of the traditional methods followed for segmentation and it is to be applied on the binary image. Detection of discontinuities is based on the points, edges, and lines. Gradient operator in edge detection is determined by Sobel, Prewitt, and Laplacian.

Gradient operator works on,

This is 3×3 matrix of a image.

A1A2A3

A4A5A6

A7A8A9

for 3×3 mask

$$H_x = (A_7 + A_8 + A_9) - (A_1 + A_2 + A_3)$$
$$H_y = (A_3 + A_6 + A_9) - (A_1 + A_4 + A_7)$$
$$\tag{5.9}$$

$$\nabla f = \begin{bmatrix} H_x \\ H_y \end{bmatrix} = \begin{bmatrix} \frac{\partial f}{\partial x} \\ \frac{\partial f}{\partial y} \end{bmatrix}$$

$$|\nabla f| = [H_x^2 + H_y^2]^{1/2}$$

$\frac{\partial f}{\partial x}$ - gradient operator in x direction

$\frac{\partial f}{\partial y}$- gradient operator in y direction

Image gradient is used for edge detection as well as feature matching in image processing. Clustering-based segmentation works based on the nearest pixel of K values.

Laplacian Operator

This operator is a function f(x, y) which may be called second-order derivative and will be characterized like

$$\nabla^2 f = \frac{\partial^2 f}{\partial x^2} + \frac{\partial^2 f}{\partial y^2} \tag{5.10}$$

The Laplacian of Gaussian (LOG) is an operation that can be determined by

$$\nabla^2 h(r) = -\left[\frac{r^2 - \sigma^2}{\sigma^4} \right] e^{-\frac{r^2}{2\sigma^2}} \tag{5.11}$$

Representation and Description

After image segmentation, the segmented pixel will move further computer processing to be represented and described for feature extraction. Image representation can have two ways in order to represent and describe them:

1. External characteristics (Shape)
2. Internal characteristics (Color and Texture)

Chain code is the best binary representation that connects the straight line and it is generally represented by 4 connectivity or 8 connectivity segments. This chain code gives better results by using binary images with shape contour. The direction of chain code is given (Figure 5.5).

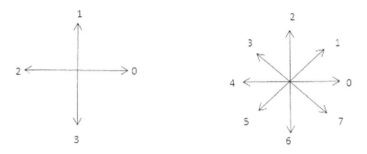

FIGURE 5.5 Four and 8 direction chain code.

Previous equation of discrete Fourier transform (DFT) will be derived by,

$$a(u) = \frac{1}{K} \sum_{k=0}^{K-1} s(k) e^{-j2\pi uk/K} \tag{5.12}$$

$$S(m) = x(m) + jy(m) \tag{5.13}$$

j. Object Recognition

Object recognition is a method applied on images to find the appearance of the real object. Object recognition can be executed by using deep machine learning and artificial intelligence. Training the model is a very important task in object recognition to detect the place accurately. Huge amounts of data are needed for training the model to obtain the accuracy where the data should be separable using class labels. Segmentation is one of the techniques to focus the object without displaying background. Feature extraction is used for object detection based on the shape, size, area, perimeter, structure, mathematical, and statistical values which is followed after segmentation. Many machine learning algorithms are available to recognize the pattern based on the data given by the programmer (Figure 5.6).

5.2 LITERATURE REVIEW

There are various algorithms available to attain good research work. Several engineers and researchers are focusing towards the best developments in the field. Here is the collection of related articles that are used for my work (Table 5.1).

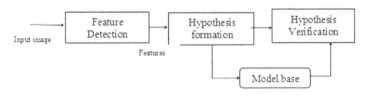

FIGURE 5.6 Object recognition system.

TABLE 5.1

Review of Various Research Articles

Author/ Date	Theme of the Work	Description of the Techniques	Source	Research Findings
PL. Chithra & P. Bhavani	Preprocessing, segmentation, feature extraction and classification	Histogram equalization, DWT (Discrete Wavelet Transform), SVM	Apple fruit, banana fruit, bay berry, and number plate data set acquired by digital camera	Found that different kind of data set and algorithm will attain the best results in classification.
Dhankhar, P & Sahu,N	Overall review on Digital Image Processing	Summary of data	Reference images have been taken from the various research engines.	Various tools and applications are presented
Ryszard S et al	Done Content Based Image Retrieval in biometric images and essential features are extracted depends on, texture, shape, and color.	CBIR and Biometric system	Iris image database, Lena in built images in the tool.	Found color moments in range of 0,45,90 degree for coke and motyl. RGB, HSV,CYR Color models were analyzeds.
Lagrid, M et al.	Taken mammographic collective data set as well as thyroid illness data set using supervised classification algorithm of coherence verification, stochastic validation, and absolute validation	CBR System	Failed to mention the source of database.	Novel case based segmentation methods introduced for the purpose of retrieving information.
Poonam Dhankhar, Neha Sahu	Review on various edge detection techniques in image processing.	Sobel, Prewitt, LoG, Roberts, and Canny edge detection.	Used original building images further.	Proved canny edge detection algorithm is better than others using

TABLE 5.1 (Continued)
Review of Various Research Articles

Author/ Date	Theme of the Work	Description of the Techniques	Source	Research Findings
				non-maximal suppression.
Sonam Saluja et al.	Used various edge detection techniques.	First and second order derivative based edge detection, Sobel, Prewitt, LoG, Roberts and Canny edge detection.	Preferred important structural properties in a image over entire paper.	Analyzed canny edge detection is best in order to edge detection. Adaptive edge detection algorithm helps to adjust a noise level.
P. Mohanaiah et al.	Feature extraction using GLCM properties. Graycomatrix function has been used to obtain the features.	Angular second moment, Inverse difference moment, correlation.	Failed to mention the source of database.	Found that GLCM is suitable for image feature extraction and DWT is suitable for video processes.
Dharun, V	Extracted texture features by GLCM using brain tumor MRI images and shape features for regions	GLCM	Brain MRI images were collected from various diagnosis centres in Kerala, India.	Connected region concepts used for texture and shape feature extraction.
Ansari, M. et al.	Implemented Edge detection techniques and quality measurements PSNR and MSE	Sobel, Prewitt, LoG, Roberts and Canny edge detection- First Order Derivative - PSNR, MSE	Acquired original sun flower and face images by the authors own.	Second order derivative gives the best result for canny edge detection.
Chandwadkar, R. et al.	Comparison between various edge detection methods	Sobel and Canny detector	General images were taken from public search engine.	Taken review on essential of edge detection algorithm
Cauchard, V et al.	CBR approach for reusing the IP (Image	CBR Module, plan based architecture ad	Biomedical images of cytology and histology	Author felt that choosing common vocabulary for

(Continued)

TABLE 5.1 (Continued)
Review of Various Research Articles

Author/ Date	Theme of the Work	Description of the Techniques	Source	Research Findings
	Processing) application	case based architecture	images from cancer research canter	IP program is difficult and also can be solved by retrieval algorithm
Izquierdo, E	Knowledge management in case of surveillance application	K-Space framework as well as cost292	Real-time CCTV surveillance video	Automatic classification and semantic-based analysis
Adnan, K, & Akbar, R.	Retrieving meaningful information from unstructured data from literature review	Information extraction techniques	Review papers	Limitation of IE techniques in the area of image, video, text, and audio.
Chen, H. et al.	Used knowledge management, text mining, and data mining in medical utilization.	Machine learning approaches are probational and statistical modeling.	Biomedical informatics	Machine learning algorithms are best handing for biomedical data to secure the information
Wang, Y. et al.	Presented clinical information extraction based on recent review papers	Discussed and analyzed machine learning algorithms from previous authors	Electronic Health Records (EHR)	Tried to analyze the gap between EHR and IE.
Mahamune, M. et al.	Knowledge Discovery Database (KDD) is data mining techniques used for the purpose of retrieving data from the healthcare database	KDD data mining techniques and KM model	Healthcare database	Found KDD is the best data-mining techniques to explore the hidden information from medical data-ware house.

TABLE 5.1 (Continued)
Review of Various Research Articles

Author/ Date	Theme of the Work	Description of the Techniques	Source	Research Findings
Adnan, K & Akbar, R	IE techniques in case of unstructured data	Review article analyzed	Multifaceted unstructured data	May consider variety of unstructured data in IE techniques

The previous table refers to the various research articles related to the data science, computation excellence, information computing, and society for feature extraction under various applications.

5.3 PROPOSED METHODOLOGY

5.3.1 DATA COLLECTIONS

The data set of cervical cancer pap smear images has been taken from the Herlev database Hospital. The collection of data set will have both normal and abnormal cancerous images. The image is called a pap smear microscopic image where test by the pathologist. The sample test sample will be go for screening process to the laboratory. Based on the size of nucleus and cytoplasm the result will be obtained. Nearly 65 images are taken for the experiment to show the result. The main objective of this chapter is information extraction by using CBR and OBIA algorithm by following segmentation and feature extraction (Table 5.2).

The previous table shows the two types of data set of normal and abnormal cell type under cervical cancer. These are the data which has been working for this whole chapter, implemented by the MATLAB Tool.

5.3.2 PREPROCESSING – IMAGE ENHANCEMENT

Image enhancement is a process of improving contrast adjustment and sharpness in the image. This is the initial step for the entire image processing application. The

TABLE 5.2
Details of Data Collection

Types of Data set	Image Type	Size of Data	Database
Normal Cell	BMP	58	Bethesda Database
Abnormal Cell	BMP	72	

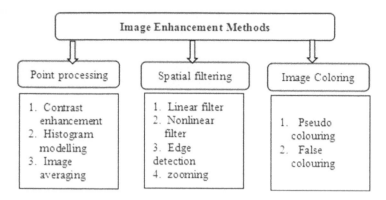

FIGURE 5.7 Image enhancement methods.

main working process is adjusting the pixel values until better visualization. The basic image enhancement methods are shown (Figure 5.7).

The general formation of brightness and contrast adjustment equation given below:

$$Brightness\,(J) = \frac{1}{XY} \sum_{i=1}^{X} \sum_{j=1}^{Y} f(i, j) \tag{5.14}$$

$$Contrast\,(C) = \sqrt{\frac{1}{MN} \sum_{i=1}^{M} \sum_{j=1}^{N} [f(i, j) - J]^2} \tag{5.15}$$

Where XY and MN denote dimension of the image, f(i,j) denotes value of gray level on the original image at ((i,j).

5.3.3 Intensity Transformation Function

In gray-scale images, if we want to improve the contrast, we need to adjust the intensity values. For example, requirement of darker or lighter images that can be applicable using intensity either black or white. Intensity transformation function may have four subparts:

1. Photographic negative function
2. Gamma transformation function
3. Logarithmic transformation function
4. Contrast stretching transformation function

a. **Photographic Negative**

Photographic negative is the easiest function in intensity transformation. While working with a gray-scale image, it assumes that 0 is black such as 1 is white. In a photographic negative, it is like a 0 become 1 and 1 become 0. It is a complement of original image. It describes true black color will be true white color as well true white color will be converted into true black color. Imcomplement() function is used on a image by using [0,L-1] with s = L-1-r as reversing the intensity value of an image:

$$G(x, y) = L - 1 - f(x, y)$$

L-1 shows the upper-limit pixel value of the object r denotes pixel value.

b. Gamma transformation

Gamma transformation is a tool for adjusting the brightness level based on different kinds of images. If the value of $_{gamma}$ is less than 1, it is to be brightened and if the value of $_{gamma}$ more than 1, it is to be darkened.

$$S = c \ r^{\gamma}, \quad c, \gamma \geq 0$$

Whereas (γ) is fractional gamma power. This is otherwise called gamma law transformation. Based on the gamma power, the intensity will be determined.

c. Logarithmic Transformation

Log transformation is substituting all pixel value using logarithm value. The aim of transformation is to enlarge the values of a dark pixel in a image when compressing higher values (Figure 5.8).

The general form of Log transformation calculation is determine by:

$$s = c \ \log(r + 1), \quad r \geq 0$$

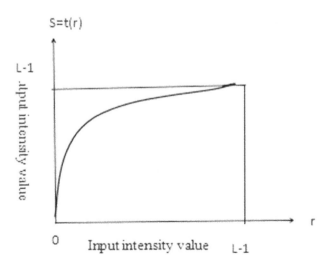

FIGURE 5.8 Narrow and wider range of input and output intensity values.

The main properties of log transformations are having two phases:

1. Gray level will be expanded where in lower extend of input image.
2. Gray level will be compresses where in higher extend of input image.

d. **Contrast Stretching**

The determination of low contrast image will be less illumination, deficiency of dynamic range, lack of camera during image acquisition. It may otherwise be called normalization. Improve the contrast level by adjusting the stretch intensity range of pixel values. Specify and fix the lower and upper limit of pixel range between 0 to 255 before stretching. Then call the function a and b; also P is each pixel value.

$$P_{out} = (P_{in-c})\left(\frac{b - a}{d - c}\right) + a \qquad (5.16)$$

The value can be given from 0 or 255. Find the lowest and highest pixel value by calling c and d whereas P is each pixel value and will be scaled by applying the equation.

Min-Max contrast stretching is defined by lower and upper pixel values of the input image. This algorithm works based on:

$$X_{new} = \frac{X_{input} - X_{min}}{X_{max} - X_{min}} \times 255 \qquad (5.17)$$

Whereas lower limit might be 0 and upper limit might be 255.

5.3.4 SEGMENTATION

Segmentation is a process grouping the correlated pixels in a image. Discontinuity and similarity are the essential properties in the image. Based on these properties, the object will be partitioned. The discontinuity of a pixel is depends on edges, lines, and points. This chapter gives edge detection techniques of Sobel, Prewitt, Canny, and Laplacian operators with the implementation of cervical cancer data set on Pap smear images.

a. **Sobel Edge Detection**

To calculate the intensity level of the image, measure the matrix.

Step 1: The first step in sobel edge detection is x and y direction of the image that can be separated initially.
Step 2: After that, combine both edges in a image.
Step 3: Conversion of image from RGB to gray-scale image.

Step 4: Kernel convolution 3×3 matrix using weighted indexes for filtering purpose.
Step 5: Scan operation has to be done across the X and Y direction of an image for scanning huge changes in the gradient.

The following equation is used to calculate the gradient:

$$G = \sqrt{Sx^2 + Sy^2} \tag{5.18}$$

Whereas,

G - Sobel operator
S_x - Horizontal Sobel gradient
S_y - Vertical Sobel gradient
The basic calculation of sobel convolution kernel is under 3×3 matrix (x,y) pixels.
The calculation of x gradient:

$$S_x = (b_2 + cb_3 + b_4) - (b_0 + cb_7 + b_6) \tag{5.19}$$

The value of c will be 2.

-1	0	1
-2	0	2
-3	0	1

G_x

+1	+2	+1
0	0	0
-1	-2	-1

G_y

The calculation of y gradient will be given:

$$S_y = (b_0 + cb_1 + b_2) - (b_6 + cb_5 + b_4) \tag{5.20}$$

b. **Prewitt Edge Detection**

Prewitt operator is based on the gradient pixel values in the image. It has two types that are horizontal edges and vertical edges. Prewitt edge detection is slightly similar to Sobel edge detection but the difference is the coefficient of the mask. The main properties of the Prewitt operator are masked, and should have opposite sign when compared to original matrix and the sum of both should be equal to 0. The horizontal and vertical derivations are:

-1	0	+1
-1	0	+1
-1	0	+1

M_x

+1	+1	+1
0	0	0
-1	-1	-1

M_v

Prewitt edge detection algorithm follows:

$$M = \sqrt{M_x^2 + M_y^2}$$

(5.21)

Pixel value is (i,j) represented by:

$$
\begin{matrix}
a_0 & a_1 & a_2 \\
a_7 & [i, j] & a_3 \\
a_6 & a_5 & a_4
\end{matrix}
$$

The derivations of M_x and M_y will be:

$$M_x = (a_2 + ca_3 + a_4) - (a_0 + ca_7 + a_6)$$

(5.22)

$$M_y = (a_6 + ca_5 + a_4) - (a_0 + ca_1 + a_2)$$

(5.23)

c. Laplacian Edge Detection

Laplacian is a second spatial derivation of a image. This algorithm used highlights the region of pixel intensity changes in the image. It may otherwise be called Laplacian of Gaussian. The primary work of Laplacian is reducing noise and makes the image smoothing. This algorithm may have Inward edges as well as Outward edges. Also, it contains two different kinds of operators which can be implemented one after another. The two operators are:

1. Positive Operator
2. Negative Operator

Positive Laplacian Operator describes that the middle element should be negative and four corners of the image should be 0. This operator considers outward edges in an image.

Negative Laplacian Operator has the center element as positive, such that the corner should be 0 and remaining element in the mask should be -1.

Laplacian can be defined as:

$$\nabla^2 = \nabla \cdot \nabla = \begin{bmatrix} \frac{\partial}{\partial x} \\ \frac{\partial}{\partial y} \end{bmatrix} \cdot \begin{bmatrix} \frac{\partial}{\partial x} \\ \frac{\partial}{\partial y} \end{bmatrix} = \frac{\partial^2}{\partial x^2} + \frac{\partial^2}{\partial y^2}$$

(5.24)

While applying on the image, the result will be:

$$\nabla^2 f = \left(\begin{bmatrix} \frac{\partial}{\partial x} \\ \frac{\partial}{\partial y} \end{bmatrix} \cdot \begin{bmatrix} \frac{\partial}{\partial x} \\ \frac{\partial}{\partial y} \end{bmatrix} \right) l = \frac{\partial^2 l}{\partial x^2} + \frac{\partial^2 l}{\partial y^2}$$

(5.25)

The result of the 2D image matrix in form of:

0	1	0
1	-2	1
0	1	0

+

0	0	0
0	-2	0
0	0	0

=

0	0	0
0	-2	0
0	0	0

Laplacian is cheaper to implement while compared to gradient and it is more sensitive to remove noise. Also it will not provide any details about edge directions and it is simply called an isotropic operator, which means all weights in equal directions.

The Laplacian of Gaussian operator will be:

$$\nabla^2 f = \frac{\partial^2 f}{\partial x^2} + \frac{\partial^2 f}{\partial y^2} \tag{5.26}$$

Gaussian filter is used; call it LoG(Laplacian of Gaussian):

$$G(x, y) = e^{-\frac{x^2+y^2}{2\sigma^2}}$$

Where σ standard deviation controls smoothing in a image.

Laplacian of Gaussian(LoG) function centers 0 with σ standard deviation of Gaussian in term of:

$$LoG(x, y) = -\frac{1}{\pi\sigma^4}\left[1 - \frac{x^2 + y^2}{2\sigma^2}\right]e^{-\frac{x^2+y^2}{2\sigma^2}} \tag{5.27}$$

Smoothing works depends on the standard deviation value which may varying at each time.

5.3.5 FEATURE EXTRACTION

Feature extraction comes under the part of dimensionality reduction. Feature extraction in digital image processing is accessing and manipulating the pixel and intensity values from extracted information in a image. The main duty of feature extraction is reducing the size of the data and also extracting shape, size, color, and texture from the image. Image processing has various feature extraction algorithms (Figure 5.9).

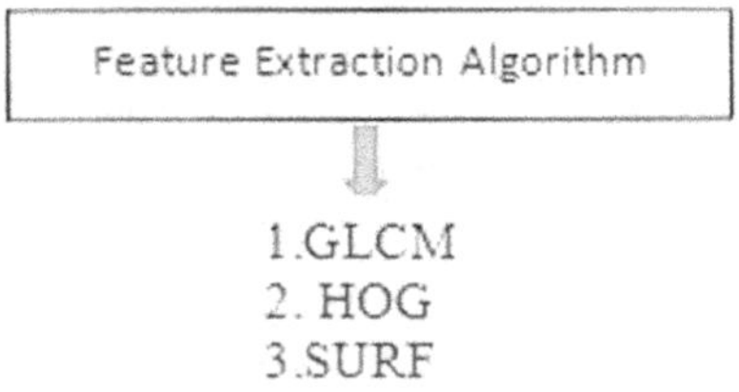

Feature Extraction Algorithm

1.GLCM
2. HOG
3.SURF

FIGURE 5.9 Three kind of Feature extraction algorithm.

a. **GLCM Feature Extraction**

GLCM feature calculates texture attributes in an image to find feature statistical properties. These are the important properties in GLCM spatial gray level dependence. Features will be taken from the segmented images to move further for classification. The statistical and numerical values are obtained after extracted information that can be load using machine learning algorithms.

The GLCM properties are given below:

$$1. \quad Contrast = \sum_{a,b=0}^{N-1} I_{a,b}(a-b)^2 \tag{5.28}$$

$$2. \quad Correlation = \sum_{a,b=0}^{N-1}\left[\frac{(a-\mu_a)(b-\mu_b)}{\sqrt{(\sigma_a^2)(\sigma_b^2)}}\right] \tag{5.29}$$

$$3. \quad Energy = \sum_{a,b=0}^{N-1} I_{a,b}^2 \tag{5.30}$$

$$4. \quad Entropy = \sum_{a,b=0}^{N-1} I_{a,b}(-\ln I_{a,b}) \tag{5.31}$$

$$5. \quad Homogeneity = \sum_{a,b=0}^{N-1} \frac{I_{a,b}}{1+(a-b)^2} \tag{5.32}$$

$$6. \quad Mean = \mu_a = \sum_{a,b=0}^{N-1} a(I_{a,b}), \mu_b = \sum_{a,b=0}^{N-1} b(I_{a,b}) \tag{5.33}$$

$$7. \ Standard\,Deviation = \sigma^2 = \frac{1}{N-1}\sum_{i=0}^{N-1}(x_i-\mu)^2, whereNisnumberofsample$$
$$\tag{5.34}$$

These are the important texture features of GLCM.

b. **HOG (Histogram of Oriented Gradients)**

Histogram of gradients generally is used to expand some information from an image. It may otherwise call as feature descriptor. The main focus of HOG is concentrated on structure and shape of the image data. It calculates localized portion which means the entire image will be broken into small regions, gradients and orientation have to be calculated to each region. The first step in a HOG feature descriptor is preprocessing the original image; re-sizing the image from 720×475 pixels to 64×128 pixels, which is the standard size of the image. Now calculate gradients with x and y direction. Gradients should be calculated for each pixel. For example, the sample pixel vale of the image:

```
120  43   23   58   98
63   87   60   187  244
04   47   65   155  167
78   54   58   95   234
40   36   93   44   101
```

Sample 5×5 matrix

The center pixel value is 65; now find gradient in x-direction, for that subtracts the value from right side to the left side value. Same as y-direction, subtract the above pixel value from the below pixel value. So the gradient of x and y direction looks like:

X-direction(G_x) = 155-47=108
Y-direction(G_y) = 60-58=2
Now calculate gradient magnitude by following:

$$\text{Totalgradientmagnitude} = \sqrt{[(G_x)^2 + (G_y)^2]} \tag{5.35}$$

G_x is 108 and G_y is 2.

Gradient of magnitude will be get by following the example.
Orientation and angle of the same pixel value calculated by:

Orientation $- >\tan(\phi) = \frac{G_y}{G_x}$ and Angle$>a\tan\left(\frac{G_y}{G_x}\right)$

c. **SURF (Speeded-Up Robust Features)**

SURF is kind of feature detector and feature descriptor. This technique can be used by various applications such as object recognition and classification model. This is slightly similar to SIFT (Scale-Invariant Feature Transform). SURF is very advanced and fast version than SIFT. To detect the object in a image, it uses determinant of Hessian blob detector, which consists of three integer operator. SURF descriptor will be used to recognize the object, like face, car, anything and it is used to track the object using points of interest. SURF has major three phases:

1. Interest point orientation
2. Nearest neighbor description components
3. Feature matching

Basically, SURF uses square shape filters by Gaussian smoothing. Square four corners are used for filtering the image, it might be very fast to filter if the input image is integral. The equation used for filter the integral image:

$$X(x, y) = \sum_{a=0}^{x} \sum_{b=0}^{y} I(a, b) \tag{5.36}$$

Depends on Hessian Matrix, blob detector will be calculated the point of interest in SURF. The original image referred as I, point shows p=(x,y) and the Hessian Matrix H(p, σ), point p as well as σ denotes scale:

$$H(p, \sigma) = \begin{pmatrix} A_{xx}(p, \sigma)A_{xy}(p, \sigma) \\ A_{yx}(p, \sigma)A_{yy}(p, \sigma) \end{pmatrix} \tag{5.37}$$

$A_{xx}(p, \sigma)$ is the structure of second-order derivative under Gaussian filter in the image I(x,y) at a point of p.

For example, the size of the box filter is 9 ×9 and the value of Gaussian $\sigma = 1.2$. The approximation of C_{xx}, C_{yy}, C_{xy} can be determined by Gaussian:

$$\det(H_{approx}) = C_{xx}C_{yy} - (wC_{xy})^2, \text{ where} w = 0.7(\text{Bay'svalue}) \tag{5.38}$$

5.4 ROLE OF KNOWLEDGE MANAGEMENT AND EFFECTIVE INFORMATION EXTRACTION IN IMAGE PROCESSING

The main objective of this chapter is to solve the complexity in image processing techniques in medical applications. Totally there are three different concepts discussed in this chapter which includes techniques in digital image processing, handling role of knowledge management and feature extraction concepts in image processing, effect of cervical cancer on pap smear images under microscopic objects. A knowledge management system is simple called as problem solving method which describes about the collection of data in a specific field like medical, finance, banking, agriculture, and many more.

Generally, a knowledge-based system is developed from the several distinct applications and also artificial intelligence is developing area which grows under a knowledge-based system. Decision-making is done based on the available information behind the database of each field. Information is very essential for future references to generate new data from the database by applying KBS. This KBS is helpful in healthcare departments where doctors can easily find out the disease state of the patient based on the stored data.

Knowledge management has many advantages that healthcare can benefit from. Knowledge management system should develop their process in order to increase the business strategy and improve decision making system. KM is able to store the data, generate new data, and recycle the data. It helps to identify the medical errors. KM technology can resolve the complexity problems in healthcare applications. Knowledge-based technology defines the information on techniques and algorithms used to improve the quality of the image in image processing. For instance, preprocessing, noise removal, image enhancement, partition the object, and feature derivation.

Cervical cancer is one of the scariest diseases and nowadays it is gradually increasing, which spoils the women's lives. Even though many technologies are available, also many scientists and doctors are developing their methodology but still it does not control the early stage. Pap smear images taken from the women's cervix captures the abnormal cell tissue form, followed by PAP and HPV test. Those microscopic images will be sent to the pathology laboratory for analyzing the cells' state of normal and abnormal tissue. Based on the size, shape, and structure of the cell, the doctor can get the patients report whether they are in the initial stage or mild attack of cervix or in final stage (severe stage).

Information extraction is the same as feature extraction of an image; consists of number of pixels and intensities. The determination of a image is depends on the pixel values and intensity values. Features will differ from each image such as to detect the tumor or abnormal state in medical image; the features may vary from other application image. The features of remote sensor images differ from medical images. So, features are very essential in order to extract the information for the purpose of obtaining accurate results in any application. Digital image processing is a vast area to develop the several applications. Many fields are getting benefits by using image processing technologies to recover from the complexity and challenges in their filed. The following methodologies can be done over image processing by removing unwanted pixels, eliminating noise, improving image quality based on pixel range, developing visuals of the entire image, detecting specific objects from the image, and recognizing and analyzing the object in the final stage to get the result.

5.5 KNOWLEDGE MANAGEMENT IN CASE-BASED REASONING METHODOLOGY ON HEALTHCARE HANDLING

CBR is widely used in medical and industry applications. CBR is a powerful tool in image processing in order to improve the quality based on image interpretation system and also it supports the decision-making framework. It solves problems based on the previous solved records from the source case which may called as case base. The primary objective of the CBR is selecting the most similar cases to solve the new problem.

The process of case-based reasoning as follows:

1. Recovering most similar cases
2. Apply existing solution which are similar to the new cases
3. Take a revision on new proposed solution
4. Retaining the solution of new problem, can be save for future references.

Knowledge management is kind of reusing the cases to improve the application. CBR technology may be applicable in various sectors such as data analytics, data design, classification and regression models, decision making, and planning design. The main goal of both KM and CBR is collection of data and reusing the data for future references. For example, in the medical sector, it has huge amounts of

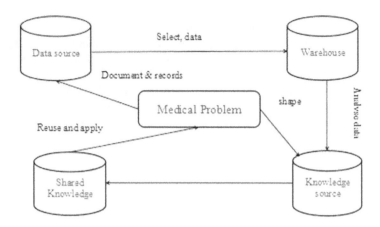

FIGURE 5.10   Process of knowledge management.

databases containing patients' records. Those records are controlled by machine learning algorithms for classification, as same as knowledge management is also a kind of information storing system that separately stores images, documents, files, and all (Figure 5.10).

5.6 OBIA (OBJECT-BASED IMAGE ANALYSIS)

The main theme of object-based image analysis is analyzing the images based on the regions. OBIA is deeply works towards the characteristics of the features in a image which are contextual and hierarchical relationship between the pixel. This is widely used in remote sensing application as well as geographic information science. Currently OBIA is used in the field of biomedical application and it is used in segmentation part to segment the image into region. It will segment the region depends on the colors, textures, intensity, and edges in a image. OBIA can have two approaches:

1. Explicit approach - Human perspective image analysis
2. Implicit approach - Machine learning algorithms for classification with training and testing data

In both approaches, features are the important role to success in segmentation. OBIA have many statistical features. They are given below:

1. Intensity features
2. Shape features
3. MBR shape features (Minimum Bounding Rectangle)
4. Texture features
5. Border relation features
6. Context features

5.7 RESULTS AND DISCUSSION

5.7.1 IMAGE SEGMENTATION

This proposed work concentrated on image analysis of image enhancement and image segmentation. In image enhancement, there are three algorithms implemented for better results. Sobel, Prewitt, and Laplacian edge detection will be shown here. The implementations were executed in MATLAB R2016a Version. The data set is taken from cervical cancer pap smear images with normal and abnormal cell types (Table 5.3).

The table shows the result of Sobel, Prewitt, and Laplacian edge detection techniques of normal and abnormal cells of the nucleus and cytoplasm.

5.7.2 FEATURE EXTRACTION

a. GLCM (Gray Level Co-Occurrence Matrix)

GLCM feature extraction is mainly used for the purpose of measuring the statistical attributes in an image. After segmentation, feature extraction is playing essential

TABLE 5.3
Result of proposed work

Image Segmentation				
	Pap smear Normal Cell		Pap smear Abnormal Cell	
Sobel Edge Detection	Input Image	Output Image	Input Image	Output Image
Prewitt Edge Detection	Input Image	Output Image	Input Image	Output Image
Laplacian Edge Detection	Input Image	Output Image	Input Image	Output Image

TABLE 5.4
GLCM Features with Their Properties

GLCM Features

Features	Normal Cell	Abnormal Cell
Contrast	2.0891	1.064
Correlation	−0.1675	−0.0465
Energy	0.0012	0.0276
Entropy	0.0564	0.9612
Homogeneity	0.0265	0.0823
Mean	0.1276	0.3185
Std.deviation	0.0587	0.5397

role to detect the images. GLCM features for normal and abnormal pap smear images given below with statistical values (Table 5.4).

b. HOG (Histogram of Oriented Gradients)

HOG is used to detect the object in a image string of gradients. Here is the result of HOG using sample cell image (Figure 5.11).

c. SURF (Speeded-Up Robust Features)

SURF is an object detector. It is to be implemented with normal and abnormal cells of pap smear microscopic images. It is shown as the result of SURF detector feature extraction. Those circles are denoting the pixel range of the image (Table 5.5).

CellSize = [2 2]	CellSize = [4 4]	CellSize = [8 8]
Length = 250992	Length = 60516	Length = 14400

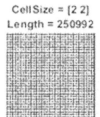 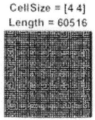 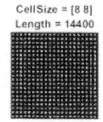

FIGURE 5.11   HOG feature descriptor.

TABLE 5.5
SURF Result of Normal and Abnormal Cells

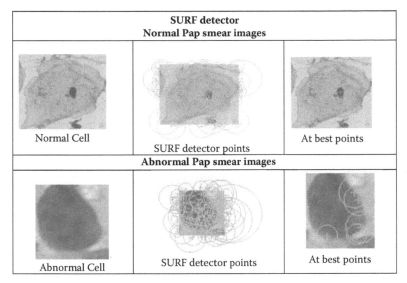

SURF detector Normal Pap smear images		
Normal Cell	SURF detector points	At best points
Abnormal Pap smear images		
Abnormal Cell	SURF detector points	At best points

5.8 CONCLUSION

This chapter presented features and information extraction of a cervical cancer data set. Edge detector is a very effective image processing method to discover the edges in a image. Segmentation is a kind of detection of the object to take out the features. Feature extraction is statistical numeric values that are calculated for further classification phase. This chapter organizes the flow of segmentation and feature extraction. Additionally, it involves knowledge management using CBR and OBIA concepts. It implements pap smear images under Sobel, Prewitt, and Laplacian edge detector in image segmentation as well as GLCM, HOG, and SURF used for feature extraction. The contribution of the work is entirely related to data science, computational excellence, information computing, and Society 5.0 under image processing for knowledge management and effective information extraction.

REFERENCES

Adnan, K., & Akbar, R. (2019). An analytical study of information extraction from unstructured and multidimensional big Data. *Journal of Big Data*, *6*(91), 1–38. https://doi.org/10.1186/s40537-019-0254-8.

Adnan, K., & Akbar, R. (2019). Limitations of information extraction methods and techniques for heterogeneous unstructured big data. *International Journal of Engineering Business Management*, *11*, 1–23. DOI: 10.1177/1847979019890771.

Ansari, M., Kurchaniya, D., & Dixit, M. (2017). A comprehensive analysis of image edge detection techniques. *International Journal of Multimedia and Ubiquitous Engineering*, *12*(11), 1–12. DOI: 10.14257/ijmue.2017.12.11.01.

Cauchard, V., Revenu, M., & Porquet, C. (1999). Knowledge management in image processing by means of case-based reasoning. *Workshop on Knowledge Management and Organizational Memories IJCAI* 99, Stockholm, Sweden (pp. 36–54).

Chen, H., Fuller, S. S., Friedman, C., Hersh, W. H., Chen, S. S., Fuller, C., Friedman, W. Hersh. (2005). Knowledge management, data mining, and text mining. (pp. 5–33). Boston, MA: Springer & Medical informatics. ISBN 978-0-387-25739-6.

Chandwadkar, R., Dhole, S., Gadewar, V., & Raut. D., & Tiwaskar, S. A. (2013). *Comparison of edge detection techniques.* Paper presented at Conference of Institute of Research and Journal,·August 2013. DOI: 10.13140/RG.2.1.5036.7123.

Chithra, P. L., & Bhavani, P. (May 2019). A study on various image processing techniques. *International Journal of Emerging Technology and Innovative Engineering, 5*(5), 316–322. ISSN 2394-6598.

Dhankhar, P., & Sahu, N. (2013). A review and research of edge detection techniques for image segmentation. *International Journal of Computer Science and Mobile Computing, 2*(7), 86–92. ISSN 2320-088X.

Dharun,V. (2016). Extraction of texture features using GLCM and shape features using connected regions. *International Journal of Engineering and Technology,* 2926–2930. ISSN (Print): 2319-8613, ISSN (Online): 0975-4024.

Izquierdo, E. (2006). Knowledge-based image processing for classification and recognition in surveillance applications, *ICIP-IEEE.* 1-4244-0481-9. IEEE, 2377–2380.

Lagrid, M., Benlabiod, L. B., Rubin, S., Tebibel, T., & Hanini. M. R. (2020). A case-based reasoning system for supervised classification problems in the medical field. *Expert Systems With Applications, 150,* 1–20. DOI: 10.1016/j.eswa.2020.113335.

Mahamune, M., Ingle S., Deo P., & Chowhan S. (2015). Healthcare knowledge management using data mining techniques. *Advances in Computational Research, 7*(1), 274–278. ISSN: 0975-3273 & E-ISSN: 0975-9085. http://www.bioinfopublication.org/jouarchive.php?opt=& jouid=BPJ0000187.

Mohanaiah, P., Sathyanarayana, P., & GuruKumar, L. (2013). Image texture feature extraction using GLCM approach. *International Journal of Scientific and Research Publications, 3*(5), 1–5. ISSN 2250-3153.

Ryszard, S. (2007). Image feature extraction techniques and their applications for CBIR and biometrics systems. *International Journal of Biology and Biomedical Engineering, 1*(1), 6–16.

Sonam Saluja, Aradhana Kumari Singh, Sonu Agrawal. (2013). A Study of Edge-Detection Methods. International Journal of Advanced Research in Computer and Communication Engineering, 2(1). ISSN (Print) : 2319-5940 & ISSN (Online) : 2278-1021.(PP. 994-999.)

Wang, Y., Wang, L., M., Mojarad, Moon, S., Shen, F., Afzal, N., … Liu, H. (2018). Clinical information extraction applications: A literature review. *Journal of Biomedical Informatics, 77,* 34–49. DOI: 10.1016/j.jbi.2017.11.011.

6 Recreating Efficient Framework for Resource-Constrained Environment: HR Analytics and Its Trends for Society 5.0

Kamakshi Malik[1], Rakesh K. Wats[2], and Aman Khera[3]

[1]Assistant Professor, Department of Management DAV College, Chandigarh, India

[2]Professor and Head, Department of Media Engineering NITTTR, Chandigarh, India

[3]Assistant Professor, University Institute of Applied Management Sciences, Panjab University, Chandigarh, India

CONTENTS

DOI: 10.1201/9781003132080-6

6.1   INTRODUCTION

It has been evidently articulated by top managements that in pursuit of attaining competitive advantage, it is not the originality of the product, the marketing strategy, or the state-of-the-art technology that matters, but it is the system of attracting and managing the best in class talent or the human resource that ultimately matters. This responsibility of attracting retaining, training, and engaging the talent within an organization lies with the Human Resource Department of an organization. Additionally, there are unparalleled challenges brought by the concomitant changes in the economy, technology, employee demography and societal trends. These trends have further created a multitude of organizational transformations viz. disruptions, changing customer and consumer demands, changing requirements of the workforce from companies, to name a few. This change in the business scenario has also transformed the role of the HR team which is no longer limited to record keeping or issuing appointment/termination letters and training employees, etc. It has gradually yet quintessentially moved from being operational contributor to strategic partner, change agent and people advocate, to achieve the greater objectives of the organization. Since organizations are operating in a resource constrained environment today, they are under immense pressure to perform and that too with competitive advantage. The organizational resources, be it time, finances, customers, and talent (employees) are valuable yet limited. And if organizations want to play a wider strategic role in organizations and overcome this pressure of achieving business objectives with limited resources, they cannot simply rely on intuition, hunches, and referrals.

Thus, there is a greater necessity of analytics in the field of human resource management. Analytics is the systematic process of exploring data with the intention of acquiring useful information that can be used to make more accurate data-centric strategic decisions and thereby enhance business results. When this mechanism is used to recruit, maintain, train, and engage the human capital of an organization, it is called HR Analytics.

While the use of HR analytics is limited to larger companies in the field of business, consultants in the area of analytics agree that the concept is here to remain and that the enterprises need to be flexible enough to accept it. In terms of value development, the advantages of analytics are apparent from the fact that if an organization can recognise a causal relationship between training investment and efficiency, it can boost the former to have a quantifiable effect on the latter. Despite its potential benefits, the emergence of talent analytics as a separate field of business analytics has been very slow (CIPD, 2013; OrgVue, 2019). According to a global survey by Deloitte (2014), organizations understand the significance of building their talent analytics capabilities, yet are not ready to embrace it holistically and effectively. Similarly, Falletta (2014), stated that only 15% of respondents asserted a key role of talent analytics in their organization.

6.2   LITERATURE REVIEW

The existing literature on HR analytics was reviewed and the articles were classified into three broad categories: the What? Why? Where? of HR analytics. Researchers in academics and consultancies have defined it in a number of ways, with no universally accepted definition as of now. Lawler, Levenson, and Boudreau (2004) defined it as a logic-based process that links HR practices to organizational performance by using statistical techniques, while Levenson and Pillans (2017) defined it more comprehensively as a mechanism of identifying employee related data leading to improved decision making and implicating it further to enhance organizational effectiveness. The literature on HR analytics have also clearly stated, with enough evidences, the reason why it must be employed by organizations. HR analytics has been directly linked to organizational effectiveness (Momin, 2014), attracting and retaining talent (Harris, Craig, & Light, 2010) etc. Further, the previous studies also demonstrated the applications of data analytics in various HR functions viz. selection and recruitment (Morrison & Abraham, 2015), training and development (Barbar, Choughri, & Soubjaki, 2019), and strategic decision making (Garcia-Arroyo & Osca, 2019). Table 6.1 gives some research findings from academic and consultant contributors related to the "what, why, and where" aspects of HR analytics.

A critical review of literature on HR analytics also led to a number of key lessons, which, India HR professionals should pay attention to while using HR analytics (Angrave, Charlwood, Kirkpatrick, Lawrence, & Stuart, 2016). The first and foremost is that the HR teams must understand the contribution of their people or the human resource towards the success of their organization. They must design organization-specific strategies rather than generic ones (Boudreau & Ramstad, 2007). Second, in order to garner logic-driven analytics with meaningful insights, analytics must be embedded in deep understanding of data and the milieu in which it is collected (Boudreau & Jesuthasan, 2011). This will help in creating useful data metrics which can further measure the cost and benefit relationship of various HR strategies. Third, these tools help in identifying those groups of employees whose performance makes the most strategic difference to the business and its performance (Boudreau & Jesuthasan, 2011). Thus, the analysis must be done using advanced statistical and econometric techniques.

TABLE 6.1

Literature Review on "What, Why, and Where" of HR Analytics

What is HR analytics?	Contribution	Contributor	Reference
	It is a logical method that connects HR activities through the use of statistical techniques to organizational results.	*Academics*	Lawler et al., 2004
	It is a knowledge-based method for proactive decision-making on the facets of an organization relevant to individuals, using common methods and technology such as HR metrics and predictive modeling.	*Consultancy*	Bassi, 2011
	A method to quantify and examine individual performance.	*Academics*	Aral, Brynjolfsson, & Wu, 2012
	An HR practice involving data management system to evaluate the effects of business and promote evidence-driven decision-making through the use of informative, graphic, and predictive data analyses pertaining to human capital, operational performance and	*Academics*	Marler & Boudreau, 2017

TABLE 6.1 (Continued)
Literature Review on "What, Why, and Where" of HR Analytics

What is HR analytics?	Contribution	Contributor	Reference
	external economic metrics.		
	It is the mechanism by which concrete trends in workforce-related data are identified, analyzed, and shared to inform decision-making and optimize results. It is not only limited to data collection, but extends to managing change, to improve organizational effectiveness.	*Consultancy*	Levenson & Pillans, 2017
Why use HR analytics?	HRA ensures organizational agility and optimization of the workforce resulting in more profitable business.	*Academics*	Van denHeuvel & Bondarouk, 2017; Kapoor & Kabra, 2014
	It aids in maximizing HR-linked data and discovering insights related to antecedent of employees' performance resulting in enhanced organizational effectiveness.	*Academics*	Momin & Mishra, 2016; Nienaber & Sewdass, 2016
	It impacts the financial performance of business entities thereby contributing to	*Academics*	Etukudo, 2019

(Continued)

TABLE 6.1 (Continued)

Literature Review on "What, Why, and Where" of HR Analytics

What is HR analytics?	Contribution	Contributor	Reference
	competitive advantage by improved preparation and execution of decisions as well as enabling forecasting in a VUCA (volatile, unpredictable, complex, and ambiguous) climate.		
	It helps to recognize and handle critical human talent (e.g., high-performance, high-potential, pivotal workers), HR planning, anticipate workforce expectations and attitudes, and customize HR practices to recruit and retain talent, to respond efficiently to changing market environments, priorities, and competitive risks.	*Academics*	Harris et al., 2010
Where is HR analytics being used?	Selection and recruitment/talent acquisition management	*Academics*	Morrison & Abraham, 2015; Illingworth, Lippstreu, & Deprez-Sims, 2015; Landers & Schmidt, 2016
	Assessment of contribution of human resources towards the organization and	*Academics*	King, 2016; Chamorro-Premuzic, Akhtar, Winsborough, & Sherman, 2017

TABLE 6.1 (Continued)
Literature Review on "What, Why, and Where" of HR Analytics

What is HR analytics?	Contribution	Contributor	Reference
	development of talent through machine learning		
	Knowledge management, decision making, learning performance (through machine learning and algorithms)	*Academics*	Roberts, 2013; Liu, Wang & Lin, 2017; Garcia-Arroyo & Osca, 2019
	Strategic decision making affecting organization's efficiency and performance	*Academics*	Garcia-Arroyo & Osca, 2019
	Managing safety and well being of employees in emergency situations through app-based-location information	*Academics*	Karlskind, 2014

6.3 CONCEPTUALIZING HR ANALYTICS

The data analytics is a tedious task, involving a number of vital steps but the results can give valuable insights into business performance and judiciously forecasting threats and opportunities. HR analytics also known as People analytics, Talent analytics or Workforce analytics is the application of statistics, simulation, and measurement of employee-related variables to improve business outcomes. It is the science of collecting, processing and reviewing data relevant to HR tasks, such as recruiting, talent acquisition, employee engagement, employee performance and retention, ensures better decision making in all these fields. It has emerged as a range of methodologies that enable companies to define trends in labour data in order to control the workforce, promote change (Guenole, Ferrar, & Feinzig, 2017), and eventually generate value (Marler & Boudreau, 2017; Pease, Byerly, & Fitz-enz, 2012).

By using various types of HR software and technology, HR departments are creating a large amount of data every day. In this process, the employee data is timely collected by the Human Resource Department of the organization and compared it with the organizational objectives. This is done so that one can measure

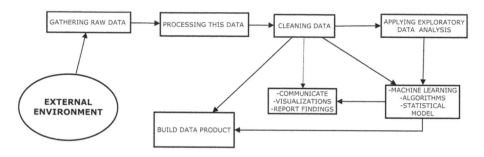

FIGURE 6.1   Data analytics process.

how HR actions are in sync with and adding to the organizational goals and strategies.

Figure 6.1 gives the steps of data analytics process where raw data is initially collected through various sources, it is processed and cleaned viz. irrelevant, duplicated, and incomplete data is eliminated. The relevant and complete data is then subjected to analysis using algorithms, statistical models, or machine learning. On the basis of the results so accrued, graphic representations and reports are formulated and communicated to the top management. Data visualization in the form of graphs and charts is frequently used to identify the valuable patterns in the data and are easy to understand. By delving deeper into the results of the analysis done, the top leaders are in a better position to take well-thought-out decisions for the profitability of the organization.

6.3.1   RELEVANCE OF HR ANALYTICS

Human Resource Analytics comprises a "series of methods to collect, clean, organize, and analyse large amounts of data from multiple sources (like internal and external stakeholders), enable predictions about performance" (Garcia-Arroyo & Osca, 2019) and support decisions related to the personnel linked to business results and organizational performance (Marler & Boudreau, 2017).

Progressive organizations such as Google, IBM, and Best Buy have increasingly learned exactly how their top performers will guarantee the optimum performance, organizational commitment, and then reinforce their achievements. In order to strengthen their competitive edge, they are increasingly implementing advanced methods of evaluating employee data. If organizations want better performances from their top employees – who are perhaps their greatest asset and largest expense – they must favor analytics over gut instincts (Davenport, Harris, & Shapiro, 2010).

Moreover, with the advancement of fast-paced technologies and networks, the data and information inputs in organizations have increased manifold. This has caused a severe necessity of enhancing how this data is converted into information which is subsequently managed effectively and efficiently to gather acumen that can help business performances. For instance, an organization receives a large amount of personal employee data in the form of resumes, the analyses of which, by a mere human is not only tardy but also subjective.

Data analytics, however, provide the ability to aid in forecasting employee behavior, promoting workforce analysis, offering evidence-based HR readiness strategies, and general market strategy. Nowadays, both options have been data-driven rather than past knowledge (Zerktouni, 2018).

Furthermore, by using data analytics in the overall HR processes, there is an increase in the quality of newly hired because of the strategic decisions of recruitment, so taken, are more analytical and evidence based (Villinova University, 2015).

The researchers of data analytics have recently associated HR analytics with organizational performance, although there is no precise explanation of the path that supports such a correlation (Braganza, Brooks, Nepelski, Ali, & Moro, 2017; Wamba et al., 2017). By allowing one to observe and assess staff behavio r, incentives for them can be aligned and thus HR analytics is associated with improved performance (Aral et al., 2012). The benefit of using analytics is attributed to the organizational level changes it creates (Lam, Sleep, Hennig-Thurau, Sridhar, & Saboo, 2016), the employee engagement levels it improves (Marler & Boudreau, 2017), and the implications it has on the overall effectiveness of organizations.

Finally, HR analytics has become popular among companies over the last five years (OrgVue, 2019) and consequently, its importance and necessity in HR departments is now better understood than ever (Andersen, 2017; Bersin, 2012).

6.3.2 Benefits of Using HR Analytics

Data analysis is becoming a ubiquitous phenomenon in the business world. Various departments like finance, customer-service, and sales and marketing have been using data extensively since a decade ago, but organizations are now seeking more analysis in the realm of human resources as well.

The department of HR is witnessing the emergence of big data, i.e. vast quantities of organised and unstructured data produced by organizations' routine operations and is being commonly used by organizations (Mayer-Schönberger & Cukier, 2013). The power of HR Analytics and big data is making companies dump their hunch-oriented decision making and taking more informed, data-based decisions now. Official decisions have become more reliable and accurate with the use of HR analytics. As per a survey by Tata Consultancy Services, business organizations believe that the single possible advantage of big data and HR analytics is that it improves employee retention (TCS, 2013). That is why companies are investing huge money into talent management software and recruiting special employees like data scientists, data miners, and data analysts to name a few.

Nowadays, CEOs are adopting evidence-based research methodology to develop insights that not only help the organizations to shift their attitudes to a wide variety of modern HR problems but also outline their performance plan and provide competitive advantage (SHRM, 2016). Researchers also believe that organizations can benefit from HR analytics as it is an important tool that can be "used to assimilate the data gathered in order to describe, analyze, predict and optimize the employee's potential" (Reddy & Lakshmikeerthi, 2017) and also enable the

FIGURE 6.2 Benefits of HR analytics.

managers to understand employee behaviors and take timely corrective actions for improvement. When integrated with wider business problems like workforce planning, diversity, and internal mobility, some organizations experienced meaningful changes (Kaur & Fink, 2017). Some other benefits of using HR analytics are depicted in Figure 6.2.

6.4 APPLICATIONS OF HR ANALYTICS FOR INDUSTRY 5.0

Given the constantly evolving nature of work and the transformation of the population, it is much more important that organizations pursue a collective research strategy. This section encompasses the application of data analytics in various HR functions through brief case studies from the global organizations. These case studies have been compiled from a global survey report like "The Strategic Workforce Analytics report by Corporate Research Forum" (Levenson & Pillans, 2017).

6.4.1 Managing the HR Through Analytics: A Case Study of Google

Google is known for its creative work culture and employee-centered HR activities as one of the leading global organizations. Google is also one of the strong contenders in the area of human analytics and, through its widespread use, has reached higher market heights. Some data-oriented ventures undertaken by the tech giants in various aspects of HR management are discussed.

A "People & Innovation Laboratory" – popularly known as the Pi-Lab, is a unique concept that conducts applied experiments within the organization to not only inculcate a "data-driven approach" to decision-making in HR management but also to determine the most effective techniques for managing people and maintaining a productive environment. The lab has even reported to improve employee health by reducing the calorie intake of its employees by relying on scientific data and experiments (by serving food in smaller sized plates [Sullivan, 2013]). Another attempt in the same arena is "Project Oxygen," which is a research-oriented project, the aim of which is to analyze internal data and identify the leadership traits of managers. It not only determined that great leaders are indispensable for employees' performance and retention, but also identified eight behavioral traits of effective managers/leaders. The said results were compiled after detailed analysis of the variety of comments given by employees in employee surveys, their expectations from managers which were based on the grievances and admirations mentioned in performance reviews, and phrases in top manager awards etc. Effective leader qualities, according to Project Oxygen are to be a good coach; inspire the team instead of controlling; show interest/concern about the performance of the team members and their personal well-being; be very productive/result-oriented; be a good communicator–listen and exchange information; support the career advancement of your team; and provide a specific vision/strategy for the team; and have technical expertise to assist or guide the team (Shrivastava, Nagdev, & Rajesh, 2018).

6.4.2   Employee Recruitment: A Case Study of JP Morgan and Rentokil Initial

Recruitment is one of vital functions of the HR department whereby the organization seeks for people best suited to their culture. Going through thousands of resumes and analysing the capabilities or skills of the candidates through those pages might not be as simple as it seems. HR analytics has come to the rescue of HR leaders as it helps in:

- Facilitating automated gathering of candidate data from multiple sources, thereby saving time.
- Attaining insight into the cultural fit and learning and development capabilities of potential candidates.
- Recognizing candidates with traits similar to those of the top performers in the organization.
- Ensuring equal opportunity for all candidates without any routine bias.
- Enabling departments to be more organized and well-informed when the need to hire arises.

The following two case studies exhibit the transformation of routine recruiting process using HR analytics, and thereby benefitting from it.

JPMorgan Chase, the well-known US-based bank, evaluated the efficacy of campus recruitment using data analytics and discovered that graduates from public universities perform at par to their counterparts from private universities. Thus, it changed the mindset of recruiters and Chief Human Resource Officer who now targeted both public and private universities while recruiting graduates.

Rentokil Initial suffered from an inefficient recruitment process, leading to deteriorating sales performance. The team applied HR analytics to analyse their data and identified the traits of its best sales people. It found six selection tests that effectively predicted those attributes. By analyzing employee behaviors of the star performers, they developed automated assessment techniques to select candidates further. This led to a 40% increase in sales and 300% return on investment.

6.4.3   Employee Retention: A Case Study of HP, IBM, and WIPRO

Organizations across the globe are trying to retain their best in class talent but despite sincere efforts, employees look for greener pastures. Employee turnover is a costly affair for organizations in this resource-constrained environment. Thus, it is imperative for organizations to look out for strategies to improve retention. HR analytics has played a significant role in doing so by

- Identifying trends and patterns exhibiting the reasons of employee turnover based on the analysis of past data turnover.
- Understanding the status of existing employees in terms of their performance, productivity, and level of engagement in the organization.
- Associating or correlating both types of data to understand the dynamics of employee turnover.
- Crafting a predictive model to identify and ensign employees who may fall into the pattern associated with employees that have quit.

The following case studies illuminate as to how analytics helped in reducing the organizations' turnover, thereby enhancing employee retention.

Hewlett-Packard (HP) applied HR analytics to forecast employee attrition by creating a "Flight Risk" score. This score predicted the prospects as well as the reasons of employees leaving the organization. Further, certain factors like higher pay-scale, promotion opportunity, and better performance ratings were found to have a negative relation with flight risk. Thus, predictive HR analytics helped HP in strategizing policies for employee retention.

IBM, too, analyzed employee turnover akin to the above stated example to predict employee retention. The analytics team used the Watson machine learning capabilities of IBM to create an "algorithm that included sources such as data on recruitment, work experience, history of promotion, performance, job responsibility, remuneration, location, job position etc." It also hypothesized that when employees think of quitting their current organization, their engagement with social media declines. Thus, the analysis of employees' sentiments measured through Social Pulse, helped the analytics team to anticipate which employees are thinking of quitting.

Wipro Ltd. has increased capabilities for training, motivating, and compensating employees, discovering new and exciting opportunities for organization by using HR analytics. Moreover, it not only provides a planned vision to the HR leaders for taking sound decisions but also predicts which employees will be star performers and which one will leave the organization (Vasudevan, 2015).

6.4.4   Employee Engagement: A Case Study of Clarks and Shell

Employee engagement is the level of passionate involvement and commitment of employees towards their job and organization, paving the way for its success.

Today, organizations do not just look out for good employees rather they seek engaged employees due to the organizational benefits associated in terms of enhanced customer satisfaction, better employee well-being, improved productivity, good corporate image, to name a few. HR analytics has not only made it more employee-centric but also data centric apart from being more accurate and reliable.

The employee engagement surveys now use data analytics and help in:

- Interpreting survey data for basic questions like "Are your employees proud to be a part of this organization?", "As an employer, are you providing them satisfactory incentives to stay motivated?", and "Are you trying to resolve the grievances of your employees?" can help organizations derive interesting insights.
- Understanding employee's emotions about their job, their supervisors, and the things they would want their employer/organization to change to bring in more transparency.
- Identifying the factors which employees like or don't like about their job and the organization and preparing relevant analytical models that will boost levels of engagement.
- Linking the employee engagement levels with the organizations' productivity and profitability and thereby quantifying engagement in terms of financial performance.

The following case studies will exemplify the previous claims.

Clarks, the global shoe retailer, tried to establish an association between employees' engagement level and the financial performance. Though the company enjoyed satisfactory levels of employee engagement, it wanted to ascertain the profits attained because of this (engagement). The HR team of Clarks, in collaboration with statisticians, analyzed data from both internal and external sources, combined with the corresponding people survey result. It was found that an improvement of 0.4% in the organization's productivity was contributed by a 1% increase in the employee's engagement.

Another such example is that of Shell, where a 1% increase in employee engagement resulted in a 4% decline of "recordable case frequency," which is considered to be a vital industrial safety standard. This safety performance parameter was directly associated with business performance. Shell's research, using HR analytics, also exhibited that when employees are engaged in their job and with the organization they perform better, thereby confirming a causal link between engagement and sales in business.

6.4.5 COMPENSATION AND BENEFITS: A CASE STUDY OF CLARKS

Compensation is the remuneration that the employer pays to his employee for his contribution to the organization. It inarguably occupies a crucial place in the life of an employee as it determines the employee's living standard, status, loyalty, and motivation to work. Every organization has a certain compensation philosophy that guides the management to ensure equity, transparency and consistency in payment. It is important for the organizations to understand the employee's sentiments with regard to the compensation they expect as it is directly linked to greater employee engagement, improved performance, and better business outcomes. Today, Compensation analytics has emerged as a modern discipline that focuses on maximizing the expense of a workforce to accelerate bottom line growth. This analytics division helps organizations in:

- Building an employer identity that expresses a winning employee value proposition efficiently. In effect, pay analytics helps HR leaders to identify and appropriately reward high-performers to improve the workplace morale, engagement, and retention,
- Comparing the characteristics of pay, bonus rates, and employee characteristics of those on the same position or in comparable roles. Equipped with this evidence, executives will make choices that are no longer dependent merely on sentiment, but on facts and figures,
- Identifying and addressing compensation bias, thereby eliminating gender, race or age discrimination when fixing and maintaining salaries and ensuring pay equity,
- Recognizing the factors that determine the right compensation or the pay scale (such as level of employee engagement, performance, recognition, career development, and workload).
- Clarks used analytics to optimize the remuneration being offered to employees. They conducted a survey and told the employees to articulate a preference between two different variables and assessed those benefits, which, employees are willing to substitute. They built a basic framework as to what people truly valued, and adjusted their compensation and pay packages accordingly (Levenson & Pillans, 2017).

6.4.6 EMPLOYEE TRAINING

Training refers to imparting knowledge, abilities, and specific skills to employees so that they can perform their job in an efficient and effective manner. The need for training and development is determined by the gaps in the expected or standard employee performance and actual performance. Providing relevant, timely, and adequate training followed by assessing the training program through feedback from employees is an important function of the HR department. Quantifying the benefits of training especially in terms of financial performance of the organization and assessing its effectiveness is a tedious task. Through analytics, the HR departments can:

- Design customized training modules according to the needs, requirements, and existing skills of employees,
- Measure effectiveness of training programs during the program itself and if need be, may be altered likewise (in case employees feel the program to be too difficult or monotonous, it may be made a little more employee - friendly and more engaging for them),

- Enable the organization to identify and eliminate the performance gaps by analyzing individual performance data,
- Improve upon the content of the training modules to be more user centric and engaging based on the feedbacks from employees undergoing it,
- Measure the benefits of employee training programs in terms of business performance by using statistical analysis.

The following case study reveals how HR analytics helped in doing so.

A fast moving consumer goods (FMCG) retailer in the Netherlands, which used HR analytics to build the relationship between investments in training and growth and market performance, announced a very inspiring outcome. The researchers found that the supermarket outlet's training program had a positive effect on its performance and that the return on training investment was more than 400%, which gives a lucid judgment of how training employees can be advantageous for the overall business performance.

6.5 RECENT TRENDS IN HR ANALYTICS

This section will briefly discuss the tools that are trending and are being increasingly used by HR analytics teams in global organizations.

6.5.1 R PROGRAMMING LANGUAGE

An open-source software, R is a kind of software whose source code is available to the general public and they are allowed to review, amend and enhance the software. R comes under the category of such open-source softwares. There is a history behind the name "R" used for the programing language. It derives its name from the two creators, Ross Ihaka and Robert Gentelman, and was named after the first letter of both the authors.

The R programing language was designed by statisticians and in today's world it is quite popular in the field of Data Science and Machine Learning. It is used primarily for statistical analysis and visualization of data and is quite apt for exploring, cleaning, and analyzing voluminous data sets distributed over unimaginable rows and columns. However, analysts require technical skills like an understanding of linear regression, non-linear modeling, time series tests, etc., to run this tool efficiently.

There are several reasons why R language is gaining popularity in the field of HR analytics:

- Open Source: As stated previously, this is one of the key reasons for its success and currently Google is also utilizing R language for performing various forms of statistics and DMLs (Data Manipulation Language) which covers inserts/updates/deletes to the data.
- A platform agnostic language: R is considered to be a platform independent language which means that a developer can use or code in R the same way if he/she is on Windows/Mac/Linux.

- Easily installable: R has a very extensive library with various packages which can be installed easily and can create visually comprehendible conclusions, after running the statistical analysis of the data.
- Implementation of the S language: There are some key differences between languages R & S. In R language, the data scientist gets a minimal set of output in the first step, stores this result in an object and the further examination is done by the R functions. Contrary to this, in S language the entire statistical analysis is performed as a series of several steps and the intermediary results are stored in the objects.
- An expression language: Another major reason for its popularity is that it is an expression language and has got a far simple syntax as compared to other programing languages.
- Usage of objects and functions: The two major artifacts of R programing language are objects that allow to store data in the memory and functions that help to automate several mundane tasks and perform complex computations.

R may be applied in the following aspects of HR management: perform employee churn analysis (for example, to assess whether tenure in an organization affect churn), data cleansing, and predictive analysis and correlation studies.

6.5.2　PYTHON

To understand Python we should first have an understanding of the three different types of programing languages viz. interpreted, interactive, and object oriented. An interpreted programing language unlike the compiler based programing language is the language which can execute the instructions directly without compiling a program; an Interactive as the name suggests, is a programing language which allows a programmer to make changes to the program whilst it is running and an object oriented programing is based on the concept of objects which comprises data and the code.

Python is considered to be an interpreted, interactive, and "Object Oriented Programming Language'" and this is the major reason for its popularity. The Python language was written by Guido van Rossum and at that time he was reading the script of Monty Python's *Flying Circus* (a BBC Comedy) and he wanted a name which was short and mysterious for the people; hence he named his language - "Python."

The key reasons why Python is considered to be a powerful language in the field of data science are as follows:

- Easy to understand: Python is considered to be a popular language for developers as it is easy to code and considering it is an interactive language, it is easy to understand as well.
- Open source: Python is an open-source software which can easily be downloaded for free from the python website.
- Availability of standard library: Python has got a large standard library consisting of several modules and functions that a developer can utilize and this helps the coder to avoid writing the code from scratch for specific use cases.

- Dynamically typed language: Python is considered to be a dynamically typed language as the developer does not have to declare the variables in advance, like integer, character, etc. and the type of variable is actually checked at run time.
- Portable language: Another reason for Python's popularity is that it is a portable language which means that the code written in python on Linux operating system will work the same way on other operating systems like Windows OS and Mac OS.
- Visualization and graphics: Python has got strong capabilities around visualization and graphics and the ability to generate charts and graphical layouts makes it a favorite amongst the data scientists.
- Wide support: It enjoys a wide variety of support and developers in the form of assortment of packages deliberated for diverse industries.

Python can be used interchangeably with R and there is a conundrum among data scientists about which of the two will become the ultimate tool of choice. While R is better at doing statistical analyses and more apposite for visualization of data, Python, on the other hand offers greater effectiveness in data mining, imaging and data flow capabilities and is easier to learn. It is also called as the "glue language" because it can be incorporated with existing software and other programing languages effortlessly. It is acclaimed for its readability, on the spot feedback, and scalability.

Python, together with methods like natural language processing and text analysis, is being used in some HR functions to conduct analysis from social media websites, employee engagement, and pulse surveys in order to gain insight on organizational networks, culture, and sentiment and examine how they drive change.

6.5.3 Business Intelligence

BI refers to infrastructure, applications, and practices for the selection, incorporation, interpretation, and display of business information. Some authors have considered it to be a "philosophy, comprising of strategies, processes, applications, data, products, technologies and technical architectures to support the collection, analysis, presentation and dissemination of business information" (Dedic & Stanier, 2016). The most important and rather interesting fact about Business Intelligence is that its usage can be different for different fields. For a user interface designer it could be related to the creation of the dashboards; for a database administrator it could be more about normalization of data. So, business intelligence can help the organizations to take proactive decisions which can help to make strategic decision making.

There are specific artifacts which fall under the umbrella of business intelligence. These are discussed as follows:

- RDBMS - RDBMS stands for Relational Database management System, which is a software used for the storage, management and administration of the Databases. In an RDBMS the data is stored in the form of tables which have columns and rows. There is also a concept of unique or primary key constraints which help to avoid duplicate data and the indexes help with the faster retrieval of data. The most popular RDBMS Software available are Oracle, Microsoft Sql Server, and MySql.

- SQL - SQL stands for Structured Query Language, which is a software program that is used to interact with the databases and helps in the retrieval of the data in the form of rows and columns from the table of the database. An example of how a sql is written is, let's say we need to retrieve all the records of a table called EMP then it is written as Select * from emp.
- ETL - ETL stands for Extract, Transform, and Load, as the name suggests, helps to copy the data from source to target and also perform transformation before loading the data into the target environment. There are some popular ETL tools that data analysts use like Informatica, Microsoft SQL Server SSIS, and Azure.
- Data Warehouse - A data warehouse is a type of database which helps to house the data that has been extracted from different sources or databases. The purpose of a data warehouse is for analytics and reporting purposes. The most popular data warehouses are Amazon Redshift, Snowflake, and Oracle Exadata.
- Dashboards - A dashboard is a GUI or Graphical User Interface which provides a view of the KPI (Key Performance Indicators) which helps to understand the progress report and is also considered to be a form of data visualization. A dashboard normally contains icons, images, drawing objects, and different types of graphs. The working of BI is shown (Figure 6.3).

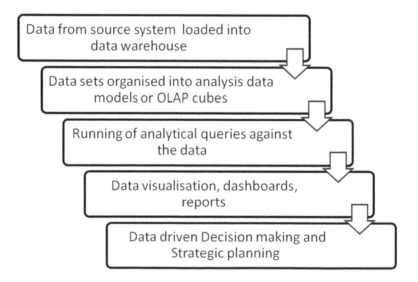

FIGURE 6.3   The working of BI.

The first step is to load the data from a data silo or several source databases and bring them into one target data warehouse environment. Once the data has been loaded into a data warehouse, the next step is to organize the tables in the form of Analysis Models or OLAP (Online Analytical Processing) cube which is a multi-dimensional array of data. It allows the user to query on multi-dimensional data. The next step is to run the analytical queries against these tables which are normally run using the SQL or Structured Query Language and that helps to select and manipulate the data. Based on the output of the query, the data can be visualized in the form of dashboards and reports which helps to provide a view organizations' KPIs which are the "Key Performance Indicators" and this helps the core management team of a company to make data driven decisions and perform proactive "Strategic Planning."

Researchers believe that BI, inarguably plays a crucial role in organizational development by providing competitive advantage in the context of achieving positive information asymmetry (Marchand & Raymond, 2008; Thamir & Poulis, 2015). Furthermore, its contribution in "optimisation of business processes and resources, capitalization on profits and improvement in pro-activity" (Olszak, Ziemba, & Koohang, 2006), and strategic decision-making (Popovič, Turk, & Jaklia, 2010) cannot be denied and is well documented. BI is a powerful tool that not only makes analytic processes easier and swifter but also helps organizations make better choices, based on information learned from the available data and alternatives. Organizations using BI can have a greater and quicker access to the main activities and procedures that must be followed by their efficient teams to accomplish their priorities and goals (Kapoor & Sherif, 2012).

6.6 CHALLENGES OF IMPLEMENTING HUMAN RESOURCE ANALYTICS

HR analytics, being a novel concept is not devoid of challenges, posed in its efficient and effective implementation in the organization. There is also an opinion about the same that there is nothing as big data analytics in HR (Cappelli, 2017) and that there are significant obstacles and intricacies in the practice (Hamilton & Sodeman, 2020). This section will answer the question – What hampers organizations in moving towards HR analytics?

The first and the foremost challenge is the very fact that analytics rely on big data that consists of 3Vs viz. large volumes of data, which are varied and come from diverse sources, produced at high velocity (Wenzel & Van Quaquebeke, 2018). Moreover, the HR systems of organizations do not "own" these data sets; instead the entire data is dispersed over numerous parts of the organization (Angrave et al., 2016; Marler & Boudreau, 2017). Routing of such disparate, dispersed and voluminous data, ascertaining its compatibility with the databases and subsequently analyzing and using it may pose as a challenge (Dahlbom, Siikanen, Sajasalo, & JarvenpÃ¤Ã¤, 2019; Waters, Streets, McFarlane, & Johnson-Murray, 2018).

Secondly, the basic marginalized positioning of the HR departments in organizations' hierarchy, too, is responsible for immobilizing the support for HR analytics. Due to their own relegation, regrettably, HR has been reduced in eminence in contrast to functions that are considered to directly influence profitability

(Anderson, 2014; Benko & Volini, 2014). Another issue with the HR departments has been that there is a lack of understanding as to how to incorporate data analytics with the organizational strategy of value creation to attain competitive advantage. Also, the focus of HR managers is more on the managerial expenditures of processing applications or on-boarding new hires than on the overall organizational profitability (Hamilton & Davison, 2018; Marler & Boudreau, 2017). Furthermore, while some authors report "that the HR function lacks the skills, knowledge and aptitude to ask the right questions from the HR data at their disposal" (CIPD, 2013), others feel that majority HR professionals are sceptical to substitute people-lead-decisions with algorithms (Hamilton & Sodeman, 2020). Thus, the architecture and the very fabric of the HR departments in various organizations globally, is posing a challenge to HR analytics.

The problem of overfit and algorithmic bias leads to misrepresented analysis and subsequent prejudiced conclusions, thus failing the very purpose of HR analytics. Over-fit happens when a model starts catching the noisy or irrelevant and inaccurate values in the data and fits more data than it actually needs. Due to these large data sets, the machine learning detects patterns that do not have content validity or are spuriously significant, as a result, the efficiency and accuracy of the model decreases. Algorithmic bias occurs when the algorithm is poorly trained and inaccurately includes or excludes data. The more categorised data an algorithm sees, the better job it performs. The trade-off to this strategy, though is that deep learning algorithms, depending on what is lacking or too abundant in the data on which they are trained, can form weak spots. For example, a photo-app developed by Google, erroneously tagged a photo of two black people as gorillas because its algorithm had not been trained with adequate images of people with dark-colored skin.

Last but not the least, the employees perceive HR analytics as an insidious tool that inevitably barges into their private lives, thus raising privacy concerns. Some interesting examples of how these systems can act like surveillance cameras come from companies like Walmart, Microsoft, and Amazon. These organizations developed certain tracking systems that could even collect audios, geo-location, accelerometer, and other employee-related data throughout their workday (Shell, 2018; Sheng, 2019), in order to identify the most productive actions and knowledge-sharing activities by individuals and teammates. Though it is done with the intention of collecting useful information yet these actions may be presumed to be invading into employee privacy. According to a 2019 Accenture survey, 64% of employees are concerned about a possible eroding of their privacy (Sheng, 2019), and as a response to this pervasiveness, they might exhibit discriminatory or biased behaviors (Tomczak, Lanzo, & Aguinis, 2018).

6.7   HR ANALYTICS FRAMEWORK: THE WAY FORWARD

There has been a substantial rise in the interest of organizations in the field of HR analytics over the past few years, with no indications of it subsiding. Keeping in view the relevance and benefits of analytics mentioned earlier in the chapter, it is pertinent for organizations to develop an efficient and effective framework for HR analytics with the following.

6.7.1 DEVELOPING LEADERS

The effectiveness of any new initiative relies on the leaders who take it forward, and HR analytics is no different. In reality, the dedication of leaders to this approach is the single most important factor determining whether it succeeds in the organizations or not. Leaders who agree that knowledge from human resources can be used to address market issues need to continually push for choices and decisions focused on facts and results instead of beliefs, hearsay, or presumption. Furthermore, any insight gained from applying analytics is valuable only if it contributes to action, and substantial change in the management. This is exactly where the role of a leader comes in play. The organizations must therefore develop such leaders who have the mindset and attitude of adopting new technologies and a clear vision of how HR analytics can be imbibed in the very fabric of the HR department. Today, as discussed earlier, the organizations are facing shortages of vital resources be it financial or non-financial (time, customers, talent) so a capable, visionary leadership can set the boat sailing in the rough seas by implementing analytics effectively.

6.7.2 ENSURING REQUISITE SKILL SETS

The backbone of HR analytics is the skills required to understand, analyse and interpret analytics results. The core skills required for HR analytics are statistics, data analytics, and research methodology. Additional skills in data visualization, machine learning and data processing also seem to be useful, but the need for these will depend on the nature of the problem. The HR teams must ensure that they have the right people to perform these analyses (Kaur & Fink, 2017). Also, it is not always necessary to use fancy data analysis tools, simple statistical tools like correlation and regression using IBM SPSS (Statistical Package for Social Sciences) can also solve the purpose. Thus, organizations must hire people who can skilfully use the employee data ethically and with confidentiality.

6.7.3 FOCUSSING ON CLARITY OF VISION AND MISSION OF ADOPTING HR ANALYTICS

Before implementing HR analytics, the HR teams must have a clear vision and mission (or purpose) as to why we need analytics? What issues shall be addressed by this analysis? Is the organization ready to take actions as a result of the analysis? Is the organization prepared to change its systems, processes or behaviors as a result of the analysis? Simply asking questions and not taking necessary actions or implementing changes will lead you to nowhere. Thus, HR analytics must be used when organizations are geared up for change.

Furthermore, it is crucial to communicate the purpose or vision of using analytics to all the key stakeholders (both internal and external) so as to make them understand its relevance and seek their support.

6.7.4   UNDERSTANDING AND CONNECTING WITH THE BUSINESS STRATEGY

It is believed that HR analytics focuses more on cleaning data, building systems, and developing numbers in an attempt to gain information, forgetting the fact that its basic intention is improving business performance. Hence, analytics must begin with understanding the business strategy and should be embedded in strategy in order to improve its implementation. Significant questions to be answered are: What are the strategic goals of my organization? What gaps are impeding us to achieve these goals? How can HR analytics address these gaps? This will help the organization in connecting HR analytics with business strategy, subsequently leading to attain organizational effectiveness.

6.7.5   COLLABORATING WITH STAKEHOLDERS

By strategically partnering with key stakeholders like line managers and employees, HR analytics can be implemented more effectively. For instance, if HR team assists a particular division in the organization that suffers from high errors or product returns, through analytics, it will be able to develop credibility among line managers. Developing this faith in analytics is very crucial in reducing the scepticism among managers. Secondly, some employees may consider data analytics as an attempt to invade their privacy. Thus, organizations must clear this fog and be transparent in collecting and using employee data. Also, the analysis should focus more on how the outcomes of data analysis will help an individual employee to perform better rather than only highlighting weaknesses. Thus, HR analytics should empower employees to understand the significance of their job and its contribution towards overall organizational performance.

The previously mentioned framework has been depicted in Figure 6.4.

6.8   DISCUSSION

The organizations are witnessing a paradigm shift due to the demographic, environmental, and technological changes taking place at a breakneck pace. This has led to the evolution and transformation of the roles and the responsibilities of the HR teams. The employees are more demanding with greater expectations from the employer, the employee–employer relationship has greater proclivity towards

FIGURE 6.4   Framework for efficient implementation of HR analytics.

the employee, the business models are changing at the blink of an eye. To remain sustainable, relevant and profitable simultaneously, in the resource constrained scenario, the organizations must step towards analytics, with agility and innovation. Though many organizations are already using it extensively, many are still in a quandary owing to the scepticism of its prospective benefits and lack of requisite skills in employees performing analytics. There is no denying the fact that keeping in view the future of analytics, its relevance and scale cannot be elbowed away. Various functions of the HR viz. recruitment, training, compensation, engagement, etc. can be managed more effectively and efficiently by using machine learning, algorithms, and artificial intelligence, thereby leading to organizational effectiveness.

This chapter proposes a framework for the efficient implementation of HR analytics which includes developing leaders who have a vision for analytics, ensuring analytical skills and basic knowledge of analytics, focus on how analytics can be aligned with the vision and mission of the organization, collaborating with the organizations' stakeholders (line manager, employees, etc.) and understanding the business strategy for implementing analytics. Whatever analytics an organization prefers to use, it must first concentrate on developing strategies around this framework so as to attain competitive advantage and be effective in this era striving for computational excellence through data science.

6.9 CONCLUSION

The purpose of this chapter was to discuss as to what encompasses HR analytics and its relevance in the resource-constrained business environment and apprise about certain recent trends like algorithms, business intelligence, etc. that organizations are adopting in the field of HR to become more strategic, data-driven partners. This chapter has talked about numerous case studies of how organizations across the globe are using HR analytics innovatively to come out with solutions of problems that have always posed a challenge for the HR departments. Data acquiring is not new to HR departments as it has always been flooded with human resource data in the form of application forms, training feedbacks, employee engagement surveys etc. but the need of the hour is to re-create effective frameworks to understand and analyze this data, aligning it with the objectives of the organization. This will not only help resource-starved organizations save on time, cost, and talent but also help them in gaining competitive advantage. Conclusively, HR analytics is essentially the potential game-changer and a success mantra for organizations that want to be global leaders and yearn to stay germane in Society 5.0.

REFERENCES

Andersen, M. K. (2017). Human capital analytics: the winding road. *Journal of Organizational Effectiveness: People and Performance, 4*, 133–136.
Anderson, C. (2014). What HR needs to do to get a seat at the table. *Harvard Business Review.* Retrieved from https://hbr.org/2014/11/what-hr-needs-to-do-to-get-a-seat-at-the-table

Angrave, D., Charlwood, A., Kirkpatrick, I., Lawrence, M., & Stuart, M. (2016). HR and analytics: Why HR is set to fail the big data challenge. *Human Resource Management Journal*, *26*(1), 1–11.

Aral, S., Brynjolfsson, E., & Wu, L. (2012). Three-way complementarities: Performance pay, human resource analytics, and information technology. *Management Science*, *58*(5), 913–931.

Barbar, K., Choughri, R., & Soubjaki, M. (2019). The impact of HR analytics on the training and development strategy-private sector case study in Lebanon. *Journal of Management and Strategy*, *10*(3), 27–36.

Bassi, L. (2011). Raging debates in HR Analytics. *People & Strategy*, *34*, 14–18.

Benko, C., & Volini, E. (2014). What it will take to fix HR? *Harvard Business Review*. Retrieved from https://hbr.org/2014/07/what-it-will-take-to-fix-hr

Bersin, J. (2012). *Big data in HR: Building a competitive talent analytics function: The four stages of maturity*. Oakland: Bersin and Associates.

Boudreau, J. W., & Jesuthasan, R. (2011). *Transformative HR: how great companies use evidence-based change for sustainable advantage*. San Fransisco: Jossey Bass.

Boudreau, J. W., & Ramstad, P. M. (2007). *Beyond HR: The new science of human capital*. Boston: Harvard Business Press.

Braganza, A., Brooks, L., Nepelski, D., Ali, M., & Moro, R. (2017). Resource management in big data initiatives: Processes and dynamic capabilities. *Journal of Business Research*, *70*, 328–337.

Cappelli, P. (2017). There's no such thing as big data in HR. *Harvard Business Review*. Retrieved from https://hbr.org/2017/06/theres-no-such-thing-as-big-data-in-hr

Chamorro-Premuzic, T., Akhtar, R., Winsborough, D., & Sherman, R. A. (2017). The datafication of talent: How technology is advancing the science of human potential at work. *Current Opinion in Behavioral Sciences*, *18*, 13–16.

CIPD. (2013). Talent analytics and big data: The challenge for HR, research report 2013. London, UK: Chartered Institute of Personnel Development.

Dahlbom, P., Siikanen, N., Sajasalo, P., & JarvenpÄ¤Ä¤, M. (2019). Big data and HR analytics in the digital era. *Baltic Journal of Management*, *15*(1), 120–138.

Davenport, T. H., Harris, J., & Shapiro, J. (2010). Competing on talent analytics. *Harvard Business Review*, *88*(10), 52–58.

Dedic, N., & Stanier, C. (2016). *An evaluation of the challenges of multilingualism in data warehouse development*. Proceedings of the 18th International Conference on Enterprise Information Systems, 96–97. Retrieved from https://scholarworks.waldenu.edu/cgi/viewcontent.cgi?article=7836&context=dissertations

Deloitte. (2014). Global Human Capital Trends 2014-Engaging the 21st-century workforce. Deloitte University Press. Retrieved from https://www2.deloitte.com/content/dam/Deloitte/ar/Documents/human-capital/arg_hc_global-human-capital-trends-2014_09062014%20(1).pdf

Etukudo, R. U. (2019). *Strategies for Using Analytics to Improve Human Resource Management*. Walden Dissertations and Doctoral Studies. 6557.

Falletta, S. (2014). In search of HR intelligence: evidence-based HR analytics practices in high performing companies. *People and Strategy*, *36*(4), 28–37.

Garcia-Arroyo, J., & Osca, A. (2019). Big data contributions to human resource management: A systematic review. *The International Journal of Human Resource Management*, 1–26. DOI: 10.1080/09585192.2019.1674357.

Guenole, N., Ferrar, J., & Feinzig, S. (2017). *The power of people: Learn how successful organizations use workforce analytics to improve business performance*. New York: Pearson Education.

Hamilton, R. H., & Davison, H. K. (2018). The search for skills: Knowledge stars and innovation in the hiring process. *Business Horizons*, *61*(3), 409–419.

Hamilton, R. H., & Sodeman, W. A. (2020). The questions we ask: Opportunities and challenges for using big data analytics to strategically manage human capital resources. *Business Horizons, 63*(1), 85–95.

Harris, J., Craig, E., & Light, D. (2010). Talent and analytics: New approaches, higher ROI. *Journal of Business Strategy, 32*, 4–13.

Illingworth, A. J., Lippstreu, M., & Deprez-Sims, A. S. (Eds.). (2015). *Big Data in talent selection and assessment.* New York: Routledge.

Kapoor, B., & Kabra, Y. (2014). Current and future trends in human resources analytics adoption. *Journal of Cases on Information Technology, 16*(1), 50–59.

Kapoor, B., & Sherif, J. (2012). Human resources in an enriched environment of business intelligence. *Kybernetes, 41*(10), 1625–1637.

Karlskind, M. (2014). How 5 companies use mobile apps to manage big data. *Material Handling & Logistics, Cleveland.* Retrieved from https://www.mhlnews.com/global-supply-chain/article/22050884/how-5-companies-use-mobile-apps-to-manage-big-data

Kaur, J., & Fink, A. A. (2017). Trends and practices in talent analytics. Society for Human Resource Management (SHRM)-Society for Industrial-Organizational Psychology (SIOP) Science of HR White Paper Series.

King, K. G. (2016). Data analytics in human resources: A case study and critical review. *Human Resource Development Review, 15*(4), 487–495.

Lam, S. K., Sleep, S., Hennig-Thurau, T., Sridhar, S., & Saboo, A. R. (2016). Leveraging frontline employees' small data and firm-level big data in frontline management: An absorptive capacity perspective. *Journal of Service Research, 20*(1), 12–28.

Landers, R. N., & Schmidt, G. B. (Eds.). (2016). *Social media in employee selection and recruitment: An overview.* London: Springer.

Lawler III, E. E., Levenson, A., & Boudreau, J. W. (2004). HR metrics and analytics: Use and Impact. *Human Resource Planning, 27*, 27–35.

Levenson, A., & Pillans, G. (2017). Strategic Workforce Analytics. OrgVue. Retrieved from https://www.orgvue.com/uploads/sites/2/2020/05/strategic-workforce-analytics-report.pdf

Liu, C. H., Wang, J. S., & Lin, C. W. (2017). The concepts of Big Data applied in personal knowledge management. *Journal of Knowledge Management, 21*(1), 213–230.

Marchand, M., & Raymond, L. (2008). Researching performance measurement systems. *International Journal of Operations & Production Management, 28*(7), 663–686.

Marler, J. H., & Boudreau, J. W. (2017). An evidence-based review of HR Analytics. *The International Journal of Human Resource Management, 28*(1), 3–26.

Mayer-Schönberger, V., & Cukier, K. (2013). *Big data: A revolution that will transform how we live, work, and think.* Boston: Houghton Mifflin Harcourt.

Momin, W. Y. M. (2014). HR analytics transforming human resource management. *International Journal of Applied Research, 1*(9), 688–692.

Momin, W. Y. M., & Mishra, K. (2016). HR analytics: Re-inventing human resource management. *International Journal of Applied Research, 2*(5), 785–790.

Morrison, J. D., & Abraham, J. D. (2015). Reasons for enthusiasm and caution regarding Big Data in applied selection research. *The Industrial-Organizational Psychologist, 52*(3), 134–139.

Nienaber, H., & Sewdass, N. (2016). A reflection and integration of workforce conceptualisations and measurements for competitive advantage. *Journal of Intelligence Studies in Business, 6*(1), 5–20.

Olszak, C. M., Ziemba, E., & Koohang, A. (2006). Business intelligence systems in the holistic infrastructure development supporting decision-making in organisations. *Interdisciplinary Journal of Information, Knowledge & Management, 1*, 47–58.

OrgVue. (2019). *Making people count: 2019 report on workforce analytics.* Retrieved from https://www.orgvue.com/resources/ebook/making-people-count-report/

Pease, G., Byerly, B., & Fitz-enz, J. (2012). *Human capital analytics: How to harness the potential of your organization's greatest asset* (Vol. 64). New York: John Wiley & Sons.

Popovič, A., Turk, T., & Jaklia, J. (2010). Conceptual model of business value of business intelligence systems. *Management: Journal of Contemporary Management Issues, 15*(1), 5–30.

Reddy, P. R., & Lakshmikeerthi, P. (2017). 'HR Analytics'—An effective evidence based HRM tool. *International Journal of Business and Management Invention, 6*(7), 23–34.

Roberts, B. (2013). The benefits of big data. *HR Magazine; Alexandria, 59*(10), 20–22, 24, 26, 28, 30.

Shell, E. R. (2018). The employer-surveillance state. *The Atlantic*, October 15.

Sheng, E. (2019). Employee privacy in the US is at stake as corporate surveillance technology monitors every move. *CNBC*. Retrieved from https://www.cnbc.com/2019/04/15/employee-privacy-is-at-stake-as-surveillance-tech-monitors-workers.html#

Shrivastava, S., Nagdev, K., & Rajesh, A. (2018). Redefining HR using people analytics: the case of Google. *Human Resource Management International Digest, 26*(2), 3–6.

SHRM. (2016). Use of workforce analytics for competitive advantage. What's next: Future global trends affecting your organization. *Society for Human Resource Management*. Retrieved from https://www.shrm.org/foundation/ourwork/initiatives/preparing-for-future-hr-trends/Documents/Workforce%20Analytics%20Report.pdf

Sullivan, J. (2013). How Google is using people analytics to completely reinvent HR. *Talent management and HR*. Retrieved from https://www.tlnt.com/how-google-is-using-people-analytics-to-completely-reinvent-hr/

TCS. (2013). The emerging big returns of big data. Tata Consultancy Services. Retrieved from https://www.scribd.com/document/153044091/TCS-Big-Data-Global-Trend-Study-2013.

Thamir, A., & Poulis, E. (2015). Business intelligence capabilities and implementation strategies. *International Journal of Global Business, 8*(1), 34–45.

Tomczak, D. L., Lanzo, L. A., & Aguinis, H. (2018). Evidence-based recommendations for employee performance monitoring. *Business Horizons, 61*(2), 251–259.

Van denHeuvel, S., & Bondarouk, T. (2017). The rise (and fall?) of HR analytics: A study into the future applications, value, structure, and system support. *Journal of Organizational Effectiveness: People and Performance, 4*(2), 157–178.

Vasudevan, S. (2015). The no pain route to analytics: Enabling speedy decision making through 'Analytics-as-a-Service'. Wipro. Retrieved from https://www.wipro.com/content/dam/nexus/en/service-lines/analytics/latest-thinking/3115_the-no-pain-route-to-analytics.pdf

Villinova University. (2015). The talent gap in data analytics. Villanova University Blog. Retrieved from: https://taxandbusinessonline.villanova.edu/blog/the-talent-gap-in-data-analytics/

Wamba, S. F., Gunasekaran, A., Akter, S., Ren, S. J., Dubey, R., & Childe, S. J. (2017). Big data analytics and firm performance: Effects of dynamic capabilities. *Journal of Business Research, 70*, 356–365.

Waters, S. D., Streets, V. N., McFarlane, L. A., & Johnson- Murray, R. (2018). *The practical guide to HR analytics*. Alexandria, VA: Society for Human Resource Management.

Wenzel, R., & Van Quaquebeke, N. (2018). The double-edged sword of big data in organizational and management research: A review of opportunities and risks. *Organizational Research Methods, 21*(3), 548–591.

Zerktouni, J. (2018). Data Science: A necessity for HR in the competitive business world. *Analytics Insight*. Retrieved from https://www.analyticsinsight.net/data-science-a-necessity-for-hr-in-the-competitive-business-world/

7 Integration of Internet of Things (IoT) in Health Care Industry: An Overview of Benefits, Challenges, and Applications

Afshan Hassan, Devendra Prasad,
Meenu Khurana, Umesh Kumar Lilhore, and
Sarita Simaiya
Chitkara University Institute of Engineering and Technology,
Chitkara University, Punjab, India

CONTENTS

7.1 INTRODUCTION

Due to advancements in the field of technology preferably in healthcare sector integration of IoT in healthcare field has become a necessity. In order to improve the efficiency, accuracy, and quality of healthcare services for the patients, IoT has been introduced as it shows great potential to ensure the productivity and safety of patients in hospitals (Alamr, Kausar, Kim, & Seo, 2018). In simple terms

DOI: 10.1201/9781003132080-7

IoT can be defined as a network consisting of different interconnected smart objects that are capable of sending and receiving data in the form of signals (Atzori, Iera, & Morabito, 2010). Before the introduction of "IoT in Healthcare," the patient-doctor communication was limited to physical visits, communication through calls or texts. However, after the application of IoT in healthcare IoT-enabled devices are capable of monitoring the patient's health remotely to ensure that the patient stays healthy and safe. The use of IoT technology ensures delivery of utmost care for the patient by the healthcare professional (Bagci, Raza, Roedig, & Voigt, 2016). This not only increases the engagement and satisfaction of patient's interaction with doctors but also increases the efficacy of delivery of healthcare services. Different types of wearable devices like bands reporting fitness or other devices that are capable of monitoring patient's blood pressure, heart rhythms, glucose levels, calorie counts, etc. enable tracking of various health conditions remotely especially for people who are living far away from their families (Chi et al., 2013).

The integration of IoT in healthcare has caused healthcare to skyrocket by allowing continuous and accurate monitoring of patient's health. The application of IoT in healthcare sector is not only beneficial for patients but it also helps physicians monitor patient's health more accurately and effectively (Choi et al., 2018), as shown in Figure 7.1. If there is any change or disturbance in any of the parameters monitored like fluctuation in blood glucose levels or blood pressure, the data will be collected by the sensors and signals will be sent to the concerned healthcare provider. The concerned professional will then connect with the patient remotely and after analyzing the data so collected, he can deliver the best treatment to the patient as per the concern. Apart from this, IoT devices are

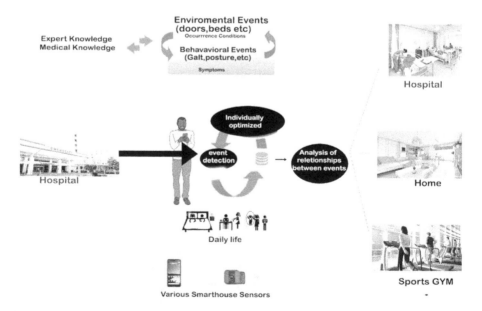

FIGURE 7.1 IoT healthcare service delivery model.

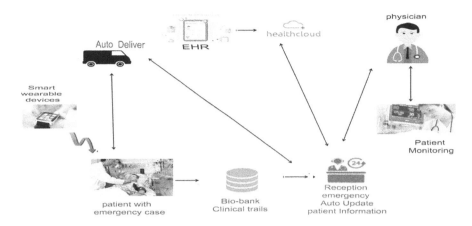

FIGURE 7.2   The expanded IoT architecture for smart inpatient system scenario.

capable of tracking the real-time location of different monitoring equipment's like wheelchairs, nebulizers, oxygen pumps etc. IoT devices enabled with hygiene monitoring can help prevent the patient from catching infection in hospitals. IoT intelligent devices can help insurance companies to detect and negate fraud claims by patient (Cvitić & Vujić, 2015). IoT achieves this by introducing transparency between insurers and customers. Insurance companies can validate their claims with the data gathered through IoT-enabled devices fitted on the patient's body. IoT has the capability to take the healthcare industry to the next level.

The adoption of IoT in the medical industry has helped not only in driving down the costs but also in improving the healthcare outcomes. Figures 7.2 and 7.3 illustrate an IoT-hospital scenario wherein a patient is fitted with a wearable sensor. The IoT-enabled sensing device has the capability to detect the nearest hospital offer the required service to the patient. After this, an ambulance from the nearest hospital is deployed to the patient's location. The ambulance then links various prescription histories of patient to help health practitioners understand the status of patient's health from his medical history saved onto the cloud, thereby treating the patient quickly, effectively, and in a timely manner in case of an emergency (Garkoti, Peddoju, & Balasubramanian, 2014).

7.1.1   ELEMENTS OF IoT

There are five fundamental elements of IoT that are explained below (Gope & Hwang, 2016), as shown in Figure 7.4.

- **Gateway:** The efficient flow of data between different networks and platforms is managed by a gateway. Gateway is responsible for processing the data collected or gathered from the sensors, and encryption of data flowing through the network; using modern encryption techniques in order to filter illegal or malicious packets to thwart any attack.

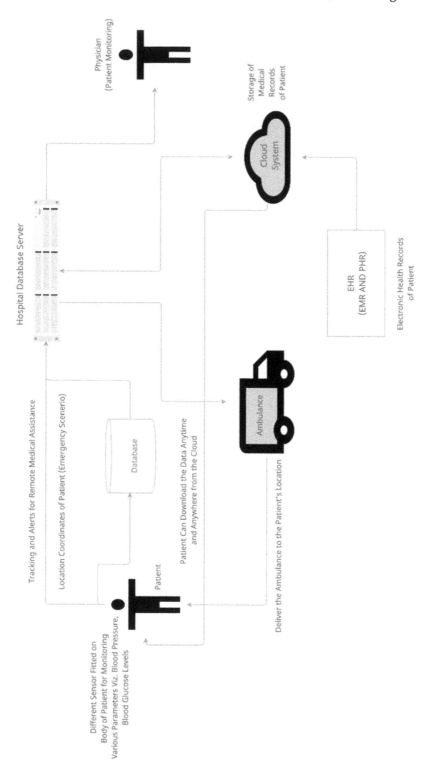

FIGURE 7.3   The expanded IoT architecture for smart inpatient system scenario.

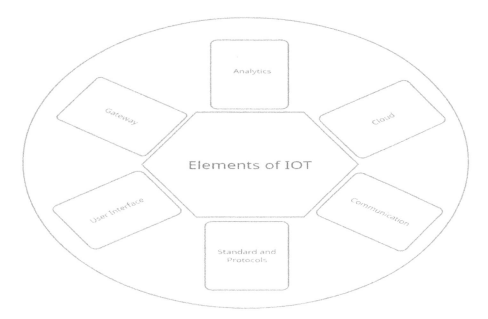

FIGURE 7.4   Elements of Internet of Things.

- **Analytics:** In an IoT-enabled solution, the sensors attached to the body of the patient gather or collect different parameters. The data so collected is sent to the caregiver, so that he can prescribe appropriate treatment depending on the scenario. The data that flows from sensors has to be converted to a form so that it would be readable to the doctor and other medical staff. This data is stored in database or on the cloud so that it can be accessed and analysed. Analysis of data in bulk helps to understand the different trends in data which can be used by business oranizations to see various future opportunities for growth, development and future prospects.
- **Cloud:** Various oranizations can gather data from the sensors and applications with the help of an IoT ecosystem. For the purpose of data processing, different tools and technologies can be used to capture, process, store and manage data efficiently and effectively in order to deliver high performance and all of this can be achieved by an IoT cloud. It is a high-speed daunting network that is capable of linking different database servers, devices, sensors and actuators to maximize the processing performance of bulk data by multiple machines at once. Data analytics and traffic monitoring also form an important part of IoT cloud.
- **User interface:** User interface determines the physical and recognizable portion of an IoT system that the user can navigate in order to avail different services. It is crucial for the programmer to construct or develop a consumer friendly interface that is accessible without putting in any additional efforts into it and that can assist in simple and easy navigation of the platform. There are several dynamic designs that can be used conveniently and that are

capable of solving any complicated query. For example; hard controls used at home earlier are continuously being replaced by touch panels or touchpads that can perform much better in terms of speed and are more convenient to use.

- **Standards and protocols:** The use of scripting languages like HTML, CSS3, PHP, etc., by the websites has made it more convenient, secure and easy to use the Internet. There are various standard protocols that has made Internet more acceptable and a reliable service to use. However Internet lacks a standard. It is therefore crucial to have an IoT standard that is same throughout in order to ensure smooth communication across different networks and platforms.

7.1.2 CHARACTERISTICS OF IoT

The following are the characteristics of IoT (Gardašević et al., 2017):

- **Interconnectivity:** IoT devices are capable of connecting anything with the universal information networks and infrastructure to guarantee efficient and smooth communication.
- **Intelligence:** IoT systems include several processing units and applications used mostly for automating the decision predicting capabilities, thereby improving the performance overall.
- **Energetical efficiency:** All the IoT-enabled devices must be capable of using recyclable energy in order to improve and boost energy efficiency, henceforth increasing the system output.
- **Data sharing:** IoT systems must have the capability of data sharing wherein all the connected devices stay updated with the current content shared by these devices in order to enhance and improve the flow of communication within the network.
- **Things related services:** IoT has the capability to provide things-related services within certain constraints viz personal privacy, semantic consistency, etc. and this will change both the technology as well as the information world for good.
- **Heterogeneity:** IoT devices are dependent on various hardware architectures and platforms. They can send and receive messages using different networks and platforms.
- **Safety:** One needs to design for security in order to ensure the protection of personal records, all the end nodes, networks as well as the data that flows through an IoT-powered network. This can be achieved by building a security framework that evolves with time to adapt to the changes.

7.2 ARCHITECTURE OF IoT IN HEALTHCARE

IoT has a four-step architecture, as shown in Figure 7.5, that is basically phases in the process (Islam, Kwak, Kabir, Hossain, & Kwak 2015):

FIGURE 7.5   IoT reference architecture in the healthcare industry.

- *Phase 1:* Networked things (wireless sensors and actuators): The first phase is concerned with the deployment of various IoT-enabled devices like detectors, fitness bands, and sensors that are capable of collecting the patient's data remotely. These devices possess the capability of intervening the physical reality (Ko & Song, 2016). For example, adjusting the room temperature, switching on/off the lights, etc. These sensors may be wired or wireless.
- *Phase 2:* Data aggregation and analog to digital data conversion: The data collected through different wearables is usually in analog form. In order to process the data, it has to be digitized or converted into a digital form. This phase is concerned with the conversion of data from analog to digital form. Also since huge data is produced by sensors, a LAN (Wi-Fi, Ethernet) or WAN (GSMs, 5G) network is required to transmit the data (Kortuem, Kawsar, Sundramoorthy, & Fitton, 2009).
- *Phase 3:* Edge IT (analytics, pre-processing): This phase deals with analysis and management of data at required levels. After the application of advanced analytics on the data gathered, one can take necessary actions and ensure improved treatment outcomes, better performance, and patient experience for healthcare providers before transferring it to the cloud (Liang et al., 2012).
- *Phase 4:* Storage of data: Since the volume of data captured is very large, therefore it is transferred to the cloud or data center in this phase. To ensure an in-depth review. Data from other sources may be included hereafter all the

7.4   IOT HEALTHCARE SECURITY CHALLENGES

The medical sector has flourished through the adoption of wide variety of applications, devices, and services that IoT offers. The smart IoT-enabled devices embedded on a patient's body are mostly connected to servers in order to collect patient's personal healthcare data globally, henceforth making the data vulnerable to attackers (Ren, Chen, & Chuah, 2012). Therefore security of patients' data is a critical parameter that needs to be take care of while implementing IoT in healthcare. In order to guarantee security in IoT-base healthcare solutions, the following security requirements need to be focused upon:

- **Confidentiality:** Confidentiality resists the revealing of patients' critical data to unauthorized parties. The database containing the critical patient information should be password protected even for medical practitioners who happen to be the legitimate users of data, in order to guarantee some degree of security.
- **Authentication:** Authentication ensures "proof of identity." It prevents a masquerader or an imposter pretending to be the peer who is communicating on the other side.
- **Integrity:** In most of IoT-enabled solutions, the sensors are responsible for collecting data related to the patients. These sensing devices are in turn connected to some centralized server which is responsible for storage of patient's data, so that it may be accessible anytime and anywhere. However, while monitoring the different patient parameters, it needs to be ensured that whatever data is being sent is neither changed in transit nor in the database i.e. integrity dictates that data is not tampered with and is same as is sent by the sender.
- **Availability:** Availability dictates that the services offered by IoT are 24/7 available to the patient, so that in case of an emergency one may not suffer due to unavailability of resources or services. One of the attacks defeating the purpose of availability is Denial of Service (DOS) attack wherein an attacker floods the network with bogus messages, which in turn prevents the legitimate users from availing the services.
- **Authorization:** Authorization ensures that IoT-enabled resources and services are available only to users who are authorized for the same.
- **Non-repudiation:** The principle of non-repudiation dictates that a sender cannot deny sending content later in time, after sending the same.
- **Data freshness:** The data detected by the sensors and stored in storage should be fresh or recent. Data freshness ensures that old data packets are not being replayed and data so received is fresh or latest in time.
- **Fault tolerance:** In the event of network fault e.g., a software glitch or a node failure, there should be a provision wherein the network continues providing the promised services in an efficient manner.
- **Resilience:** If a node gets compromised within the network, the connected devices should still be able to operate without any compromise of the services.

7.5   DESIGN CONSIDERATIONS

- **Computational limitations:** The processors embedded inside IoT-enabled devices are not capable of performing expensive complex computations. This is due to the fact that these processors are less powerful in terms of speed and henceforth cannot display high performance. Therefore, finding a solutions maximing the resource utilization is a tedious task (Ren et al., 2012).
- **Memory limitations:** The IoT-powered devices have system software and operating system (OS) embedded inside them, which is responsible for the activation of these devices to perform the required tasks. However these devices suffer from low on-device memory which prevents them from performing well in complex computational tasks.
- **Mobility:** The IoT sensors are attached to the body of the patient for monitoring different body parameters viz heart rate, body temperature, blood glucose levels, etc. These parameters are reported to the caregiver to keep him updated about the patient's condition. These network devices are mobile in nature. The user might be at his home/in office or any other location and in all the scenarios the network settings and configurations need adjustment. Therefore a mobility-centric algorithm needs to be developed which is difficult to built.
- **Scalability:** IoT has taken the healthcare sector by storm. Every day thousands of devices are connected to the global network. Hence there is a need of developing a protocol or a scheme to ensure that these networks are as scalable as they can be, which is quite a challenging task.
- **Tamper resistance:** An attacker can attack the IoT network, extract the key, modify the source code or load malicious code or content by accessing various files from the database and create havoc. Hence there has to be mechanism or scheme that guarantees physical tamper resistance, which is quite difficult to implement in practise.
- **Energy limitations:** The IoT sensors have limited battery power. These devices are switched to power-saving mode when sensors are idle (not reading data). The CPU speed is also lowered down in absence of processing the data. Hence energy constraint of IoT devices is one of the grave concerns to be focused upon.
- **Dynamic security updates:** IoT networks are vulnerable to a number of attacks like phishing, pharming, etc. Therefore, these networks need a security protocol, which stays up-to-date with the current security definitions or updates to mitigate vulnerabilities. Hence, a security mechanism needs to be built which is capable of installing updated security patches dynamically.

7.6   APPLICATIONS OF IOT IN HEALTHCARE

IoT offers a huge potential for healthcare industry. IoT in healthcare has its own alias viz; Internet of Medical Things (IoMT). Below are 10 examples of IoT-enabled applications in the medical space (Shen, Liang, Shen, Lin, & Luo, 2013):

- **QardioCore:** It is a device that can be worn by users when in gym or at work, etc. This device is capable of monitoring various health conditions such as high blood pressure, fluctuations in glucose levels, etc. QardioCore is capable of sending signals to health center in case of any emergency. For example: fluctuations in glucose levels of a patient suffering from diabetes can be detrimental from a health perspective in case he is unable to contact or reach hospital. The number of physical visits to the hospital are also reduced (Suciu et al., 2015).
- **Zanthion:** It is an alert system which alerts the healthcare professionals by sending signals reporting abnormal behavior; for example, in case of situations where a body stays motionless for a long time (Touati & Tabish 2013). This device also monitors the health status of the wearer.
- **Up by Jawbone**: It is an advanced fitness tracker that is capable of not only keeping a track of calories of a person but it also monitors the sleep patterns, diet, and activity, so that the user can make better health-related decisions.
- **Swallowable sensors:** These sensors can be swallowed by a patient and they can easily detect conditions like irritable bowel syndrome or colon or intestinal cancers (Wang, Wu, & Chen, 2016). This eliminates the need of doing expensive surgeries to fix the various stomach-related diseases.
- **Propeller's Breezhaler device:** This device helps to better manage asthma. The asthma pump has the Propeller's Breezhaler sensor attached at the top which can monitor data each time it is used. The data can be collected on triggers which can in turn alert doctors to manage everything from a mobile application (Yang, Liu, & Deng, 2018).
- **UroSense:** It is a transmitter that is capable of monitoring the urine output in a patient with reproductive problems. It also helps detect prostate cancers and nib certain infections right away at the start which helps in better prevention and control of the disease by the doctor or physician (Zeadally, Isaac, & Baig, 2016).
- **Chrono Therapeutics:** It produces patches that possess the capability to deliver drugs through skin. This makes the delivery of drugs more personalized than "one-size-fits-all" approach. In case a dose is missed by a patient, the doctor is informed through signals this helps in creation of better treatment plans and prevention.
- **AwarePoint:** These systems are capable of tracking every nuance related to caregiving process. This helps in improving satisfaction of patient, patient flow, etc.
- **NHS test beds**: These are smart, connected beds that help monitor, track, and report patient's data. This not only saves time and money but also enables elderly patients or patient's with long-term illness to monitor their progress and issues with more efficacy.
- **Smart Thermometer:** It is capable of detecting the illness of a person and later on mapping all the data related to a patient to the hospital through sensors or wearables that are capable of remotely accessing the health of a patient and hence provide proper treatment after the evaluation of the issue.

The various types of devices discussed may be used for monitoring different health conditions of a patient. The data collected is monitored regularly and remotely to ensure 24/7 service to a patient who is facing a medical emergency. The healthcare professional then analyzes the data collected and delivers the best treatment to the patient after a thorough investigation and analysis of the problem.

7.7   IOT USE CASES

- **Remote patient care:** In case of people residing in far-flung areas, it gets difficult to reach the hospital for treatment in case of emergency situations due to lack of transport and other facilities and services. Healthcare professionals also find it quite difficult to reach out to patients residing in cut-off areas in case of emergencies depicted in Figure 7.3. This can have serious implications and may even lead to loss of life in adverse cases. This issue with time-consuming commute can however be solved with remote patient care powered by IoT. The sensors fitted onto a patient's body may monitor different types of parameters which can be sent to the doctors anytime and anywhere in real time, so that he can prescribe medicines thereby aiding in patient's treatment remotely. In certain cases, face-to-face interaction can be provided with the help of IoT-enabled solutions, which will help the healthcare providers to prepare plans and get the equipment ready, until the patient reaches the hospital. Different types of medical reports in form of images like X-rays, CT-scan reports, etc. can be shared with the medical practitioners through video/audio streaming, which can further help the doctor in treating the patient at a distance.
- **Augmenting surgeries:** IoT has penetrated all the dimensions of medical field including operation theatres. Robotic devices in the form of robot-assisted surgeons powered by machine intelligence are capable of performing surgeries in precise and an efficient manner. Real-time tracking of various activities and procedures before and after the operating stages can be done through various types of IoT sensors in order to prevent complications.
- **Wearables:** In the form of wearables, IoT devices may help care providers to gather numerous data points on the circadian rhythms, movement, oxygen levels, pulse, etc. of the patient. These devices can provide updates to nurses and patients in real time. Consider a scenario wherein a heart patient gets an abnormal heart rate. The device may automatically relay the alert to nurses thereby encouraging them to provide urgent and prompt aid to the patient. This will also assist inremote health observation of aged patients outside of the grounds of the hospital but require continuous supervision. There is considerable demand for IoT in healthcare and it will prove to be highly useful for patients and healthcare providers. It is built to improve healthcare and business efficiency. The power of IoT has also been leveraged by many hospitals around the world under their smart hospital programs.
- **Staff and inventory tracking:** IoT sensors can be used to track the location of staff and patients in real time. Different types of IoT-powered devices like

RFID cards or others may be used to track the inventory and locate the staff in case of emergency. This will help in driving down the operational costs and increasing the efficiency of oranization overall. Real-time monitoring of staff, available resources, or other hardware equipment with the help of IoT-enabled devices will help dedicate more time to patient care.

- **Remote hardware monitoring and maintenance:** Hardware and software both are required equally for all the IoT-powered healthcare-related services to function. Power failures, equipment failures, etc. are considered detrimental from a healthcare perspective and may even lead to loss of life. In such scenarios IoT devices come to rescue. e-Alert by Phillips is one such solution which is capable of tracking and reporting glitches in software or any hardware failures. If any issue is found in the medical hardware, an alert is sent to the hospital staff, so that appropriate measures can be taken to ensure availability of services without any intervention.
- **Pharmacy management:** The drug industry is worth millions of dollars and is very complex. As there are multiple steps in moving and handling drugs from a factory to a medical storage facility, there are a variety of concerns related to the safety of drugs. IoT will help integrate the best security techniques with the recent advancements to promote quicker drug delivery, better procedures, and improved patient outcome. Take for example, smart refrigerators that might be used to preserve vaccines and protect them from being destroyed while storing, transportation, or transition. IoT-enabled hospitals will greatly enhance working performance and efficacy, glitch free hospital care and safety resulting in improved patient experience.
- **Healthcare automation and error reduction:** Routine operational and administrative tasks can be managed well by IoT-based healthcare solutions. IoT devices are capable of handling huge data sets and generating multiple metrics related to client's health conditions. This computerized or sensor-based statistical analysis excludes the chance of human error in diagnosis due to less human intervention, thereby promoting error reduction.

7.8 CONCLUSION

The integration of IoT in healthcare has resulted in the transformation of conventional healthcare procedures. Smart sensors embedded in the form of wearables in the patient's body are capable of providing immediate assistance in case of emergency situations. Not only this, introduction of IoT in healthcare has improved the efficiency and reduced the cost and effort that needed to be invested on physical visits to hospitals. IoT has improved the healthcare outcomes to a remarkable and realizable extent. The sensors fitted onto a patient's body may monitor different types of parameters which can be sent to the doctors anytime and anywhere in real time, so that he can prescribe medicines thereby aiding in patient's treatment remotely. In certain cases, face-to-face interaction can be provided with the help of IoT-enabled solutions, which will help the healthcare providers to prepare plans and get the equipment ready until the patient reaches the hospital. IoT sensors can be used to track the location of staff and patients in real time. Different types of IoT-powered devices like RFID cards or others may be used to track the inventory and locate the staff in case of emergency. This will help in driving down the operational costs and increasing the efficiency of organizations overall. In a nutshell we can say that IoT has and will continue to revolutionize the healthcare sector to a measurable extent.

REFERENCES

Alamr, A. A., Kausar, F., Kim, J., & Seo, C. (2018). A secure ECC-based RFID mutual authentication protocol for internet of things. *The Journal of Supercomputing, 74*(9), 4281–4294.

Atzori, L., Iera, A., & Morabito, G. (2010). The internet of things: A survey. *Computer Networks, 54*(15), 2787–2805.

Bagci, I. E., Raza, S., Roedig, U., & Voigt, T. (2016). Fusion: coalesced confidential storage and communication framework for the IoT. *Security and Communication Networks, 9*(15), 2656–2673.

Chi, L., Hu, L., Li, H. T., Sun, Y., Yuan, W., & Chu, J. F. (2013). Improved energy-efficient access control scheme for wireless sensor networks based on elliptic curve cryptography. *Sensor Letters, 11*(5), 953–957.

Choi, J., In, Y., Park, C., Seok, S., Seo, H., & Kim, H. (2018). Secure IoT framework and 2D architecture for End-To-End security. *The Journal of Supercomputing, 74*(8), 3521–3535.

Cvitić, I., & Vujić, M. (2015). Classification of security risks In the IoT environment. *Annals of DAAAM & Proceedings, 26*(1), 5–9.

Gardašević, G., Veletić, M., Maletić, N., Vasiljević, D., Radusinović, I., Tomović, S., &Radonjić, M. (2017). The IoT architectural framework, design issues and application domains. *Wireless Personal Communications, 92*(1), 127–148.

Garkoti, G., Peddoju, S. K., & Balasubramanian, R. (2014). Detection of insider attacks in cloud based e-healthcare environment. *2014 International Conference on Information Technology* (pp. 195–200). IEEE.

Gope, P., & Hwang, T. L. (2016). BSN-care: A secure IoT-based modern healthcare system using body sensor network. *IEEE Sensors Journal, 16*(5), 1368–1376.

Islam, S. R., Kwak, D., Kabir, M. H., Hossain, M., & Kwak, K. S. (2015). The internet of things for health care: A comprehensive survey. *IEEE Access, 3*, 678–708.

Ko, H., & Song, M. (2016). A study on the secure user profiling structure and procedure for home healthcare systems. *Journal of Medical Systems, 40*(1), 1–9.

Kortuem, G., Kawsar, F., Sundramoorthy, V., & Fitton, D. (2009). Smart objects as building blocks for the internet of things. *IEEE Internet Computing, 14*(1), 44–51.

Liang, X., Barua, M., Chen, L., Lu, R., Shen, X., Li, X., & Luo, H. Y. (2012). Enabling pervasive healthcare through continuous remote health monitoring. *IEEE Wireless Communications, 19*(6), 10–18.

Liang, X., Li, X., Shen, Q., Lu, R., Lin, X., Shen, X., & Zhuang, W. (2012, March). Exploiting prediction to enable secure and reliable routing in wireless body area networks. *2012 Proceedings IEEE INFOCOM* (pp. 388–396). IEEE.

Lu, R., Lin, X., & Shen, X. (2012). SPOC: A secure and privacy-preserving opportunistic computing framework for mobile-healthcare emergency. *IEEE Transactions on Parallel and Distributed Systems, 24*(3), 614–624.

Muralidharan, S., Roy, A., & Saxena, N. (2016). An exhaustive review on Internet of things from Korea's perspective. *Wireless Personal Communications, 90*(3), 1463–1486.

Paschou, M., Sakkopoulos, E., Sourla, E., & Tsakalidis, A. (2013). Health internet of things: Metrics and methods for efficient data transfer. *Simulation Modelling Practice and Theory, 34*, 186–199.

Raza, S., Shafagh, H., Hewage, K., Hummen, R., & Voigt, T. (2013). Lithe: Lightweight secure CoAP for the internet of things. *IEEE Sensors Journal, 13*(10), 3711–3720.

Ren, Y., Chen, Y., & Chuah, M. C. (2012). Social closeness based clone attack detection for mobile healthcare system. *2012 IEEE 9th International Conference on Mobile Ad-Hoc and Sensor Systems (MASS 2012)* (pp. 191–199). IEEE.

Shen, Q., Liang, X., Shen, X., Lin, X., & Luo, H. Y. (2013). Exploiting geo-distributed clouds for a e-health monitoring system with minimum service delay and privacy preservation. *IEEE Journal of Biomedical and Health Informatics, 18*(2), 430–439.

Suciu, G., Suciu, V., Martian, A., Craciunescu, R., Vulpe, A., Marcu, I., & Fratu, O. (2015). Big data, internet of things and cloud convergence–an architecture for secure e-health applications. *Journal of Medical Systems, 39*(11), 1–8.

Touati, F. & Tabish, R. (2013). U-healthcare system: State-of-the-art review and challenges. *Journal of Medical Systems, 37*(3), 1–20.

Wang, Y., Wu, X., & Chen, H. (2016). An intrusion detection method for wireless sensor network based on mathematical morphology. *Security and Communication Networks, 9*(15), 2744–2751.

Yang, Y., Liu, X., & Deng, R. H. (2018). Lightweight break-glass access control system for healthcare internet-of-things. *IEEE Transactions on Industrial Informatics, 14*(8), 3610.

Zeadally, S., Isaac, J. T., & Baig, Z. (2016). Security attacks and solutions in electronic health (e-health) systems. *Journal of Medical Systems, 40*(12), 1–12.

8 Cloud, Edge, and Fog Computing: Trends and Case Studies

Eng Lieh Ouh[1], Stanislaw Jarzabek[2],
Geok Shan Lim[1], and Ogawa Masayoshi[1]
[1]Singapore Management University, Singapore
[2]Bialystok University of Technology, Poland

CONTENTS

DOI: 10.1201/9781003132080-8

8.1   INTRODUCTION

Cloud computing technology allows software providers to sell their software to more customers than with traditional software deployment methods. With an economy of scale reaped by service providers, consumers are also able to pay a lower price for the service. With the growing popularity of cloud hosting, application developers and software vendors have been favoring multi-tenanted over traditional single-tenanted cloud service architectures. Multi-tenancy is an architecture style where multiple clients use a single cloud application instance. Different service architectures isolate tenant's data to a varying degree. At the same time, a service architecture must accommodate the highest possible degree of service variability to allow a service provider to offer a cloud service to tenants with a wide range of varying requirements for service functionality. This case study describes formalized evaluation of profitability of cloud service offering strategies.

Whether adopting a specific cloud service solution will be profitable or not is of prime concern to a service provider, but the decision is not without challenges. The following discussion illustrates this point: Service profitability (Ouh & Jarzabek, 2016) depends on the "cost of engineering a service for a given customer base, on service provisioning cost, and on the revenue gained from selling the service to that customer base." There are many choices of service architecture models for service providers to select depending on the required level of sharing and isolation among tenants (Ouh & Jarzabek, 2014). A service provider aims to maximize the service profitability by trying to select the most economically viable service architecture, while catering for a possibly wide range of tenant-specific requirements. "As the unfit costs for the tenant decrease, the relative economic advantage of the service business model increases" (Ma, 2007). Suppose service provider understands that her tenants have moderate varying requirements for service functionalities, Then, a service provider can choose a service architecture that allows all the tenants to use the same service code and service instance. However, such a resource sharing solution may not scale to the situations when tenants' requirement for service vary drastically. The degree of sharing of resources limits the extent to which we can engineer the variations of the service. Then, a service providers might consider service architecture model with a higher degree of isolation, such as a dedicated service instance for each tenant. In this case, the provisioning cost of a cloud service is higher. However, with isolated resources for each tenant, this decision opens up more options for engineering varying requirements to a service. There are degrees of resource sharing between the Fully Shared and No-Shared architectures, offered by so-called hybrid service architectures.

To decide on service architecture, a profit-driven service provider evaluates the engineering and provisioning costs to offer a service against the amount of revenue earned when the tenants pay a fee to use this service. The many service-related factors and constraints make manual analysis of service profitability undependable and not scalable. Informed analysis of this matter is vital to service provider as the primary target of offering a service is to achieve profitability. Although service profitability has been discussed before in the literature, these discussions have been primarily focused on deriving more types of pricing models, cost-effective engineering, and provisioning methods in independent studies, giving a potentially incomplete analysis of service profitability. To the best of our understanding, no formal evaluation of service profitability taking into account all the important factors and constraints has been proposed. We fill the gap by paving the ground for a more formal and quantitative treatment of service profitability problem.

As the first step towards that goal, we define a conceptual model of various factors that must be taken into account in evaluating cloud service profitability. Some of those factors the reader can spot in the above discussion of service profitability tradeoffs. We further elaborate on those factors in Section 8.2. We then formalize dependencies among those factors that determine service profitability as a step towards quantitative analysis, which is discussed in Section 8.3. With an extended conceptual model, we show how service provider's decisions on service architecture, dynamic versus static service adaptation techniques, or the profile of the tenants affect service profitability. Finally, we propose a simulation process and software tool to further to aid service providers in the analysis of cloud service profitability.

8.2   OVERVIEW OF THE MULTI-TENANCY CLOUD SERVICE MODELS

A number of architecture models for multi-tenancy cloud service have been proposed (Fiaidhi, Bojanova, Zhang & Zhang, 2012). In Figure 8.1, the leftmost design shows a dedicated app instance with a dedicated database for each tenant. As for the second design, different tenants access a shared application instance but access different database servers. The third model features a shared design where tenants access a shared application instance and shared database. These three architecture models differ in how a service is engineered, packaged, and hosted.

8.3   ENGINEERING OF CLOUD SERVICES

A service needs to be engineered to handle multiple tenants' requirements. Tenant-specific requirements are bounded to a relevant service either statically or

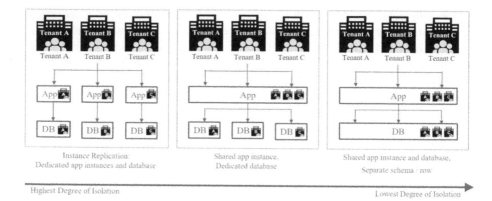

FIGURE 8.1   Overview of multi-tenancy cloud service models.

dynamically. The space of the tenant-specific requirements for all tenants is referred to as variant requirements.

8.3.1 STATIC BINDING VARIATION TECHNIQUES

Static binding variation techniques (SBVT) bind tenant-specific requirements at the service construction time. That means that such service requirements cannot change dynamically, during service execution. When a compiler builds the code to produce a custom service, one or more of the variation techniques are applied to bind tenant-specific requirements to the multiple variation locations in service code. Some of the frequently used techniques include macros for preprocessing, design patterns, Java conditional compilation, templates and parametrization. These techniques can be challenging to implement, so there are also high-level variability management tools (e.g., XVCL/ART (Jarzabek, Bassett, Zhang, & Zhang, 2003; Khue, Ouh, & Jarzabek, 2015) that extend the idea of macros and provide better support to manage the variants. Software Product Line Engineering (SPLE) method uses static binding variation techniques extensively to assemble software systems from reusable parts.

8.3.2 DYNAMIC BINDING VARIATION TECHNIQUES

Besides SBVT, another form of variation techniques are being referred to as dynamic binding (DBVT). Services built using these dynamic binding techniques can be adaptedto varying tenant requirements at runtime. Aspect-Oriented Programming (AOP) is one of the frequently used dynamic binding technique. With AOP, tenant-specific code, are injected at specified pointcuts of the service code that is common to all tenants.

Another technique of choice is the registry lookup method commonly found in Service-Oriented Service Architecture (SOA). With SOA, service variants are first registered in the registry. When application components require invoking these services, they look up the registry at runtime and the registry returns the right variant for the application components to dynamically bind to it.

8.4   PACKAGING OF CLOUD SERVICES

The encapsulation level of the service components refers to how service components of a service are packaged together to serve the tenants. When new tenants are on-boarded or when existing tenants change their service requirements, the service provider needs to adapt the service to accommodate such evolving needs with the consideration that existing tenants should have minimum impact. The packaging of the service if implemented correctly can minimize the impact on existing tenants.

8.4.1   SERVICE LEVEL

A service encapsulates a set of service components. Some of these service components can be shared among services for reuse. When there is a change required to these shared service components, the service provider needs to identify the service impact and the tenants who are using these services can be temporarily affected because the availability of the service may be interrupted due to the change. The change to the shared service components may not be required by some of these tenants but will inadvertently be affected as these service components are shared across services. This degree of encapsulation has a clear set of service components for each service, but it is not clear in terms of separation between tenants.

8.4.2   TENANT LEVEL

In this case, a tenant uses a set of functionalities, which are encapsulated with a tenant-specific set of service components. For each service, there are a distinct set of service components that are developed and used by each tenant. When there is a change required for the tenant, only the tenant-specific service components require modifications. Other tenants are not affected in terms of availability when the change is implemented to this tenant. This degree of encapsulation has a clear set of service components for each tenant.

8.5   HOSTING CLOUD SERVICES

A service engineering and encapsulation levels can differ, as can the hosting. Hosting of the service refers to how the service is executed in an application instance. An application instance comprises of software processes running on hardware or virtualized platform in a network infrastructure that can execute the service components. The application instance can be shared among tenants or dedicated to each tenant.

8.5.1   SHARED INSTANCE

When a common application instance is executing services for many tenants, we termed it as a shared instance. The service provider typically uses this option to utilize infrastructure platform resources better.

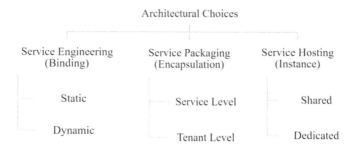

FIGURE 8.2 Architectural choices.

8.5.2 DEDICATED INSTANCE

When a tenant accesses a service on an application instance without sharing platform resources with other tenants, we term it as dedicated instance. A service provider typically uses this option when there is a need to comply with service level agreements or industry regulations or due to high variation in tenants' requirements. Figure 8.2 shows a summary of these architectural choices.

8.6 DISCUSSION ON ARCHITECTURE CHOICES

A service provider has to make these architecture choices of service engineering, packaging, and hosting that determine the degree of service variability and impact the costs and benefits to design for service variability.

8.6.1 CLOUD SERVICE VARIABILITY

"The degree of service variability is the extent to which variations in service requirements can be handled" (Ouh & Jarzabek, 2016). A service offered by service provider comprises a set of service components that are implemented to support the service. To illustrate this extent of service variability, we use an example of the boundary-controller-entity pattern that categorizes each service component into the boundary or controller or entity components.

Boundary components comprise of the interfaces created to interact with the caller of the service. It can be visual such as web or mobile user interfaces or non-visual such as a method call to invoke the service. In both cases, tenant's requirements for interface components can vary, and we termed that as Interface Variability. Controller components comprise of the compute or processing logic required of the service. Each controller can also comprise a single component or a composition of many diverse components to compute the processing logic. Similarly to the boundary component, the tenant's requirements for controller components can also vary among tenants. We termed that Logic and Composition Variability, respectively. Entity components comprise the data required to be processed and stored for the service. Each entity component can persist that survives reboot (durable) or non-durable. The data can persist in memory or file-based or a form of database such as relational or non-relational. Data variability can also

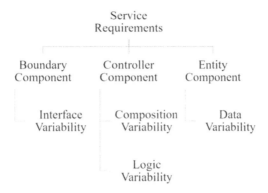

FIGURE 8.3 Variations in service requirements.

happen among tenants. Take for an example of a relational database, data from multiple tenants can persist in one table or multiple tables in one database or one database per tenant depending on the design decision made by the service provider.

The extent of variations a service can realize takes into account all the possible variations of the boundary, controller and entity components for that service. When service provider offers a service with a high degree of service variability, the service will be able to address extensive variations of service requirements from the tenants. It has more potential to on-board more new tenants with varying service requirements, resulting in higher revenue from offering a service. Figure 8.3 summarizes these variations.

8.6.2   Costs and Benefits of Designing Service Variability

The cost to design service variability includes the effort to engineer and host the service to support varying requirements of the tenants. If service provider chooses to apply static binding, encapsulating at the tenant-level and using dedicated instance, this design decision can better support a high degree of service variability. However, this design decision requires higher design costs and more computing resources. On the other hand, the service provider can increase her revenue gained by designing for service variability to widening the tenants' base. Based on the context of the service provider, she therefore needs to balance the need to reduce the cost to engineer the service and increase the revenue earned when the tenants pay for using this service.

8.6.3   Cloud Service Architecture Models

We model into three service architectural models, namely NoShared, Fully Shared, and Partially Shared, defined three different ways for how the service code is designed, packaged, and hosted. These models mainly differ in the extent of sharing or isolation of the resources among tenants. When service provider hosts the service on a common executing platform or instance, she is deciding to share the instance among tenants. With this decision in service hosting, a service provider can have the flexibility of choices in service packaging (service or tenant level) and service engineering (static or dynamic

Models	Service Engineering	Service Packaging	Service Hosting
Fully-Shared	Dynamic Binding	Service Level Encapsulation	Shared Instance
Partially-Shared	Static or Dynamic Binding	Service or Tenant Level Encapsulation	Shared Instance
No-Shared	Static or Dynamic Binding	Service or Tenant Level Encapsulation	Dedicated Instance

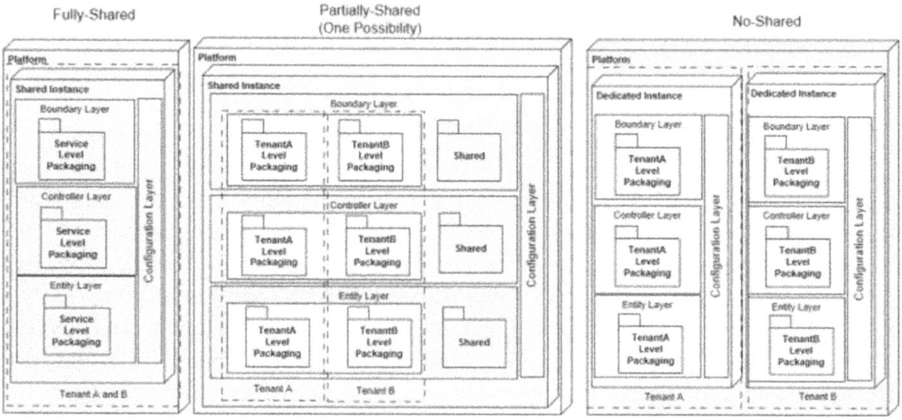

FIGURE 8.4 Cloud service architectural models.

binding). If a service provider chooses to share resources fully for service packaging (service level) and engineering (dynamic binding), these decisions resulted in the Fully Shared model. If service provider also wishes to have a certain level of isolation for service packaging (tenant-level) or service engineering (static binding), these decisions resulted in a Partially Shared model.

On the other hand, a service provider may hosts the service on an isolated platform or instance per tenant. With this decision in service hosting, a service provider can also have the flexibility of choices in service packaging (service or tenant level) and service engineering (static or dynamic binding). Whatever the decisions made for service packaging or engineering, the decision to host the service on a dedicated instance isolated the resources among tenants and resulted in the No-Shared model. Figure 8.4 summarizes these relationships.

8.7 VARIABILITY SCENARIOS – SERVICE PROVIDER PERSPECTIVE

For a profit-driven company, the service provider wishes to implement and sell a service solution that maximizes service profitability. To do that, a service provider needs to decide on a service solution that can serve as many tenants as possible with minimum cost. Below are possible scenarios that might be encountered by service provider and illustrate the complexity of the multiple factors and constraints that the service provider has to consider.

8.7.1   Cost Considerations

Suppose the service provider wishes to incur the lowest cost in their offering of a service. In that case, the service provider has to choose the service architecture model that maximize sharing of resources to reduce cost and still supports a given degree of service variability. Based on the three types of service architectures discussed earlier, a service provider may decide on the Fully Shared Model. The trade-off to this decision is the potentially lower degree of service variability that is supported by the Fully Shared Model, implying that some tenants with significantly varying requirements cannot be on-board.

8.7.2   Revenue Considerations

Suppose a service provider wishes to maximize revenue gained when offering a service. In this case, the service provider has to choose the architecture model that supports a greater extent or variation of the tenant's requirements. With more tenants paying for the service, a service provider can achieve higher revenue. Based on the three types of service architectures discussed earlier, the service provider may decide on a No-Shared Model. The trade-off to this decision is that the higher cost that service provider incurred might be transferred to the tenants as a higher fee.

8.7.3   Tenant Profile

A service provider might also wish to design and make decisions based on the tenant's profile. The service provider has to weigh the benefit of the ability to on-board these tenants to increase revenue against customizing the service to possibly varying requirements that will also increase the costs. There are two likely scenarios.

In the first scenario, we assume an initial number of tenants (e.g. 50 tenants) that service provider needs to on-board when starting to offer the service. With the assumption that service provider understands the varying requirements of these initial tenants, a service provider can design the required variability into the service at the start. The service provider can then choose the Fully Shared model to minimize the cost of hosting the service. Let the shared hosting cost of supporting 50 tenants be HostSharedCost(50), and the engineering cost for dynamic binding be DVTDesignCost(50), then the cost of offering the service can be defined as HostSharedCost(50)+ DVTDesignCost(50).

In the second scenario, we continue to assume there are 50 existing tenants on-boarded in our Fully Shared model. A service provider is now aware that there are now 30 new tenants with diverse requirements to be on-boarded. One choice that a service provider can choose is to host them in a No-Shared model as the current Fully Shared model cannot accommodate the tenant's significantly varying requirements. Let the engineering cost to design for static and dynamic binding for the 30 future tenants be SDVTDesignCost(30) and the dedicated hosting cost of supporting the 30 future tenants be HostDedicatedCost(30), then the cost equation to support both the 50 existing tenants and 30 new tenants can be defined as HostSharedCost(50) + DVTDesignCost(50) + HostDedicatedCost(30) + SDVTDesignCost(30).

In the third scenario, we again assume there are 50 existing tenants on-board in our Fully Shared model, and aservice provider can also choose to host the 30 new tenants in a Partially Shared model to accommodate the tenant's varying require-ments. If the cost to re-design for the new tenants be DVTRedesignCost(30), then the cost equation to support both the 50 existing tenants and 30 future tenants can be defined asHostSharedCost(50) + DVTDesignCost(50) + HostSharedCost(30) + DVTRedesignCost (30).

For the second and third possible scenarios, a service provider can evaluate the costs of both choices quantitatively as follows to make a decision:

Min (HostSharedCost(30) + DVTRedesignCost(30), HostDedicatedCost(30) + SDVTDesignCost(30))

8.7.4 Market Share Considerations

The service provider might also wish to on-board as many tenants as possible in order to capture the market share and create a barrier to entry to newcomers. Based on the earlier scenario to support 50 existing tenants and 30 future tenants having diverse service requirements, the service provider can choose the No-Shared model to support the 50 tenants when they start offering the service and continue the same model for the 30 new tenants. If the cost to host 80 tenants on isolated instances be HostDedicatedCost(80) and the engineering cost for static and dynamic binding be SDVTDesignCost(80), then the cost equation can be defined as HostDedicatedCost(80) + SDVTDesignCost(80). For the same scenario of 50 existing tenants and 30 new delta tenants, a service provider can now have another choice of a No-Shared model to assess quantitively and make a decision on the service architecture.

8.7.5 Service Isolation

Both external and internal factors can constrain service provider offering a service in a specific environment or context. One external constraint is the need to provide a sufficient level of isolation so that the service can comply with the laws and regulations where it is being offered. A service provider can choose either the No-Shared or Partially Shared models to provide a certain degree of isolation of the resources. For a lesser degree of resource isolation, a service provider can provide isolation of entity components while still share the boundary and controller components among tenants. This variant of the Partially Shared model allows isolation of the data among tenants but allows sharing of the compute and interface components. Other variants of the Partially Shared model involves isolating the boundary or the controller components. If the te-nants have even more stringent requirements on resource isolation, the No-Shared model with isolation at the service engineering, packaging, and hosting will be a better choice.

8.7.6 BUDGET CONSTRAINTS

Besides having external constraint, a service provider can also have constraints internally imposed with the organization. One possible such constraint is the upfront budget set to offer the service. A service is usually offered for multiple years and a budget is set aside for this purpose. A service provider has to work within this budget. In this case, it is challenging to decide on the service architecture model with the complex inter-related set of factors that may influence the outcomes. The analysis requires many combinations of computations of which manual analysis is going to be time-consuming. We will propose a method to evaluate this scenario with an optimization method to perform this analysis automatically.

8.8 CLOUD SERVICE PROFITABILITY MODEL

Given that many factors collectively determine the profitability of service offering, we introduce a concept map to summarize these factors and their inter-dependencies. This concept map illustrates the dependencies between the service cost, service revenue, tenants, and service architectures for an overview of these factors for service profitability. Below are the preliminary concepts and definitions of the concept map.

8.8.1 SERVICE TENANTS

A service provider designs a service based on a set of initial tenants' requirements. This set of initial tenants and their requirements is defined as the **Tenant Base (TB)** of a given service. During the period service is being offered by the service provider or the service period, more tenants may wish to subscribe and pay to access the service. This set of tenants a service provider expects to on-board onto the existing service architecture along with the initial tenants is defined as the **Delta Tenant Base (DTB)**.

8.8.2 RANGE OF SERVICE VARIABILITY

A service provider offering a service needs to adapt the service to the tenants, which may vary to a certain degree. We define the extent of this adaptation as the **Range of Service Variability (RSV)**. A service provider wishes to accommodate a wider range of RSV to serve more tenants, which results in higher revenue for the service provider. The **Service Architecture (SA)** chosen by the service provider determines the RSV that can be supported.

During the service period, the service provider needs to further adapt the service to more tenants with existing tenants already on-boarded. We define the degree of these further adaptations as **Delta Variability (DV)**.

8.8.3 SERVICE COSTS

When a service provider develops the service, there are **Service Engineering Costs (SEC)** incurred to engineer for the required functionality and efforts to design the variability. These sub-costs are termed as **Service Functionality Engineering Cost (SFEC)** and **Service Variability Engineering Cost (SVEC).**

Let RSV be a given range of service variability and SA be the service architecture, then Service Variability Engineering Cost (SFEC) can be defined as a function of SA and RSV. SVEC is also determined by the **variation technique (VT)** selected to either statically or dynamically bind the variants to the service. Let VT be the variation technique, then Service Variability Engineering Cost (SVEC) can be defined as a function of SA, RSV, and VT.

To host a service, a service provider also incurs **Service Provisioning Costs (SPC)** to provide the hosting and network resources to support the service. Service Provisioning Cost(SPC) is based on the choice of service architecture and can be defined as a function of SA and RSV.

For the **Total Service Costs (TSC)** when offering a service, a service provider takes into account both the SEC and SPC incurred to develop and host the service.

During the service period, there are also **Delta Cost (DC)** incurred to re-engineer (**Service Engineering Delta Cost or SEDC**) and to re-provision (**Service Provisioning Delta Cost or SPDC**) the service to support **Delta Variability (DV)** of new tenants not on-boarded initially. Similarly to SVEC and SPC, SEDC can be defined as a function of SA, DV and VT, and SPDC can be defined as a function of SA and DV.

8.8.4 Service Revenue

The **Service Revenue (TSR)** is the accumulated fee collected from tenants in a given TB who pay a service price for the usage of the service initially offered. **Delta Revenue (DR)** is the accumulated fee collected from tenants in a given DTB who are subsequently on-boarded. The price paid from these two groups of tenants may differ depending on the pricing strategy.

8.8.5 Service Profits

The **Service Profits (SP)** is the total profits in terms of monetary benefits from offering a Service. SP is defined as a function of TSC, TSR, DC, and DR. A service provider offers a service usually for many years, and we need to incorporate the discounted present value of money. When performing analysis, we use the Net Present Value (NPV) (Frakes & Terry, 1996; Mili, Chmiel, Gottumukkala & Zhang, 2000). Let Y be an investment cycle measured in the number of years, and d refers to the discount rate, then the equation for SP is:

$$\text{Service Profitability(SP)} = \text{Total Revenue} - \text{Total Costs}$$

$$SP = -(SEC_{(TB)} + SPC_{(TB)}) + \sum_{y \in Y}^{Y} \frac{TSR_{(y)} - SPC_{Op(y)} + DR_{(y)} - DC_{(y)}}{(1+d)^y} \qquad (8.1)$$

Intuitively, these equations calculate service profits by subtracting the service costs at the initial offering of the service with the yearly revenue gained. To accurately illustrate service profits over the investment years, we use the expected Return On Investment (ROI) and can be measured by:

$$\text{ROI} = \frac{SP}{\left(SEC_{(TB)} + SPC_{(TB)} \right)} \qquad (8.2)$$

Figure 8.5 shows the above model concepts in a concept map.

8.9 ANALYZING SERVICE PROFITABILITY BASED ON CONCEPT MAP

In this section, we describe a step-by-step simulation process and tooling to analyze service profitability with references to the factors identified in the concept map in the earlier section. This simulation process allows service provider to provide her estimations according to her scenario or context. Depending on the tenants' profiles or the budget constraint, the process and tool output the possible service profitability outcomes according to the types of service architectures.

8.9.1 OVERVIEW OF SIMULATION PROCESS

The first key step in the simulation process is the tenants' information. The service provider needs to provide information about the tenants to be initially on-board and potential on-board during the service period. For delta tenant base or tenants that are potentially on-board during the service period, the service provider can estimate based on parameters to simulate distributions of tenant's requirements for a specified set of tenants.

For the service cost and delta cost management, the service provider provides the costings for offering the service. These costings differ for the tenants in the tenant base that is initially on-boarded and the delta tenants that are on-boarded during the service period. Both the service engineering and service provisioning costs are to be considered. For service engineering, the service provider needs to consider which of the variation technique to adopt. For cost estimations of service engineering, existing effort cost estimation models (e.g. COCOMO II) can be used. The provisioning costs can differ in terms of the upfront and recurring costs if the service provider decides to provision the architecture on-premise or off-premises on the cloud. For cost estimations for service provisioning on the cloud, existing cloud providers do provide cost calculators in using their cloud services (e.g. AWS Cost Calculator). For delta tenants, both the service engineering delta and service provisioning delta costs are to be considered. These costs may differ from the SEC and SPC for the initial tenants as there are additional efforts to re-engineer for service variability based on the variation of requirements from the new tenants. We can still use the existing effort costs estimation models (e.g. COCOMO II) and cloud cost calculators such as AWS Cost Calculator to measure these costs quantitatively.

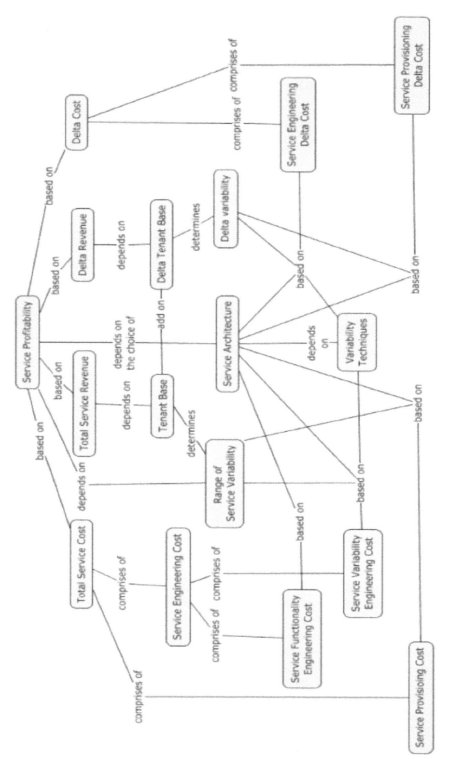

FIGURE 8.5　Service profitability concept map.

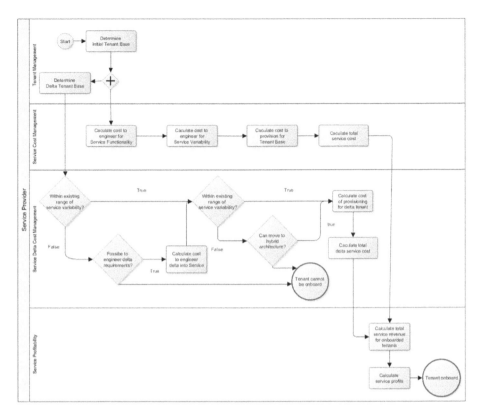

FIGURE 8.6 Simulation process for service profitability.

For the service profitability process step, the costs (as described earlier) and revenue received from the on-boarded tenant based on the type of pricing models determine the service profits. The process composed of the above process steps is summarized in Figure 8.6.

8.9.2 SIMULATING WITH CONSTRAINTS

In the earlier section, we define the concepts map of the service profitability and a process to simulate the model to analyze service profitability. On the other hand, a service provider may have a budget target and wishes to determine the service architecture based on the budget constraint. For example, a service provider may wish to know how to maximize revenue and minimize costs to achieve higher service profits within an initial budget constraint set. For this section, we describe how we enhance the simulation process with multi-swarm optimization (MSO) technique to handle this scenario. Specifically, we want to maximize service profitability by modelling the costs and revenue of providing the service within the budget constraint. The combinations of possible solutions are extensive, and manual analysis is not scalable. To effectively measure service profitability with constraints, we adopt an optimization method to automate this analysis.

Multi-swarm optimization (MSO) is one optimization method that is popular in recent years, and it is a variant of particle swarm optimization (PSO). PSO is a population-based method that got its inspiration from the information exchange behaviour of the birds in a swarm (Kennedy & Eberhart, 1995). In PSO, individuals in a swarm are called particles in a search space. The swarm initialize each particle with a random initial position and speed. At each iteration of the particle's movement, each particle moves and adapts its velocity based on the position and velocity of its neighbours, seeking its best and overall best particle positions and best fitness score. In our case, each particle is one possible solution to the costs incurred and revenue gained, and the fitness score is the service profits that can be achieved within the budget constraint. Multi-swarm optimization (MSO) extends the PSO concepts with the multiple sub-swarms instead of one standard swarm as in PSO (Liang & Suganthan, 2005; Ostadrahimi, Mariño, & Afshar, 2012; Solomon, Thulasiraman, & Thulasiram, 2011) and is a better fit for multi-modal problems. Similarly to one swarm, a multi-swam keep track of a collection of swarms and the best particle position found among all the swarms. MSO is especially fitted for optimization problems with multiple local optima. The downside of MSO versus PSO is the longer computation time required.

We first map the service profitability concepts to MSO concepts at the particle level, which is the base construct of MSO. For our problem, we model Service Engineering Costs (SEC) and Service Provisioning Costs (SPC) as two dimensions of the particle location. A service provider can implement the Fully Shared architecture (e.g. using microservices) that incurs a moderate level of SEC and SPC upfront initially. However, to subsequently on-board more tenants, the SEC cost is higher as compared to a No-Shared architecture due to the degree of resources shared. For the No-Shared architecture, a service provider incurs a high level of SEC and SPC due to the dedicated instances and engineering effort (e.g. based on software product line techniques). With a limited service budget, service providers also need to balance the SEC and SPC for a given service architecture.

8.9.3   TOOLING FOR THE SIMULATION PROCESS

The simulation process can be challenging to execute with many factors that can affect the analysis. We develop a tool to automate this process, taking into account factors that affect the service costs, revenues goals and constraints. We termed this tool Service Profitability Analyzer (SPA). SPA implements the simulation process and the concepts of the Service Profitability model in an intuitive graphical user interface. We hope that the service provider can use this tool with ease to evaluate service profitability. Figure 8.7 shows a screenshot of the SPA user interface.

A service provider can specify an expected number of tenants for the tenant base and delta tenant base. The possible delta variability values for these delta tenants for variability engineering and variability provisioning are low, medium, high and random, indicating the degree of changes required to be re-engineered for existing service. We introduce weighted cost coefficients α and β in our

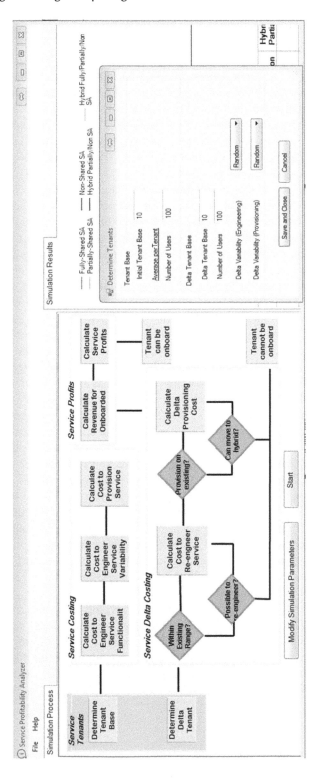

FIGURE 8.7   Service profitability analyzer.

calculations and the values of these cost coefficients vary with the values of the delta variability values (low, medium, high, random).

For estimation of the engineering costs in SPA, a service provider can input custom values or based on a specific model. The SPA interface allows service provider to compute the SFEC and SVEC based on COCOMO II (Madachy, 2015). For estimation of the provisioning costs in SPA, a service provider can input custom values or based on specific cloud implementations. The SPA interface allows for cost inputs for the essential web services based on Amazon Web Services or AWS (Amazon, 2007). With earlier Equations 8.1–8.2, SPA can calculate the service profits and ROI for a given set of inputs.

If a service provider wishes to simulate further based on a budget constraint, the tool can evaluate based on multi-swarm optimization or MSO for service profitability within the constraint. We set the upper bound of the position and speed to be the service budget since both SEC and SPC cannot exceed this budget. This setting can also reduce the computational power and time needed to run the algorithm. A fundamental operation in MSO is to compute particle velocity. For a given particle, a new velocity is influenced by several factors, namely the present velocity and position, the best-known position of the particle, the best-known position of any particle in any swarm, and the assignment of inertia weight. The inertia weight boosts a particle to move in its present direction. We use reasonable values in existing research (Liang & Suganthan, 2005) for the inertia weight (0.729) and other constants cognitive (1.49445), social (1.49445), and global weights (0.3645). Another critical function of MSO is the fitness score that is used to summarize, how close a given design solution is to achieving our aim. In our case, we want to find out how close are the values of SEC and SPC in a particle to maximize service profitability within the budget set. We introduce some constraints in the model of our context. One constraint is that SEC and SPC values are to be positive. If either value is negative, the fitness score is the negative value of the cost to signify a wrong fit. If both values are negative, we magnify the negative fitness score with a multiplication of both values. We also impose the constraint that the total costs of both SEC and SPC must be within the budget. If not, we return the difference between the total costs to the budget as a negative value. The computation of the fitness score in our context is based on the total revenue minus the total costs over the years with a given service variability, as shown in Equation (8.1). The high-level pseudo-code for our adaptation of the MSO algorithm for service profitability of multitenant service is shown in Figure 8.8.

We design the SPA to be modular and develop the tool using Visual Basic.Net. SPA allows the service provider to plug in her model or third-party models to calculate the engineering and provisioning costs. These plugins can be loaded dynamically within the tool execution without recompilation of the base codes. This option allows service provider to use other cost estimation models besides COCOMO II or AWS without modifying and building of the base codes.

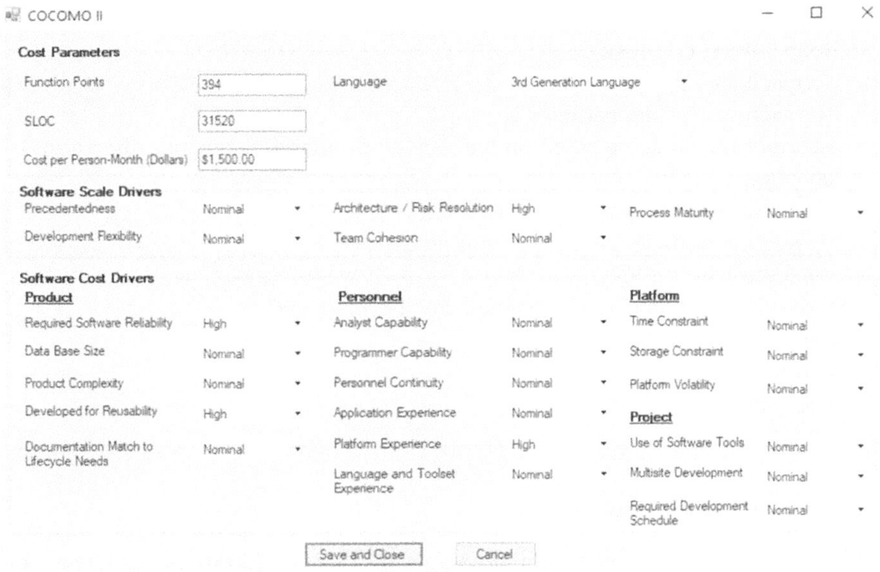

FIGURE 8.8 COCOMO II settings in SPA.

8.10 EXPERIMENTS AND EVALUATIONS

8.10.1 EXPERIMENTS OVERVIEW

To illustrate the usage of the concept map and tool, we evaluate scenarios as a service provider offering services on an open-source package Apache OfBiz (Apache, 2006) and determine the type of service architectures that maximize profits. The five types of service architectures in our evaluations are Fully Shared service architecture (SA_{FS}), Partially Shared service architecture (SA_{PS}), Non-Shared service architecture (SA_{NS}), and Hybrids service architectures (SA_{FS+PS} and $SA_{FS+PS+NS}$).

We need to manage the variability of tenants' requirements across all these service architectures in order to support multiple tenants with varying requirements. For SA_{FS}, the architecture design is based on first registering service components in a service registry. Application components can look up the registry and invoke the variant or the right service component returned during runtime for a specific tenant when required. For SA_{NS}, the architecture design style is based on product-line variability management technique (Jarzabek et al., 2003; Khue et al., 2015). The service components are statically bound to the application components during compile or build time for a specific tenant. For hybrid architectures SA_{PS}, SA_{FS+PS} and $SA_{FS+PS+NS}$, we apply each the two above architecture designs for different parts of the service variations. Below is the MSO algorithm implemented in SPA.

```
loop - maxLoop times
for each one of the swarms
for each one of the particles
measure fitness score based on the service profitability function shown in
equation (8.1)E
if a new best position has been found,
update the particle, swarm and multi-swarm
compute new velocity
use velocity to update the position of SEC and SPC
end for each particle
end for each swam
end loop
```

8.10.2 Performing Simulations

In these simulations, we set 10 initial tenants of average 100 users per tenant for the tenant base parameter and 50 delta tenants of average 100 users per tenant for the delta tenant base parameter. The investment for the service provider is over five years.

We first estimate the service costs of SA_{FS} based on COCOMO II and then use this value as a baseline to calculate the service costs for the rest of the architectures. The COCOMO II estimations are based on 692 unadjusted function points, 31520 lines of code. The Software Scale and Software Cost Drivers of COCOMO II are also available for modification in our tool. We use the default nominal values. For our experiments, we set the COCOMO II settings shown in Figure 8.8.

We use the value of SFEC to extrapolate the SVEC for each of our service architectures. A study by Poulin and Himler (2006) indicate building components based on a dynamic binding method such as using architecture design style such as registry lookup in a service-oriented architecture requires an approximate additional 20% engineering cost as compared to developing for one-time use. In this case, SVEC for $SA_{PS}$is increased by 20% over SA_{FS}. Another study of cost estimation in Software Product Lines (Nolan & Abrahão, 2010) in applying a static binding method, a higher cost is required to engineer the variation for SA_{NS}. Also based on our own experiences in applying static binding variation techniques such as the adaptive reuse technique in software product line projects, this cost is likely higher than the increase in cost for dynamic binding. These additional costs are to account the effort to develop the product line assets, implement and test the variants using the selected variation technique. In this case, SVEC for SA_{NS} is increased by 30% over SA_{FS}. For SA_{PS}, SA_{PS+NS} and $SA_{FS+PS+NS}$ due to the usage of both dynamic and static types of binding techniques, we estimated an additional 40% over SA_{FS} in terms of SVEC.

Although the initial costs to implement static binding is higher than dynamic binding, it allows for greater flexibility and lesser effort to modify due to subsequent changes. Based on this reasoning, we set the service engineering delta cost coefficient α_{NS} to be lower than α_{FS}. On the other hand, there is increased effort to design and manage both types of binding methods. We need to account for that in SA_{PS} and the rest of the hybrid architectures. Based on this reasoning, our assumption is $\alpha_{NS} < \alpha_{FS} < \alpha_{PS} < \alpha_{PS+NS}$, $\alpha_{FS+PS+NS}$. For our experiments, we assume these cost-coefficient α values for service engineering as: $(\alpha_{NS} = 0.2) < (\alpha_{FS} = 0.3) < (\alpha_{PS} = 0.4) < (\alpha_{PS+NS}, \alpha_{FS+PS+NS} = 0.5)$.

A service provider is also required to estimate the provisioning costs. The tool provides the user interface to estimate the costs when using Amazon EC2 services. A service provider can input their usage estimations such as the Amazon EC2 instances and managed services such as S3, Route 53, CloudFront, RDS, and DynamoDB. The tool uses the same calculations as on the Amazon AWS pricing calculator (Amazon, 2007). For our experiments, we set the AWS settings shown in Figure 8.9.

We need to set the value of β to denote the cost coefficience of service provisioning for the five types of service architectures. The value of the cost-coefficient value is between 0 and 1. This value is low when there are sharing of provisioned resources. On the other hand, this value is high when the provisioned resources are isolated among tenants. In our cases, the value of β_{FS} is set lower due to resource sharing, and the value of β_{NS} is 1 due to resource isolation. β_{PS} is set between β_{FS} and β_{NS} due to the partial sharing of resources. Based on the level of resources sharing, we assumed that $\beta_{FS} < \beta_{PS} < \beta_{NS}$. The weighted cost-coefficient β values are summarized as follows. For our experiments, we assume these cost-coefficient α values for service provisioning as $(\beta_{FS} = 0.2) < (\beta_{PS} = 0.5) < (\beta_{NS} = 1)$. For hybrid service architectures, we set $\beta_{PS+NS} = 0.5$ or 1 and $\beta_{FS+PS+NS} = 0.2$, 0.5, or 1 depending on the degree of sharing and isolation.

For service revenue gained by the service provider, we use the subscription-based method for the tenant to pay a fee to use the service. The fee paid depends on her varying requirements (e.g., subscribing to more features), the tenant may have to pay a higher price.

As shown in Figure 8.10, the simulation process showing animations for each step of the process on the left side of the SPA user interface, together with the simulation results on the right side. SPA execute each simulation 100 times, and the simulation results are shown in both a graph and table format. These results are divided into the five service architectures, and for each architecture, the service profits, expected ROI, and annualized profits are shown.

For the first case, we assume no tenants are on-boarded yet for initial verification of a base case. We manually calculated the functionality engineering cost based on COCOMO II and functionality variability engineering costs based on AWS Cost Calculator for each of the service architecture and compared with the tool output as are shown in Figure 8.10. The values tally between the tool output and our calculations. With no tenants on-boarded, the values are negative with costs incurred, but no revenue gained yet.

For the next three cases, we perform the simulation experiments by varying the delta variability. As mentioned in our experiments, we simulate 10 tenants for tenant base and 50 tenants for delta tenant base. For the second case, we set the delta variability to low for both the delta variability for engineering and provisioning, as shown in Figure 8.11. The SPA simulation results are shown in Figure 8.12. For this

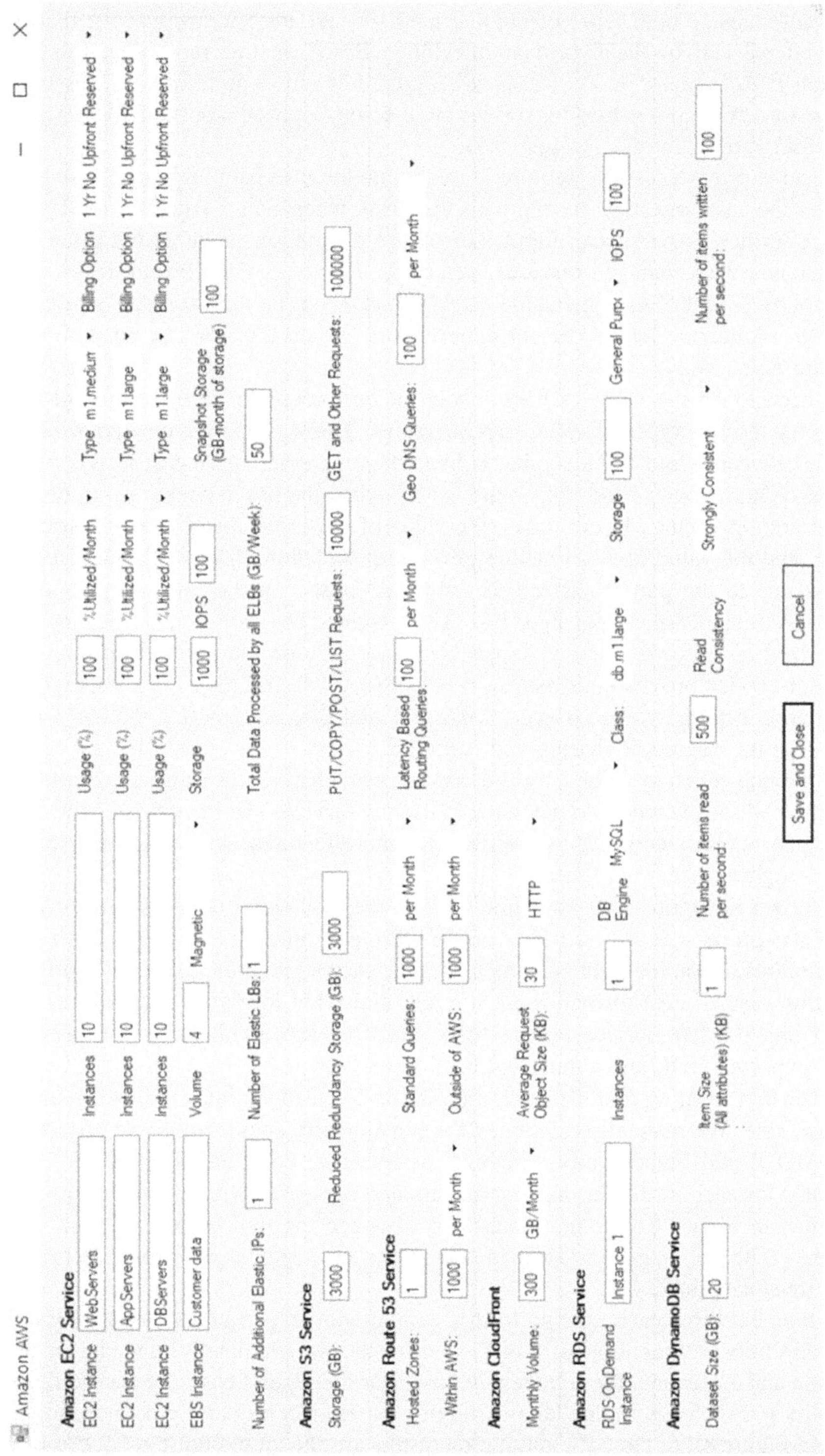

FIGURE 8.9 AWS settings in SPA.

Number of Simulations : 10   Number of Investment Years : 5
Delta Variability (Engineering) : Low   Delta Variability (Provisioning) : Low

SA	Fully Shared	Partially Shared	Non Shared	Hybrid Partially/Non	Hybrid Fully/Partially/Non
Service Profits	-$772,371	-$849,608	-$898,840	-$960,858	-$960,858
Expected ROI	-322%	-322%	-346%	-357%	-357%
Annualized ROI	-64%	-64%	-69%	-71%	-71%

FIGURE 8.10 Service profits (no tenant).

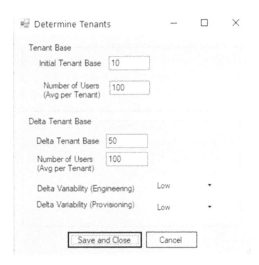

FIGURE 8.11 Tenant settings.

Number of Simulations : 10   Number of Investment Years : 5
Delta Variability (Engineering) : Low   Delta Variability (Provisioning) : Low

SA	Fully Shared	Partially Shared	Non Shared	Hybrid Partially/Non	Hybrid Fully/Partially/Non
Service Profits	$2,169,831	$1,840,815	$1,861,269	$1,880,665	$1,981,344
Expected ROI	904%	697%	716%	700%	737%
Annualized ROI	181%	139%	143%	140%	147%

FIGURE 8.12 Low delta variability.

case, service profits (SP) of the service provider adopting SA_{FS} is ($1,112k). Among the service architectures, the Annualized Return On Investment (ROI) of SA_{FS} is the highest at 86%. This high ROI can be attributed to the extensive sharing of service components for engineering and provisioning, resulting in maximizing of the service profits.

For the third case, we modify the delta variability setting to high while keeping the rest of the factors the same. The SPA simulation results for the third are shown in Figure 8.13. In this case, a service provider that adopts SA_{FS} is not a wise decision as compared to adopting SA_{NS}. The service profits for SA_{FS} is only $703,988 while for SA_{NS}, the service profits are $1,870,891. Due to this, the expected ROI and annualized ROI are also higher for SA_{NS}. As a service provider cannot on-board these tenants, there is a reduction in the revenue that can be gained from these tenants. As compared to the second case, a service provider needs to review carefully on the assumption that tenants are having low variation of requirements. The level of service profitability can vary widely based on the accuracy of this assumption.

For the fourth case, we modify the delta variability setting to random while keeping the rest of the factors the same. The SPA simulation results for the third are shown in Figure 8.14. If a service provider adopts SA_{NS}, the service profits are $1,867,829. Even for the hybrids, the service profits are $1,823,505 for SA_{PS+NS} and $1,880,206 for $SA_{FS+PS+NS}$. A service provider makes these architecture decisions can achieve better results as compared to SA_{FS} with service profits of $1,329,246. If service provider cannot determine the variation of tenant's requirements confidently, using SPA with a random distribution across the types of service architectures can help service provider to make a more informed decision about the type of service architecture to adopt. This insight also goes well with the understanding that although managing variability incurs an initial high engineering cost, the benefits of managing variability are substantial over the long run.

Number of Simulations : 10   Number of Investment Years : 5
Delta Variability (Engineering) : High   Delta Variability (Provisioning) : High

SA	Fully Shared	Partially Shared	Non Shared	Hybrid Partially/Non	Hybrid Fully/ Partially/Non
Service Profits	$703,988	$1,553,718	$1,870,891	$1,768,340	$1,789,885
Expected ROI	293%	588%	719%	658%	666%
Annualized ROI	59%	118%	144%	132%	133%

FIGURE 8.13   High delta variability.

Number of Simulations : 10   Number of Investment Years : 5
Delta Variability (Engineering) : Random   Delta Variability (Provisioning) : Random

SA	Fully Shared	Partially Shared	Non Shared	Hybrid Partially/Non	Hybrid Fully/ Partially/Non
Service Profits	$1,329,246	$1,655,422	$1,867,829	$1,823,505	$1,880,206
Expected ROI	554%	627%	718%	678%	699%
Annualized ROI	111%	125%	144%	136%	140%

FIGURE 8.14   Random delta variability.

8.10.3 BUDGET CONSTRAINT SCENARIO

Based on a budget of $400,000 for 10 initial tenants and 50 delta tenants over five years similarly to earlier cases, the tool simulates to search for an optimal solution based on the multi-swarm algorithm mentioned earlier. For each particle in each iteration, the tool calculates the fitness score based on the current two positions values of the particle. These two positions values denote the units of service engineering and service provisioning costs. If either or both positions values become negative, we reposition the particle with a negate of their current position values to make it positive. If both positions are positive, we continue to simulate the service profitability over five years that involves calculations of the engineering and provisioning costs and revenue obtained from the initial tenants and delta costs and revenue for the delta tenants. The fitness function returns the service profitability by subtracting the total revenue gained against the total costs incurred. We set 50 swarms of 1,000 particles in each swarm and loop the process 1,000 times for the model to reach a stable state. The tool seeks to achieve this stable state for the particles where the two units are optimized depending on the fitness function and conform within the budget constraint. Each simulation takes less than a minute to complete.

One application scenario is for service provider having a budget constraint and need to evaluate which service architecture can maximize service profitability. Take for example, with a Fully Shared service architecture with SEC (initial) ($258,000) and SPC (initial) ($58,000) for a random range of service variability and, the service profitability works out to be $695854. With a Partially Shared service architecture with SEC (initial) ($283,000) and SPC (initial) ($63,711) and the rest of the factors stay constant, the service profitability works out to be $ 890934. With a No-Shared service architecture with SEC ($289,499) and SPC ($69,503) and the rest of the factors stay constant, the service profitability works out to be $1,106,920. In this case, a service providers may wish to consider the higher initial costs with a No-Shared service architecture that works out to be more profitable over the years.

Another application scenario is for service providers to decide on the right balance of both the service engineering and service provisioning costs for a given budget. Take for example of a No-Shared service architecture as per the last scenario, the units generated based on the simulations are 9981 and 599 for service engineering and service provisioning costs respectively, indicating that with a given budget, a service provider should consider incurring 87717/ 88213 of the budget on engineering and 496/88213 on provisioning related costs.

8.11 RELATED WORKS

Many existing studies related to service profitability are analyzed from the perspective of novel business and pricing models. Ma (2007) and Ma and Seidmann (2008) present an analytical model to study the competition between the SaaS (Software as a Service) and the traditional COTS (Commercial off-the-shelf) software to analyze the pricing strategy in a competitive setting. They conclude that SaaS could gradually take over the whole market even when its quality is inferior.

Xu and Li (2013) propose a revenue management framework and formulate revenue maximization problem with dynamic pricing to characterize optimal conditions. Laatikainen and Ojala (2014) use multi-case research to understand the relationship between architectural and pricing characteristics of SaaS software. They conclude that flexible and well-designed architecture with scalability and a high level of modularity enables different pricing models while poorly designed architecture limits also the pricing. Our study further shows that with the right service architecture for the tenants, it also enables higher service profitability.

Another perspective of studies in service profitability focuses on how to adapt services effectively, and efficiently allocation of resources and placement of tenants. Ju, Sengupta, and Roychoudhury (2012) propose a formal model that maximize profits of the service vendor and commonality of the tenant's functional requirements, motivated as a bi-objective optimization problem. Liu, Hacıgümüş, Moon, Chi, and Hsiung (2013) propose service level agreements (SLAs) profit-aware model and approximation algorithms for database tenant placement based on expected penalty computation for multitenant servers.

There are many existing recent works on using particle swarm and multi-swarm optimization for service multi-tenancy. Chitra, Madhusudhanan, Sakthidharan, and Saravanan (2014) propose to use Particle Swarm Optimization (PSO) technique to locate and optimize a suitable workflow schedule. The study also shows increased effectiveness when there are more workflow tasks. In another work by Mezni et al. (2018), the authors propose to use a variant of MSO to develop a service placement method incorporating security concerns. In our work, we seek to use MSO to evaluate our model for service profitability. Besides the provisioning costs that are covered in many existing works, the model also takes into account the engineering costs for service profitability.

8.12   THREATS TO VALIDITY

For our simulations, the values of the parameters set for COCOMO II and AWS Cost Calculator, and cost coefficients such as α and β values can affect the simulation results. We acknowledged this early and sought to perform more simulations by varying these parameter values and cost coefficients. We observed similar results as per our earlier observations.

These experiments were conducted based on a single implementation of an open-source package. We also used only one type of cost estimation models and provisioning model. We do acknowledge that more extensive validation is required. One possible study is to use the tool to simulate these values and compare them with actual values of the costs and revenue gained to offer the service to real tenants. We also seek to open source the tool implementation so that others can review and provide suggestions to improve on its usability and performance.

8.13   CONCLUSION

In this work, we proposed an evaluation of cloud service profitability that formalizes an interplay among multiple service-related factors and business constraints. Our model takes into account factors such as the tenant profile, tenant's variation in requirements, service engineering choices, and service provisioning choices and business constraints such as the service provider's budget. Our conceptual model shows how these factors can result in multiple types of service architectures that affect service cost, service revenue, and ultimately service profits. We further proposed a process to evaluate service profitability by using simulations. We developed a Service Profitability Analyzer (SPA) tool that automates the analysis of service profitability, enabling service provider to understand the possible solutions to make her decisions that affect service profitability. The process and tool also show how business constraints such as budget can be evaluated in the simulations using multi-swarm optimization. We conducted experiments on offering services of an open-source package using our profitability model and SPA. The simulations and quantitative results go well with the understanding that though managing variability incurs an initial high engineering cost, the benefits of managing variability are substantial over the long run. We hope our discussions of our model, process, and tool experiments are insightful for service providers to make informed decisions on the type of service architectures and build a convincing business case that maximizes service profitability.

REFERENCES

Amazon. (2007, June, 27). *AWS 3-Tier web application solution*. Retrieved from http://calculator.s3.amazonaws.com/index.html#key=calc-LargeWebApp-140323.

Apache. (2006, July, 3). *Apache OFBiz*. Retrieved from: https://ofbiz.apache.org/.

Chitra, S., Madhusudhanan, B., Sakthidharan, G. R., & Saravanan, P. (2014). Local minima jump PSO for workflow scheduling in cloud computing environments. *Advances in Computer Science and Its Applications* (pp. 1225–1234). Berlin:Springer.

Fiaidhi, J., Bojanova, I., Zhang, J., & Zhang, L. J. (2012). Enforcing multitenancy for cloud computing environments. *IT Professional Magazine*, 14(1), 16.

Frakes, W., & Terry, C. (1996). Software reuse: Metrics and models. *ACM Computing Surveys (CSUR)*, 28(2), 415–435.

Jarzabek, S., Bassett, P., Zhang, H., & Zhang, W. (2003, May). XVCL: XML-based variant configuration language. *25th International Conference on Software Engineering, 2003. Proceedings* (pp. 810–811). IEEE.

Ju, L., Sengupta, B., & Roychoudhury, A. (2012, June). Tenant on-boarding in evolving multitenant software-as-a-service systems. *2012 IEEE 19th International Conference on Web Services* (pp. 415–422). IEEE.

Khue, L. M., Ouh, E. L., & Jarzabek, S. (2015, October). Mood self-assessment on smartphones. *Proceedings of the Conference on Wireless Health* (pp. 1–8).

Laatikainen, G., & Ojala, A. (2014, June). SaaS architecture and pricing models. *2014 IEEE International Conference on Services Computing* (pp. 597–604). IEEE.

Liu, Z., Hacıgümüş, H., Moon, H. J., Chi, Y., & Hsiung, W. P. (2013, March). PMAX: Tenant placement in multitenant databases for profit maximization. *Proceedings of the 16th International Conference on Extending Database Technology* (pp. 442–453).

Ma, D. (2007, July). The business model of "software-as-a-service". *IEEE International Conference on Services Computing (SCC 2007)* (pp. 701–702). IEEE.

Ma, D., & Seidmann, A. (2008, August). The pricing strategy analysis for the "software-as-a-service" business model. *International Workshop on Grid Economics and Business Models* (pp. 103–112). Berlin: Springer.

Madachy, R. (2015, June, 15). COCOMO II—Constructive Cost Model. Retrieved from http://softwarecost.org/tools/COCOMO/.

Mezni, H., Sellami, M., & Kouki, J. (2018). Security-aware SaaS placement using swarm intelligence. *Journal of Software: Evolution and Process*, 30(8), e1932.

Mili, A., Chmiel, S. F., Gottumukkala, R., & Zhang, L. (2000, June). An integrated cost model for software reuse. *Proceedings of the 22nd International Conference on Software Engineering* (pp. 157–166). https://doi.org/10.1145/337180.337199

Nolan, A. J., & Abrahão, S. (2010, September). Dealing with cost estimation in software product lines: Experiences and future directions. *International Conference on Software Product Lines* (pp. 121–135). Springer, Berlin.

Ouh, E. L., & Jarzabek, S. (2016, June). An adaptability-driven model and tool for analysis of service profitability. *International Conference on Advanced Information Systems Engineering* (pp. 393–408). Springer, Cham, Switzerland.

Ouh, E. L., & Jarzabek, S. (2014). Understanding service variability for profitable software as a service: service providerproviders' perspective. *CAiSE (Forum/Doctoral Consortium)* (pp. 9–16). http://ceur-ws.org/Vol-1164/ with ISSN 1613-0073

Kennedy, J., & Eberhart, R. (1995, November). Particle swarm optimization. *Proceedings of ICNN'95—International Conference on Neural Networks* (Vol. 4, pp. 1942–1948). IEEE.

Liang, J. J., & Suganthan, P. N. (2005, September). Dynamic multi-swarm particle swarm optimizer with local search. In *2005 IEEE Congress on Evolutionary Computation* (Vol. 1, pp. 522–528). IEEE.

Ostadrahimi, L., Mariño, M. A., & Afshar, A. (2012). Multi-reservoir operation rules: Multi-swarm PSO-based optimization approach. *Water Resources Management*, 26(2), 407–427.

Poulin, J., & Himler, A. (2006). *The ROI of SOA based on traditional component reuse.* LogicLibrary Inc. White Paper, 31 August 2006.

Solomon, S., Thulasiraman, P., & Thulasiram, R. (2011, July). Collaborative multi-swarm PSO for task matching using graphics processing units. *Proceedings of the 13th Annual Conference on Genetic and Evolutionary Computation* (pp. 1563–1570). https://doi.org/10.1145/2001576.2001787

Xu, H., & Li, B. (2013). Dynamic cloud pricing for revenue maximization. *IEEE Transactions on Cloud Computing*, 1(2), 158–171.

9 A Paradigm Shift for Computational Excellence from Traditional Machine Learning to Modern Deep Learning-Based Image Steganalysis

Neelam Swarnkar[1], Arpana Rawal[1], and Gulab Patel[1,2]

[1]Department of Computer Science and EngineeringBhilai Institute of Technology, Durg, CSVTU, Chhattisgarh, India
[2]Department of Mathematical Sciences IIT (BHU), Uttar Pradesh, India

CONTENTS

DOI: 10.1201/9781003132080-9

9.1 INTRODUCTION

The overwhelming advancements in network technologies and the presence of enormous data volumes in Internet communication channels in the recent decades has made information more prone to be eavesdropped, impersonated, illegally accessed, or tampered with and hence its secure communication has become a threat across digital communication channels. Steganography refers to the technique of hiding data in multimedia files (text, image, audio, and video) in order to conceal its very existence. Unlike steganography, steganalysis is the practice of detecting the presence of the secret (hidden) data inside the aforementioned multimedia files. These two techniques have their real-time applications in commercial communications for controlling copyright using watermarking (Katzenbeisser & Petitcolas, 1), to prevent illegal copying of content, to detect fraudulent identity cards by identifying the tampered images, to prevent leakage of confidential information in public and private sector to secure patients data in medical sciences, and are also used by security agencies (Balu, Babu, & Amudha, 2018) for transmitting confidential information. It has also been used by terrorists for intra-group communication like the 9/11 attack after which US officials claimed that Al-Qaeda used steganography for covert communication (Avcibas, Memon, & Sankur, 3), terrorists of Boko-haram sect in Nigeria used steganographic schemes for transferring secret information (Katzenbeisser & Petitcolas, 1), and an Al-Qaeda suspect in 2011 was arrested with a chip having porn video inside with 141 text files consisting of their invasions were hidden using steganography (Kolade, Olayinka, Sunday, Adesoji, & Olubusola, 4) to mention a few.

Steganography techniques are categorized in two domains: First, the spatial domain in which the secret message is hidden in the image by altering its pixel value(s) directly. Currently, the data hiding in a spatial domain is done by identifying complex regions of an image and then embedding of secret message is performed in these identified regions (also termed content adaptive data hiding) so that any change in the image pixel goes undetected by the steganographer. There have been several attempts to design content adaptive algorithms for spatial domain steganography among which Highly Undetectable steGO (HUGO) (Pevný, Filler & Bas, 5), Wavelet Obtained Weights (WOW) (Holub & Fridrich, 6), Spatial-Universal Wavelet Relative Distortion (S-UNIWARD) (Holub, Fridrich & Denemark, 7), High Pass Low-pass and Low-pass (HILL) (Li, Wang, Huang, & Li, 8), and Minimizing the Power of the Optimal Detector (MiPOD) (Sedighi, Cogranne & Fridrich, 9) are some promising algorithms. Second is the transform domain in which the Discrete Cosine Transform (DCT) coefficients of the image are altered for embedding the secret message. Most popularly used transform domain algorithms are J-UNIWARD (Holub, Fridrich & Denemark, 7), UED (Guo, Ni, and Shi 10),and F5 (Westfeld, 11). Some developing trends of modern steganography can be sketched as:

  i. Adaptive selection of pixels in an image for embedding secret message.
 ii. Minimize the embedding impact in the cover image so as to reduce the deviation of its statistics.
iii. Embedding data during the process of image creation.

Image steganalysis techniques are broadly classified into two categories: Specific or Targeted Steganalysis, which aims to break one particular steganography algorithm and Universal or Blind Steganalysis, which attacks more than one type of steganography algorithm at the same time to discriminate the image as cover (original image with no hidden data) or stego (image with hidden data). In this chapter, we discuss universal steganalyzers developed using convolution neural networks.

9.1.1 MOTIVATION

Image steganalysis is broadly considered as a binary classification problem which aims to distinguish an image as a cover or a stegoimage. In the past decades, researchers have achieved significant progress in designing image steganalyzers using the two-step ML approach which relied on the extraction and design of hand-crafted features such as Image Quality Measures (IQMs), to mention a few. Hand-engineered feature designing heavily depends on experts deep domain knowledge of both steganography and steganalysis algorithms and hence is time consuming and also not scalable in practice. The major challenge in this approach lies in the ability of designing efficient feature representations which can capture the deviations caused due to embedding operations in the image. Also, the two steps of the ML approach-feature extraction and classification are performed independently so these cannot be optimized simultaneously. Most of the ML-based steganalyzers employed rich models for feature extraction and Ensemble Classifier (EC) for classification task. Despite development of many such ML-based steganalyzer models, these could not function with fairly high accuracies as the complex image sources required feature representation in very high dimensionality. This demanded the learning of a large number of complex features automatically by designing new features rather than hand-crafting them.

The introduction of Deep Learning (DL) theory addressed these problems well as they yield (facilitate the construction) models capable of automatically learning the significant features with both semi-supervised and unsupervised techniques. DL is a subset of ML, but unlike ML, works on extracting hand-crafted features followed by classification;DL unifies both feature extraction and classification task under single architecture and utilizes the heuristics of classification stage to optimize all the parameters jointly in both the modules via back propagation. It learns the hierarchy of underlying features directly from the raw data without explicitly executing a specific set of operations. It is also capable of automatically learning representations from raw data that are unstructured and unlabeled to complex representations as the data flows through the model. These characteristics of the DL models make it well suited for developing feature extraction based universal steganalyzer. Some of the prominent CNN-based steganalyzers designed so far in chronological order are QianNet/GNCNN (Qian, Dong, Wang, & Tan, 12), XuNet (Xu, Wu & Shi, 13). YeNet (Ye, Ni & Yi, 14),

Yedroudj-Net (Yedroudj, Comby, & Chaumont, 15), Zhu-Net (Zhang, Zhu, Liu & Liu, 16), and SRNet (Boroumand, Chen, & Fridrich, 17). Though the foundations of DL like gradient descent, sigmoid activations, etc. were developed in mid-20th century but were not put into practice because they needed huge amounts of data, more numbers of neural network layers, and high processing speeds. These requirements were fulfilled with the availability of big data which was needed for better learning of the neural network, the introduction of ReLU activation favored the use of deep network, and the significant increase in the speed and availability of Graphics Processing Units (GPUs) has enabled us to train deep networks in less time. These developments in technology and Internet data facilitated the use of DL in computer vision tasks. Recently, the introduction of open-source toolboxes like Tensorflow, Keras, Python, and Pytorch have helped to build, train, and deploy the neural networks in solving very complex pattern-classification problems.

With the availability big data in 2010, so much digital content became too much to handle using traditional technologies. This big data made it possible to train machines with a data-driven approach rather than knowledge-driven approach. It also opened a world of possibilities to find useful information from the abundant data using parallel computing technologies like Mapreduce, Hadoop, and Spark. The rise of big data in 2010 sparked the rise of data science to draw insights from massive unstructured data sets. Data science is the application of scientific techniques such as regression and cluster analysis or machine learning techniques to extract useful information from the data. Data science is almost everything that has something to do with data such as collecting, analyzing, and modeling them to extract valuable information in the form of insights, product recommendations, predicting behavior, and decision support systems by designing predictive models, data visualizations etc. Handling of massive unstructured image data sets available in a current scenario is very well facilitated by deep learning models and hence in this chapter we have dealt image steganalysis using deep learning techniques. Due to the presence of large image data sets available across the Internet in the past decades, a universal steganalyzer needs to extract a large number of features (high dimensional features) from a large number of images as it is unaware of the stego algorithm and at the same time it requires increased number of iterations to minimize the loss function in order to accurately discriminate an image as a stego or cover. These requirements demand high computational complexity, which is addressed by computational excellence achieved using transfer learning and parameter sharing facilitated by DL models.

9.1.2 KEY CONTRIBUTIONS

The key learnings in the design and development of CNN-based steganalyzer are outlined as below:

- We have discussed the impact of image pre-processing on the training time and the detection accuracy of the model and have found that pre-processing operations such as application of relevant filters, image calibration, and capturing the most complex regions of an image should be performed before feeding the input to the CNN model.

- We observed that learning rate should neither be too small nor too large but should be adaptive to the data set characteristics as the training progresses, thus resulting in speedy movement of model parameters towards the minima.
- We have analyzed the gradient descent optimizers in terms of parameter update during backpropagation and have found that dynamic learning rate works better for sparse data set, training in mini batches gives good results but are unable to address the problem of non-convex error function which converges to a local minima and saddle points which is further resolved by momentum-based optimizers.
- We propose to use transfer learning and data augmentation strategies to be used for training a model when there is an insufficient amount of input data.

The rest of the chapter is organized as follows. In Section 9.2, we introduce preliminary knowledge about Universal Image steganalyzers, CNN-based models with their rationality for being used for steganalysis tasks and the most popularly used image pre-processing operations, activation functions, and gradient descent optimizers. In Section 9.3, we discuss transfer learning, data augmentation, and Generative Adversarial Network for achieving computational excellence with respect to existing state-of-the-art CNN-based steganalytic models. In Section 9.4, we enlist the open research issues and the opportunities for the researchers to achieve an improvement in the performance of steganalytic models. This chapter is finally closed with the conclusion in Section 9.5.

9.2 UNIVERSAL IMAGE STEGANALYSIS PRELIMINARIES

In the last decade, universal steganalyzers models have been designed using Machine Learning (ML) approaches. ML is a sub-realm of Artificial Intelligence (AI) that provides problem-oriented heuristics in order to design algorithms for machine-assisted automation tasks by making machine learning a set of patterns from the data feeds without explicitly stating the set of rules. ML is a two-staged process comprised of feature extraction in which features are extracted from the input data and classification that classifies a given input under different labels. The block diagram illustrated in Figure 9.1 portrays a ML framework for universal steganalysis as a binary classification problem implemented via training and testing

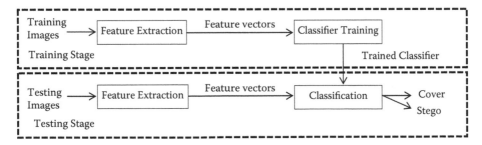

FIGURE 9.1 Two stages (training and testing) of ML approach for universal image steganalysis.

phases. In the training phase, the model is trained for several iterations by extracting recognizable and meaningful patterns from the input images. The classification task labels the input data (each image) either as a stego image or a cover image. Once the training is completed, then the performance of the model is checked by feeding an unseen input to check whether the model is able to categorize it correctly or not. If the model classifies the images very close to or correct (expected) class labels, then the model is said to be ready for implementation, else training is further extended until the predicted label and the expected label becomes nearly the same. The effectiveness of any steganalysis technique classifying the given image as a stego or cover image highly depends upon following feature design criteria:

- the ability of the features to extract the complex representations in an image;
- significant feature selection i.e. no redundant features or features with correlation be extracted;
- optimal feature selection of by ranking them as per their relevance;
- high dimensional representation for more accurate detection;
- feature dimension reduction in case of high dimensional features.

Some examples of features extracted from image data sets using ML are inter- and intra-block based 486 markov features and Subtractive Pixel Adjacency Matrix (SPAM) features from spatial domain and different statistical features like global histogram, co-occurrence, and blockiness from transform domain.

Over the past couple of years, DL has revolutionized many aspects of research and industry including things like autonomous vehicles, reinforcement learning, computer vision, motion detection, image recognition, image classification, generative modeling, robotics, Natural Language Processing (NLP), finance, and forensics. An efficient DL model is the type which is invariant of viewpoint, illumination, scale, deformation, and occlusion and also is sensitive to interclass variation (sensitive to the variation between classes but invariant to the variations within a single class). Here, designing of hand-crafted features become highly challenging due to high complexity of image data. The DL approach meets this challenge by facilitating automated feature learning. A wise attempt has been made to highlight the comparison between traditional ML and modern DL models for image steganalysis in Figure 9.2.

9.2.1 THE CONCEPTUALIZATION

CNN-based steganalyzers have been able to capture complex statistical dependencies from input images and can also learn significant feature representation signaling the presence of hidden data in them automatically, thus helping in discriminating them as cover and stego images. Some pioneering works on the development of CNN-based steganalyzers are discussed below. Care has been taken to discuss these state-of-art models under three broad criteria: either on the basis of architectural settings or image pre-processing or both the criteria inclusive.

The foremost architectural arrangement of DL for steganalysis using an unsupervised learning approach was developed by using stacked convolutional

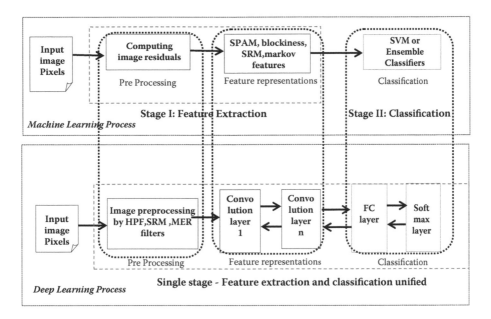

FIGURE 9.2 Comparison of traditional ML and modern DL architectures for universal image steganalysis.

auto-encoders comprising of three convolutional layers with pooling, sigmoid activation, and two fully connected layers which showed improved performance as compared to Subtractive Pixel Adjacency Matrix (SPAM) method but poor for Spatial Rich Model (SRM) method (Tan & Li, 18).

The first CNN-based steganalyzer employing supervised learning was developed by using a high-pass filter for image pre-processing, five convolution layers, Gaussian activation function, and average pooling before feeding the inputs to fully connected layers (Qian et al., 12). The model achieved comparable performance with respect to the SRM feature set which proved that a CNN model can effectively compete with the traditional hand-crafted feature-based steganalytic models.

The data in different layers follow different distributions during the training phase, which causes internal covariate shift problem resulting in slow learning rate. A customized CNN architecture was introduced to resolve the stated problem by employing appropriate parameter initialization and normalizing the layer inputs in mini batches (Ioffe & Szegedy, 19).

A CNN similar to Qian's network with five convolutional layers, an absolute values layer, tanh activation, and a 1x1 convolutional kernel was proposed, which achieved competitive performance with rich models and prevented overfitting (Xu et al., 13).

Another customized CNN model with a smaller depth and a larger width, an image pre-processing layer using High Pass Filter (HPF), two convolution layers, ReLU activation, and no pooling layers was introduced (Pibre, Pasquet, Ienco, & Chaumont, 20). The removal of a pooling layer was done to preserve any information loss caused due to pooling operation.

A transfer learning based CNN was proposed where the feature representations learned by a pre-trained CNN with a high payload is transferred to another CNN architecture aimed at detecting the same steganographic algorithm with a low payload showed better performance by improved feature learning (Qian et al., 21).

A very deep neural CNN model with larger depth and residual connections preserving the weak stego signal which gets generated during the embedding operations in the image was proposed (Wu, Zhong & Liu, 22). It achieved lower detection error rates, provided the cover and stego images are paired during training.

A CNN-based steganalyzer model with a Truncated Linear Unit as activation in addition to Rectified Linear Unit, use of selection channel knowledge, and data augmentation achieved better performance as compared to the previous SRM model on cropped and resampled images was introduced (Ye et al., 14).

A novel CNN-based steganalyzer comprising of a convolution layer, a shared normalization layer, ReLUactivation, and a pooling layer was designed (Wu et al., 23). Shared normalization enables learning discriminative patterns between cover-stego pairs, thus improving the generalization ability and can capture statistical dependencies from the training set without the need of cover-stego pairing during training and testing.

A transfer learning based CNN architecture that uses the AdamW Optimization method for model training was proposed (Ozcan & Mustacoglu, 24).

A new CNN incorporating the best features of the previously designed XuNet and YeNet architecture comprising of SRM features being extracted in the image pre-processing layer, five convolutional layers, batch normalization, TLU activation, and data augmentation was developed, which achieved better results in comparison to the existing state-of-the-art steganalyzers (Yedroudj et al.15).

An improved network architecture with optimization of the kernels in the preprocessing layer, separate convolution block to extract spatial and channel correlation and spatial pyramid pooling (SPP) to handle arbitrary size of input images was introduced, which achieved significant improvement in comparison to the existing CNN-based steganalyzers (Zhang et al., 16).

A model termed Deep Digital Steganography Purifier (DDSP) based on Generative Adversarial Attacks (GAN) networks was proposed, which was designed to remove the steganographic contents from the image, along with maintaining the visual quality of the image (Corley, Lwowski, & Hoffman, 25).

A CNN-based steganalyzer using data augmentation which combined images of data set BOWS2 and traditional BOSS-based data set both were proposed (Yedroudj, Chaumont, Comby, Amara & Bas, 26). It achieved better performance, proving that enriching the data set helps in better learning and consequently improves the performanceof the model.

The above developed deep CNN models shows that the performance of CNN-based steganalyzers have continuously improved with increasing optimization in diverse network architectures.

9.2.2 RATIONALE FOR USING CNN MODELS

It has been seen that there exists a strong correlation between the pixels of a natural image which gets changed while performing spatial domain steganography i.e when data is hidden in the pixels directly. The convolution operation which involves the product of the input image and the kernel resulting in a feature map as output has the ability to capture such changes in the pixel correlation.

In recent years, CNN models have achieved significant progress in discriminative learning such as image classification, image denoising, and generative learning such as real image generation and texture synthesis. This shows the ability of CNN to extract effective features and provide an efficient feature representation of natural images. Due to these abilities, it is preferable to use CNN models for image steganalysis to discriminate natural images (called covers) from unnatural images (called stegos).

To achieve better performance in CNN-based steganalyzers, the input images are pre-processed instead of directly being fed to CNN for classification. To begin with, we hereby discuss about the most popularly used image pre-processing operations performed on input images before feeding them to the CNN.

9.2.2.1 Image Pre-Processing Layer

The images from the data set are pre-processed before feeding them to the neural network. In most of the CNN-based steganalysis tasks this pre-processing operation is generally performed by applying a high-pass filter on the images obtained from the data set which to strengthen the weak stego signal and supress the impact of image content. Following are the commonly used operations in image pre-processing step:

a. **Filter operation:** In this method, a predefined HPF called KV kernel is multiplied with the input image matrix to produce a feature map. Mathematically the operation is expressed as:

$$R = K * I \tag{9.1}$$

here, I is input image, R is residual image (i.e image after high-pass filtering operation), * denotes convolution operation, K is shift invariant finite impulse response linear filter. K is generally chosen as KV kernel in image steganalysis tasks. The values of KV kernel is shown below:

$$K_{kv} = \frac{1}{12} \begin{bmatrix} -1 & 2 & -2 & 2 & -1 \\ 2 & -6 & 8 & -6 & 2 \\ -2 & 8 & -12 & 8 & -2 \\ 2 & -6 & 8 & -6 & 2 \\ -1 & 2 & -2 & 2 & -1 \end{bmatrix} \tag{9.2}$$

b. **Computing most effective region (MER):** The MERs of an image are the regions which have high probably of data hiding. It is determined by computing the texture complexity (C) using the contrast (G) and homogeneity (H) values obtained from the gray-level co-occurrence matrix of the input image. The texture complexity equation is given by:

$$C = \log\left(G + \frac{1}{H}\right) \qquad (9.3)$$

High values of contrast for image pixels denotes more complex image regions whereas large values of homogeneity means lack of data hiding in the image.

c. **SRM filters:** In traditional hand-crafted feature extraction based steganalyzer models the SRM feature set which comprised of 30 HPF to compute residual maps of the original image, produced the best detection rate. These 30 residual maps thus generated were then stacked together and passed to the activation function for further processing. This SRM feature set extract is used as a pre-processing step.

9.2.2.2 Convolutional Layer

A neuron in a hidden layer takes input from a patch of the image and computes the weighted sum of the product of weight and input features from the selected patch by applying convolution operationand then adds the bias term to it. The convolution output is then passed to a non-linear activation function which adds non-linearity to the input and bounds the output to lie within a certain range depending on the type of activation function being applied. The choice of activation function depends on the nature of data and the distribution of the target variables. In most of the image steganalysis tasks ReLU activation is used as it facilitates the use of large number of hidden layers and shows better performance in comparison to other activations. The output of the activation is then passed through pooling layer which is aimed at dimensionality reduction along with preserving important information and discarding the irrelevant ones. In this way the input flows through several convolution layers in the network and the output of the last convolution layer is flattened before feeding to the fully connected layers which then passes it to softmax layer which computes the probabilities to categorize the input into predicted class labels. In this way, the extracted features flow through the network via three sequential steps convolution, non-linearity, and pooling. The basic architecture of a CNN-based steganalyzer is shown in Figure 9.3.

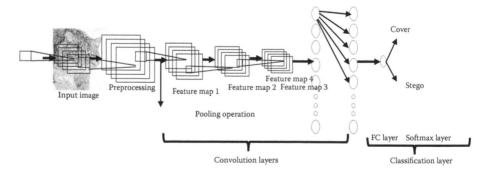

FIGURE 9.3   The basic architecture of convolution neural network.

Over the past couple of years, DL has revolutionized many aspects of research and industry including things like autonomous vehicles, reinforcement learning, computer vision, motion detection, image recognition, image classification, generative modeling, robotics, Natural Language Processing (NLP), finance, and forensics. An efficient DL model is the type which is invariant of viewpoint, illumination, scale, deformation, and occlusion and also is sensitive to interclass variation (sensitive to the variation between classes but invariant to the variations within a single class). Here, designing of handcrafted features become highly challenging due to high complexity of image data. The DL approach meets this challenge by facilitating automated feature learning.

In DL-based steganalyzers the concept of "feature learning" is introduced as a brand-new paradigm. In the feature learning process the network learns from low (simple features) to high (complex features) level feature representation in the top layers of the network. This is done by the convolution layers of the network, where each layer performs the three basic operations, namely convolution, activation, and pooling. We explain these three operations in this section.

Convolution operation: In this layer, the input image as shown in the Figure 9.4 is convolved with one or more filters generally a square matrix of odd order (common order is 3x3) extracts a feature and generates a feature map for subsequent processing. Kernel values are learnable parameters and are generally randomly initialized. When the input matrix of the order n_*n (here n=5) is convolved with an $f * f$ kernel (here f=3), then the size of the output matrix will be of the order $(n - f + 1) * (n - f + 1)$ (here 3x3).

Padding: The convolution operation outputs an image with a reduced dimension and if we keep on convolving then after a few layers the image size will become too small to extract features any further. Also during convolution the kernel slides over the input images i.e. the pixels on the corners and the edges are used much less than those in the middle and hence will be unable to preserve information stored in the border if any. Padding solves these problems by adding a layer of zeroes on the borders, as shown in the Figure 9.5.

So if an input is $n * n$ matrix convolved with an $f * f$ kernel with padding p, then the size of the output matrix will be of order $(n + 2p - f + 1) * (n + 2p - f + 1)$. In the above example, $p = 1$.

In this way, padding prevents shrinking and increases the contribution of the border pixels of image by bringing them into the middle of the padded image. There are two common choices for padding:

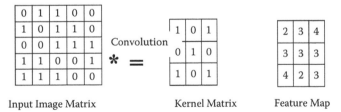

Input Image Matrix   Kernel Matrix   Feature Map

FIGURE 9.4 Convolution operation between input image and kernel matrix and production of feature map.

0	0	0	0	0
0	1	2	1	0
0	3	4	6	0
0	4	7	8	0
0	0	0	0	0

0	1
1	3

*

=

3	7	5	1
10	17	23	6
15	29	37	8
4	7	8	0

Padded Matrix　　　　　Kernel Matrix　　　　　Feature Map

FIGURE 9.5　Padding operation: A zero layer is added on all four sides of input image.

i. **Valid:** It means no padding. In case of *valid padding*, the output matrix will be of the order $(n - f + 1) * (n - f + 1)$.

ii. **Same:** In case of the *same padding*, the size of the input and output image is same i.e., $n + 2p - f + 1 = n$ and $p = \frac{(f-1)}{2}$. Padding helps in retaining information of the boundary and also preserves the size of the image.

Stride: Stride is the number of pixels shifted over the input matrix in the horizontal and vertical directions separately. For input size $n * n$, filter size $f * f$, padding p, and stride s, our output image dimension will be of the order $\left[\frac{(n + 2p - f)}{s} + 1 \right] * \left[\frac{(n + 2p - f)}{s} + 1 \right]$. Stride is used to reduce the size of the image, a particularly useful feature.

9.2.2.3　Non Linear Mapping Layer

In this section, we outline the nonlinear activation functions used for image steganalysis task. The choice of activation function functions in deep networks has a significant impact on the training time of the model and performance of the application. It restricts the output within a particular range to reduce computational complexity. It also adds non-linearity in the data set, which enables it to learn any continuous function. The activation functions are broadly classified as linear and nonlinear functions. Modern deep learning models uses nonlinear activation functions as they are capable to learn the complex dependencies from the data sets having high dimensionality. Unlike linear functions, the non-linear functions allow computing derivatives and stacking of many intermediate layers which are capable to learn complex data sets with higher accuracy. We hereby discuss the most widely used activation functions for image steganalysis tasks:

i. **Sigmoid/Logistic:** It is an S-shaped continuous and differentiable function, which limits the output between 0 and 1. Because of such range of its output, sigmoid is used in the soft classification problems (i.e. where output is predicted in the form of probability values). It is mathematically expressed as:

$$\text{sigmoid}(x) = \frac{1}{1 + e^x} \tag{9.4}$$

It has been observed that the gradient becomes almost zero if the values of input are either very high or very low thus resulting in no change of the predicted output i.e. the model ceases to learn. This leads to a vanishing gradient problem, which eventually causes slow convergence of the network.

$$\tanh(z) = \frac{e^z - e^{-z}}{e^z + e^{-z}} \tag{9.5}$$

ii. **Hyperbolic tangent (tanh):** It is a zero centred non-linear activation function with output ranging between -1 and +1. It is mathematically expressed as:

It suffers from a vanishing gradient problem and is computationally expensive due to the presence of term e^z.

Modern deep learning systems are designed with larger depth which enables the model to capture more complex statistical dependencies among a hierarchy of features and employs nonlinear activation such as Rectified Linear Unit (ReLU) and its variants in place of the saturated counterpart like sigmoid and tanh because such functions are capable to solve the vanishing/exploding gradient problems and also leads to faster convergence of the network. Most of the researchers have used RELU activation for CNN-based image steganalysis tasks. This section discusses the variants of RELU.

In all of these non-saturated activation functions, the most widely used activation for image steganalysis tasks is Rectified Linear Unit (ReLU) (Nair & Hinton, 27). In this section, we introduce the variants of Rectified Linear Units: Leaky Rectified Linear (Leaky ReLU) (Maas, 28), Parametric Rectified Linear (PReLU) (He, Zhang, Ren, & Sun, 29), and Randomized Rectified Linear (RReLU) (Srivastava, Hinton, Krizhevsky, Sutskever, & Salakhutdinov, 30), Exponential Linear Unit (ELU) (Clevert, Unterthiner, & Hochreiter, 31), Scaled Exponential Linear Unit (SELU) (Madasu & Rao, 32), and Softplus (Dugas, Bengio, Bélisle, Nadeau, & Garcia, 33).

$$\text{Leaky ReLU}(x) = \begin{cases} x & if \ x \geq 0 \\ \frac{x}{a_i} & if \ x < 0 \end{cases} \tag{9.7}$$

here a_i is a constant in the range $(1,+\infty)$. The limitation of Leaky ReLU is that it does not provide consistent predictions for negative input values.

$$\text{PReLU}(x) = \begin{cases} x & if \ x \geq 0 \\ ax & if \ x < 0 \end{cases} \tag{9.8}$$

Here, a is a trainable parameter learned by the network itself for faster convergence, but when a is fixed to 0.01, the function performs as a Leaky ReLU function. It is used when the leaky ReLU function fails to solve the problem of dead neurons.

$$RReLU(x) = \begin{cases} x & if \ x \geq 0 \\ ax & if \ x < 0 \end{cases} \tag{9.9}$$

where a is randomly sampled from U(l,u) and $l < u$ and l, u $\in$ [0, 1].

vii. **Softplus Function:** It is a smooth version of the ReLU with output ranging between 0 to +infinity. It helps in faster convergence of the network in less epochs. It is given by:

$$f(x) = \log(1 + e^x) \tag{9.10}$$

viii. **Exponential Linear Unit (ELU):** It allows for negative input values which centres the mean of the activation outputs very close to zero thus enabling faster training of the network by bringing the gradient closer to the natural gradient. It saturates to a negative value as the input gets smaller and smaller and hence reduces the bias shift effect. It is mathematically expressed as:

$$ELU(x) = \begin{cases} x & if \ x > 0 \\ a(e^x - 1) & if \ x \leq 0 \end{cases} \tag{9.11}$$

iii. **Rectified Linear Unit (ReLU):** It is a nonlinear function which maps the negative input values to zero and retains the positive part. In this way, the neuron with negative weight does not get activated thus creating sparsity in the network; hence allowing for faster convergence of the network It is mathematically expressed as:

$$ReLU(x) = \begin{cases} x & if \ x \geq 0 \\ 0 & if \ x < 0 \end{cases} \tag{9.6}$$

In practice, it has been observed that ReLU converges faster than sigmoid and tanh activations. But ReLU suffers from the following problems:

a. **The dying neuron problem:** For the negative values of input, the gradient of the loss function becomes zero and hence prevents the network from learning any further.

b. **The exploding gradient problem:** For larger gradients the weight updation occurs in such a way that some neurons do not get activated on certain input values during the training thus leading to no learning by the neurons.

iv. **Leaky ReLU:** For positive input values it functions the same as ReLU but for negative inputs it assigns a small positive non-zero slope in the negative region thus resolving the dying neurons problem of ReLU. As an improvement to ReLU, it allows for the flow of small amount of information for negative input values. It is defined as,

v. **Parametric rectified linear (PReLU):** It functions the same as leaky ReLU with the exception that the extremely small value a is not kept fixed but is learned automatically from the data setduring training phase. It is defined as,

vi. **Randomized Leaky Rectified Linear Unit (RReLU):** It is a randomized version of leaky ReLU which selects the slope of negative inputs from a uniform distribution $U(l, u)$ lying between the lower bound l and the upper bound u during the training and keeps it fixed during the testing phase of the neural network. This random value of the slope reduces the chances of overfitting. It is mathematically expressed as,

The term $a(e^x-1)$ defines a log curve which ensures that at least a small gradient will flow through the network even when the inputs are negative. The value of a is usually initialized between 0.1 and 0.3 and the term e^x makes the function computationally expensive.

$$SELU(x) = \lambda \begin{cases} x & \text{if } x > 0 \\ \propto e^x - \propto & \text{if } x \leq 0 \end{cases} \qquad (9.12)$$

ix. **Scaled Exponential Linear Unit (SELU):** This activation works with both negative and positive input values which helps in controlling the mean, has saturation region (i.e. derivatives approaching to zero) which helps in reducing high variance and slope greater than one to increase variance if its value is too small, all these properties makes SELU work better as against ELU. It helps in self normalizing the neural network. It maintains normalized mean and variance when the input are normalized. This internal normalization works faster than batch normalization (i.e. external normalization) leading to faster convergence of the network. It is mathematically expressed as,

While using SELU activation in a network, the weights are required to be initialized to LeCun Normal (Madasu & Rao, 32) which initializes the parameters in such a way that they follow a normal distribution and if regularization is done (if any) using the dropout technique, then a variant of dropout named Alphadropout is to be used.

$$\textbf{Swish}(\textbf{x}) = \textbf{x}_*\textbf{sigmoid}(\textbf{x}) \qquad (9.13)$$

x. **Swish:** It is a smooth, non-monotonic, and self-gated function defined as

The smooth curve of this function can be interpreted in the sense that it is differentiable at all points and there is no abrupt change in direction of the function contributing to the optimization of the model. The non-monotonicity conveys that the function may increase and decrease on different intervals of the input values and self-gated feature states that the function requires a single scalar input only. It is unbounded above – very large output values do not saturate to the maximum value and bounded below – highly negative weights become zero, thus resolving both vanishing and exploding gradient problems. It is developed by Google's Brain team (Ramachandran, Zoph, & Le, 34) who tested its performance by replacing ReLU with swish in very deep neural networks Inception-ResNet-v2 and Mobile NASNetA achieved an improved accuracy by 0.6% and 0.9%, respectively. As an improvement over ReLU, it gives output for small negative values, thus helping in capturing the underlying pattern in the data set.

$$maxout(x) = \max(w_1^T x + b_1, \quad w_2^T x + b_2) \qquad (9.14)$$

Here, w and b are the learnable parameters. This activation incurs high computational cost because the number of input parameters to a neuron gets doubled.

    xi. **Maxout**: It was first proposed by Goodfellow in 2013 and is a generalization of ReLU and Leaky ReLU activation. It outputs the maximum of the inputs so obtained and does not suffer saturation and dying neurons problem. It is mathematically expressed as,

The activation function is of significant importance for a CNN-based steganalyzer as the extraction of efficient and discriminative features heavily depends on the activation employed. In earlier steganalysis tasks sigmoid and tanhwere popularly used but could not scale to deep networks which was later addressed by the introduction of RELU activation. RELU worked better with deep layers of the network but suffered dying neuron problem which was later resolved by the introduction of Leaky RELU. Later, CNN-based steganalystic models used TLU and combination of different activations as well. Some newly introduced activation such as SELU, Swish, and maxout have been discussed previously and can be explored for achieving faster network convergence and better detection accuracies in cover-stego image classification.

9.2.2.4 Pooling Layer

The number of trainable parameters in a neural network is directly proportional to the dimension of the input image. More, the input image size more will be the number of trainable parameters thus leading to increased computational complexity. Pooling addresses this problem by reducing the dimension of the feature map obtained after convolution operation between input image and filter. Some popularly used pooling methods in image steganalysis are:

    a. **Max/Global Max Pooling**: Max pooling captures the maximum value from the pooling region. There are four pooling regions partitioned by vertical and horizontal solid black lines (as shown in Figure 9.6(a)) shown as matrices of the order 2x2 each and the maximum of our elements captured from each of these regions (matrices) are 16,20,35 and 22, respectively. Figure 9.6(b) depicts global max pooling, which captures a maximum value from the entire feature map.

    b. **Min/Global Min Pooling**: Min pooling captures the minimum value from the pooling region. There are four pooling regions partitioned by vertical and horizontal solid black lines (as shown in Figure 9.7(a)) as matrices of the

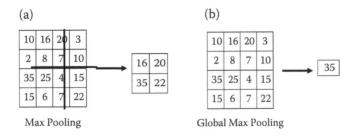

FIGURE 9.6   (a) Max pooling,; (b) Ggobal max pooling.

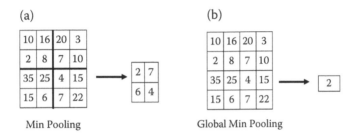

FIGURE 9.7   (a) Min pooling; (b) global min pooling.

order 2x2 each and the minimum of four elements captured from each of these regions (matrices) are 2, 3, 5, and 4, respectively. Figure 9.7(b) depicts global min pooling, which captures a minimum value 2 from the entire feature map.

c. **Average/Global Average Pooling:** Average pooling captures the average value from the pooling region. There are four pooling regions depicted in Figure 9.8(a), shown as matrices of order 2x2 each and the average values calculated from these regions (matrices) are 9, 10, 20, and 12, respectively. Figure 9.8(b) depicts Global Average pooling, which calculates the average as 12.75 using all the values of the feature map at once.

d. **Stochastic Pooling:** The stochastic pooling resolves the drawbacks of max and average pooling operation. In average pooling, the average of all the elements of the pooling region are considered, even if many pixel values are very small in magnitude. The application of ReLU activation down weights

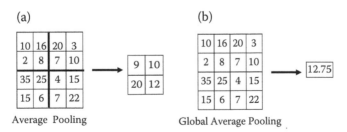

FIGURE 9.8   (a) Average pooling; (b) global average pooling.

the strong activations and therefore many zero elements are included in the average. The application of tanh activation has even worse effect as it averages the strong positive and negative activations which ultimately cancel each other out, leading to small pooled responses. Whereas max pooling does not suffer from these drawbacks, however, it reveals some other drawbacks like low bias but high variance (these are the characteristics of over-fitting in learning models) thus making it hard to generalize well on the test data set. Stochastic pooling addresses these problems and helps to prevent over-fitting. In this method, we first compute the probabilities p for each pooling region k by normalizing the activations within that region, R_k is the pooling region, the mathematical expression for calculating probability is given by:

$$p_i = \frac{a_i}{\sum_{l \in R_k} a_l}.$$

(9.15)

We now sample from the multinomial distribution based on p to pick a point from the location l within the pooling region. The pooled activation value is now a_j:

$$s_k = a_j \text{ where } j \sim P(p_1, \ldots, p_{|R_k|})$$

(9.16)

Here s_k is the selected value after application of stochastic pooling on the pooling region. This process is depicted in Figure 9.9. For each training sample in each layer for each pooling region the pooled value is drawn independently of one another. During the back propagation, the same selected location l is used to direct the gradient back through the pooling region.

As in the previous section, we discussed that the max pooling captures only the strongest activation from the pooled region. While in the stochastic pooling, probability is assigned to each location so in the same pooling region all the activations should be taken into account i.e. stochastic pooling ensures that the non-maximal activations will also be utilized.

Figure 9.9(a), (b), and (c) shows that despite of having less probability value, the activation value 1.8 with probability value 0.6 is selected as against the activation value 2.2 with probability 0.8 in the same pooling region. Hence, stochastic pooling facilitates with an opportunity to all the activations for being selected.

e. **Spatial Pyramid Pooling (SPP):** It is a pooling technique which enables handling multi sized/scaled input images by changing a 2D arbitrary size

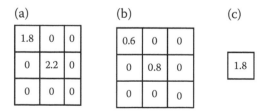

FIGURE 9.9   (a) Activations, a_i; (b) probabilities, p_i; (c) sampled activation, s.

input to fixed dimension output. In real-world scenario, the input image size is not necessarily fixed and can be of arbitrary size/scale. Resizing the input images to equal dimension does not impact classification accuracy in image classification tasks but has a significant impact on steganalysis and causes validation error leading to low test accuracy.

Unlike convolution layers, the fully connected layer requires fixed length vectors which is not possible when the input image has variable size/scale. SPP solves this problem by generating fixed length representations regardless of input sizes/scales. Figure 9.10 depicts a 3-level SPP to turn multi-sized input image pixel values into a fixed dimension output. Suppose the last convolution layer of the CNN has 32 feature maps. Then at SPP layer, each feature map is pooled to 01 then 04 and after that 16 values as a result of which 1x32-d vector, 4x32-d vector and 16x32-d vector value is obtained. These three vectors so obtained are concatenated to form a final 1-d vector which is then fed to the fully connected layer. SPP layer is implemented in a CNN by replacing the output of the last convolution layer by SPP layer, as shown in Figure 9.11. The output of all the pooling layers is flattened and concatenated to give an output of a fixed dimension irrespective of input size to the fully connected layer of the CNN. SPP performs pooling on the the last convolution layer's output and produces a N*B dimensional vector (where N=Number of filters in the convolution layer, B= Number of Bins). The vector is in turn fed to the FC layer. The number of bins is a constant value. Therefore, the vector dimension remains constant irrespective of the input image size. Average pooling is popularly used in image steganalysis tasks because the steganographic noise generated during embedding is very weak signal and application of this pooling operation propagate and preserves such weak signal which is not possible in max pooling.

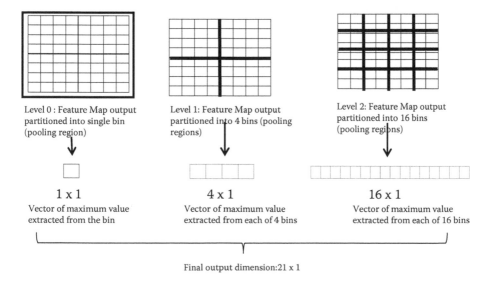

Level 0 : Feature Map output partitioned into single bin (pooling region)

Level 1: Feature Map output partitioned into 4 bins (pooling regions)

Level 2: Feature Map output partitioned into 16 bins (pooling regions)

1 x 1

Vector of maximum value extracted from the bin

4 x 1

Vector of maximum value extracted from each of 4 bins

16 x 1

Vector of maximum value extracted from each of 16 bins

Final output dimension:21 x 1

FIGURE 9.10   A three-level spatial pyramid pooling.

9.2.2.5   Batch Normalization

It has been observed that the input features in the data set have huge variations in the range of values they have, thus leading to longer training time of the neural network. For example, let us consider a five layer CNN then during back-propagation layer-2 updates its weights and biases to minimize the output error but due to this readjustment the output of the second layer (which is the input of the third layer) is changed for same initial input so the third layer has to learn again to produce the output for the same data. Due to change in the input of previous layers, the distribution of input value for current layer also changes thus forcing it to learn from new input distribution. This result in a problem called "internal covariate shift" which is the change in the distribution of network activations due to the change in network parameters such as weights and biases during training.

To overcome this problem, Batch Normalization (BN) (also known as batch-norm layer) is performed. Under this, a BN layer is added after each convolution layer and before the activation function. It converts the distribution of input in such a way that it has mean equals to 0 and standard deviation as 1. BN normalizes all inputs between -1 and +1 so the distribution of inputs remains the same for each layer irrespective of the changes in the previous layers. Therefore, every layer will not have to restart learning from scratch because the input they receive are in the same range so learning process will be less and hence hastens up the training process.

BN standardizes the input of a layer for each input layer before feeding them to activation function. A scaling factor gamma and shifting factor beta is used to keep the distribution away from zero. As training progresses, along with the weights, γ and β also learns through back propagation to improve accuracy of neural network. Generally, γ is set to 1 and β to 0 during training. The mathematical formulation of batch normalization is given as follows:

$$\mu_{batch} = \frac{1}{k} \sum_{j=1}^{k} f_j \tag{9.17}$$

$$\sigma_{batch} = \frac{1}{k} \sum_{j=1}^{k} (f_j - \mu_{batch})^2 \tag{9.18}$$

$$f_{\tilde{j}} = \frac{(f_j - E[f_j])}{\sqrt{(\sigma_{batch}^2 + \varepsilon)}} \tag{9.19}$$

$$f_j^0 = \beta f_{\tilde{j}} + \gamma \tag{9.20}$$

where k is the number of samples, μ_{batch} and σ_{batch} denotes the mini batch mean and mini batch standard deviation of the mini batch comprising $\{f_1, f_2, f_3 \ldots \ldots f_n\}$, ε is a small constant to avoid zero division, f_i denotes the ith feature of the mini batch, γ is the scaling factor, β shifting factor, γ and β are learnable parameters, and E denotes the expectation.

9.2.2.6 Classification Layer

The classification layer of the convolution neural network comprise of several Fully Connected (FC) layers followed by a Softmax layer.

i. **Fully connected layer:** The learned features from the last convolution layers are fed to the FC layer. It represents the feature vector for the input and once the network gets trained then this feature vector is used for classification tasks.

The success of a model is measured by its generalizing ability on given data sets. But due to the problems of overfitting (low bias and high variance) Figure 9.12(a) and underfitting (high bias and high variance), Figure 9.12(b) becomes unable to generalize well on training and test data sets leading to poor predictions on new (unseen) data sets. So, a model with low bias and low variance is shown in Figure 9.12(c) is considered to give better performance which is achieved by regularization which is achieved by dropout and early stopping methods.

Regularization helps in generalizing data not only in training but also in testing set (unseen data) as well. The techniques such as L1, L2, Dropout, and early

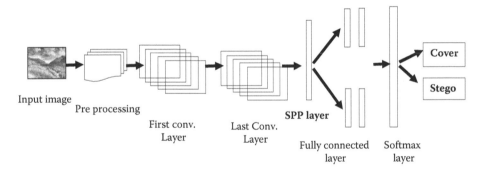

FIGURE 9.11   The output of the last convolution layer replaced with SPP layer in a CNN.

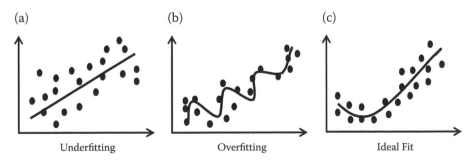

FIGURE 9.12   (a) Underfitting; (b) overfitting; (c) ideal fit.

stopping are applied for regularizing FC layers. Dropout gives efficient regularization in the CNN-based steganalyzer.

ii. Softmax layer:

In the CNN-based steganalyzer the classification of an image as being a cover or stego is performed by a two-way softmax layer. It maps the FC vector output to a probability distribution ranging from 0 to 1 for each of the classification labels being predicted by the model. The softmax function is computed by:

$$f(x_i) = \frac{e^{a_i}}{\sum_j e^{a_j}} \tag{9.21}$$

Here a denotes the vector values.

9.2.2.7   Optimization of Gradient Descent

In this section, we discuss the gradient descent optimizers which aim to achieve faster learning and better performance of the model.

In the optimization method, the gradient of the cost function U is computed with respect to the weights (or parameters l) that we want to improve. The parameters (such as weights and biases) are updated periodically in the negative direction of the gradient until the optimal weight and bias values, which gives better predictions are obtained. The model parameter update equation is given by:

$$l_{new} = l_{old} - \alpha \frac{\partial U}{\partial l_{old}} \tag{9.22}$$

Here, l_{new} is new value of parameter, l_{old} is the old value of parameter, α is the learning rate, and $\frac{\partial U}{\partial l_{old}}$ is the partial derivative of the cost function with respect to parameter l_{old}. The parameters (such as weight) are updated by subtracting the current value of the same parameter (i.e. weight) by a factor α times the gradient of the cost function.

On the basis of number of data points (examples) of the data set used to compute the gradient of the cost function three variants of GD exist. They are batch gradient descent (compute complete training data set all at once), Stochastic gradient descent(compute gradient for one training example at one time) and Mini Batch gradient descent(MBGD) (computes gradient for small batches).Of these three the MBGD is mostly employed in image steganalysis task.We discuss here the reason for doing so.GD encounters problem when the size of training data set is too big to fit in the computer memory. So in MBGD, the entire training data set is divided into small mini-batches of sizes 16, 32 and 64 for training the model iteratively. Hence the batch size becomes a hyper parameter and depends on the application. The advantage of MBGD is that the learning starts as soon as we traverse the first mini batch and also the training can be done efficiently for large data sets as well. Its parameter update equation is given by:

$$l_{new} = l_{old} - \alpha \frac{\partial U(l, x^{i:i+batch}, y^{i:i+batch})}{\partial l_{old}} \tag{9.23}$$

Here *batch* is the size of the minibatch.

The above-mentioned gradient descent does not ensure faster convergence and suffers problems like having a proper learning rate since too small or too large learning rate causes slow convergence or causes the loss function to oscillate around the minima or even to diverge. This problem is addressed by learning rate scheduler which is adaptive to the data set characteristics and performs an operation called annealing (reducing the learning rate as per a fixed schedule or when the epoch crosses a certain threshold). In practice, learning rate is kept fixed for all the model parameter updates but if the data set is sparse and there is a high variation in the frequency of different features then they are not required to be updated by the same extent but the sparsely occurring features is required to be updated by a large amount. Also, we need to ensure that the non-convex error function should not get stuck in the local minima and saddle points thus prevent the gradient to move ahead. These issues are addressed by the use of GD optimizers in the CNN architecture. In the following, we outline the most widely used GD optimization algorithms to achieve faster converge of the network.

i. **Stochastic Gradient Descent with Momentum (SGDM)**: The GD method encounters the problem of local minima (where the gradient becomes zero) as shown in Figure 9.13(a) and saddle points (where the loss function becomes flat) as shown in Figure 9.13(b), which prevents the algorithm from updating the parameters and stops the entire learning process.

Adding a momentum to the parameter update equation given below solves the previous two problems. This momentum accelerates the SGD towards the relevant direction and dampens the oscillations of the weights. Momentum creates a velocity v during training which is the running sum of gradients weighted by a factor (friction term) γ which slows down the velocity as per need.

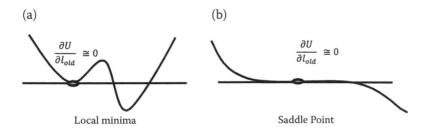

(a)   (b)

$\frac{\partial U}{\partial l_{old}} \cong 0$ $\frac{\partial U}{\partial l_{old}} \cong 0$

Local minima   Saddle Point

FIGURE 9.13   (a) Local minima; (b) saddle point.

$$v_t = \gamma \ v_{t-1} + \alpha \frac{\partial U \ (l, \ x^i, \ y^i)}{\partial l_{old}} \tag{9.24}$$

$$l_{new} = l_{old} - v_t \tag{9.25}$$

Here, v_t is the velocity at time t, γ is a constant term, and $\gamma \ v_{t-1}$ is the momentum term. The learning rate/step size towards the global minima is computed by considering not only the gradient of the loss function at the current point, but also the velocity that has built up over time. The momentum term γ is usually set to 0.9 for better performance.

$$l_{t+1} = l_t - \frac{\alpha}{\sqrt{G_t + \epsilon}} \odot g_t \tag{9.26}$$

Here, α is learning rate, which is generally initialized to 0.01, G_t is a diagonal matrix with the diagonal elements represent the sum of squares of the gradients up to time step t, and ϵ is a smoothing term to avoid division by zero, g_t is the gradient of the cost function U.

ii. **Adaptive Gradient (Adagrad):** It is an adaptive learning rate method which eliminates the need of manually tuning the learning rate α by modifying α for given parameter *l* at a given time *t* based on the previous gradients calculated for given parameter *l*. Its parameter update equation is given by:

For parameters associated with frequently occurring features, it executes smaller updates (i.e. low learning rates) and for parameters with infrequent features it performs larger updates (i.e. high learning rates). Therefore, it is well suited for addressing sparse data issues for this purpose.Its weakness is the accumulation of the squared gradients in the denominator. With every iteration the sum of the squared gradients keep on increasing which makes the learning rate infinitesimally small thus, preventing the algorithm from learning fast.

$$l_{t+1} = l_t - \frac{RMS \, [\Delta l]_{t-1}}{RMS \, [g]_t} \cdot g_t \tag{9.27}$$

Here, $RMS \, [\Delta l]_{t-1}$ is the root mean square of parameter updates at time instant t-1 and $RMS \, [g]_t$ is the root mean square (RMS) of the gradient at time instant t.

iii. **Adadelta:** It solves the problem of radically diminishing learning rates of 'Adagrad' optimizer by recursively defining the sum of squared gradients as a decaying average of all past squared gradients. The running average $E \, [g^2]_t$ at time step γ then depends only on the previous average and the current gradient. Its parameter update equation is given by:

It can be seen that Adadelta has no learning rate parameter; instead, it uses the rate of change in the parameters itself to adapt the learning rate. It uses leaky averages to keep a running estimate of the appropriate statistics.

$$E[\Delta l^2]_t = \gamma \, E[\Delta l^2]_{t-1} + (1 - \gamma)\Delta l_t^2 \qquad (9.28)$$

$$l_{t+1} = l_t - \frac{\alpha}{\sqrt{E[g^2]_t + \epsilon}} \cdot g_t \qquad (9.29)$$

iv. **Root-Mean-Square Propagation (RMSprop):** RMSprop divides the learning rate by an exponentially decaying average of squared gradients. It is identical to the first update vector of Adadelta. Its parameter update equation is given by:

Most implementations use a default value of the decay parameter γ to be set to 0.9 and a good default value of the learning rate α is 0.001. In this way, adaptive learning rate prevents the learning rate decay from decreasing too slowly or too fast.

$$l_{t+1} = l_t - \frac{\alpha}{\sqrt{\tilde{v}_t + \varepsilon}} \cdot \tilde{m}_t \qquad (9.30)$$

Here, m_t and v_t are estimates of the first moment (the mean) and the second moment (the uncentered variance) of the gradients, respectively.

$$l_t = l_{t-1} - \eta_t \left(\frac{\alpha \, \tilde{m}_t}{\sqrt{\tilde{v}_t + \varepsilon}} \right) + \gamma l_{t-1} \qquad (9.31)$$

v. **Adaptive Moment Estimation (Adam):** It was presented as a method of stochastic optimization by combining the ideas of both AdaGrad and RMSProp. It performs a form of learning rate annealing (reducing the learning rate) for each parameter by estimating the first and second moment of gradients. Its parameter update equation is given by,

vi. **AdamW:** As an improvement to Adam optimizer, instead of computing weight decay to optimize the learning rate in Adamw the hyper parameters weight decay γ and learning rate α can be optimized independently which contributes to improved generalization performance of the deep learning model. Its parameter update equation is given by,

Amongst the previously discussed gradient descent optimizers, Adamw as an improvement to Adam, which gives better performance in image steganalysis task. If the data set is sparse, the dynamic learning rate gives good results. If training is to be done in batches, then mini-batch gradient descent gives best results.

9.3 RECENT ADVANCEMENTS

In this section, we outline the recent developments in the external environment of the CNN models, which contribute to achieving better performance in terms of training efficiency.

i. **Transfer Learning:** In this technique, knowledge from a pre-trained deep model developed for an existing task is being transferred to another similar related task. It saves time i.e. learning process becomes faster, needs less training data and gives better performance. The three commonly used transfer learning strategies are:

   a. Use a pre-trained architecture directly and then apply random weights for training the model.
   b. Remove the output layer of the new architecture and use the pre-trained architecture to extract the feature from the new data set.
   c. Apply partial training i.e. freeze the weights of the initial layers and retrain the higher layers only.

It was first utilized for image steganalysis task (Qian et al., 21) in which the knowledge (parameters learned) of the features learned by a CNN-based steganalyzer designed for a high payload were transferred to a low payload, this eventually showed improved performance. In this method, a neural network is first trained with training images comprising of both stego and their respective covers with embedding done at high payload (0.4 bits per pixel (bpp) and then the learned feature representations are transferred to newly developed model with low payload (0.1 bpp) stego images for image steganalysis task. In this way, the method of transfer learning yields improvement in the detection error accuracy and faster training of the deep model.

ii. **Data Augmentation:** The performance of a CNN-based steganalyzer depends upon the size of the data set. In practise we lack sufficient sized data set to train the model for improving the efficiency of the steganalysis. So to address this problem data augmentation is used as an method to increase data set size. The data set is enriched by performing operations such as translate, scale, flip and rotate on the existing data set.

iii. **Generative Adversarial Networks (GAN):** GAN are generative models which are capable of learning to generate new data instances from random input variables with same characteristics/statistics as the original one in the training data set. It comprises of two networks called the Generator (G) and the Discriminator (D), as shown in Figure 9.14, competing against each other. GAN was introduced by Ian Goodfellow in 2014, later standardized by Alec Radford (Radford, Metz, & Chintala, 35) as Deep Convolutional Generative Adversarial Nets (DCGAN). The framework of GAN model is shown in Figure 9.14: The G network takes a sample (noise) z from the input data set and generates fake data instances (images, audio, video, etc.) which

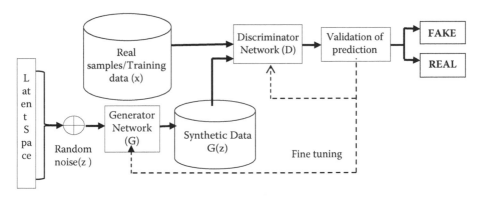

FIGURE 9.14 The Generative Adversarial Networks (GAN) architecture.

are then fed to D. The D network identifies whether data is generated or taken from the real sample. The G network analyses the distribution of the data in such a way that it maximizes the probability of making mistake in identification by the D. GANs operate in a competitive environment where D attempts to minimize its loss V(D,G) as against G which tries to maximize D's loss. The entire minmax game of GAN between D and G is mathematically expressed as:

$$\min_{G} \max_{D} V(D, G)$$

The objective function is expressed by Equation 9.32.

$$V(D, G) = E_{x \sim pdata(x)}[\log D(x)] + E_{z \sim p_z(z)}[\log(1 - D(G(z)))] \quad (9.32)$$

here E[.] denotes the expectation operator, pdata(x) is the distribution of real data, D(x) denotes the Discriminator's output, p(z) is the distribution of generator, x denotes sample from real data, z is sample from generator, and G(z) represents Generator's output.

An integration of steganalysis and steganography is achieved in GANs. The discriminator simulate the steganalysis process and generator generates the stego images (based on noise vector) resembling the cover image and tries to confuse steganalysis discriminative model. The feedback path of the D to the G helps in enhancing the quality and security of the generated stego image in the iterative training process. The stronger steganalysis model will stimulate the improvement of steganography model, and vice versa. GAN-based steganography broadly spans across three techniques of covert communication which includes modifications in the pixels of cover image, selection of cover images with more complex regions and generation of images from random noise. Another GAN-based steganography method called DCGAN employs a unique feature of Steganography without

Embedding (SWE) in which stego images are generated by the generator without performing alteration in the images (Zhou, Sun, Harit, Chen & Sun, 36). In image steganalysis, Deep Digital Steganography Purifier (DDSP), implemented using GAN pre-trained auto encoder as generator network is trained in such a manner that it eliminates maximum amount of steganographic content from the image under consideration while keeping the image quality intact.

9.4   OPEN RESEARCH ISSUES AND OPPORTUNITIES

In this section, we outline the research issues of the designing a CNN-based steganalytic model and the opportunities to overcome them.

Issue 1: How should the weight should be initialized to avoid vanishing and exploding gradient problems and achieve faster network convergence?

Scope: A good variance among the weight values avoids the above-stated problems. The weight initialization methods such as random initialization are found to perform well with all activations, Xavier/glorot initialization with tanh and the initialization with ReLU activation and hence are preferred by the research community for initializing the weights for better learning of the network for images steganalysis task.

Issue 2: A CNN model with multiple BN layers does not generalize the test data when trained with paired learning.

Scope: BN images are tested unpaired with batch statistics, which gets modified with each batch. It detects the hidden data when the cover-stego images are paired and batch statistics normalizes the data but when this condition is violated it fails to do so. The use of Shared Normalization (SN) addresses this problem. SN uses large-sized batches initially and computes the statistics which are updated over several iterations during model training. Unlike BN, SN uses shared statistics which do not change batchwise and hence is preferred to normalize input batches during model training. In BN, the feature maps learned during training are different than feature maps used during testing when inaccurate statistics are used to normalize the input data, whereas SN learns consistent statistical properties comprising of large-sized batches from the training data. The cover-stego pair is not automatically separated after SN making the model insensitive to normalization statistics. In this way, by sharing consistent statistics from training sample SN layers learn discriminative feature to distinguish cover from stego images.

Issue 3: Universal image steganalysis task suffers due lack of data available to train the DL model preventing it from achieving better classification accuracy.

Scope: In the real-world scenario, the image steganalysis suffers from lack of data set and low computation power problem leading poor performance both in terms of detection error rate and training time of the model. Transfer learning resolves the aforementioned problems by transferring the knowledge of a pre-trained model to the newly proposed model performing the same related task. Additionally, the data set can also be enriched by using data augmentation methods such as translation, rotation, flipping, scaling, merging two or more related data sets and using GAN networks. A majority of the research community utilizes BoSSBASE data set as a benchmark data set for training the CNN-based steganalytic models containing only 10,000 gray-scale images of size 512x512 of the portable gray map (PGM) format for training a deep data hungry model. So, we can use transfer learning, data augmentation or can enrich the data set by including more images from different data set such as BOWS2 and ImageNet images for better training of the model.

Issue 4: How do you handle arbitrary image size of the data set for model training?

Scope: In real world scenario, the image data set consist of multi-size images, thus making CNN-based steganalysis difficult which is addressed by SPP concept. SPP, a multilevel pooling strategy capable to obtain multi-level features, maps the feature map to a fixed dimension which is prerequisite of fully connected layer and enables better detection of tampering in the images by enhancing representation ability of features with an improvement in detection accuracy.

9.5 CONCLUSION

In this chapter, we have discussed the design and development of universal steganalyzers from traditional ML to modern DL approaches. The building blocks of CNN with their functions and the impact of customizing model parameters to achieve optimal performance of steganalysis tasks have also been covered. With much of development in the DL based network architecture for image steganalysis, there is still a scope of improvement in both the internal settings of the CNN models i.e. optimizing the model parameters such as weights, activations, learning rate; normalizing the activation outputs and external settings like image pre-processing, transfer learning, data augmentation, and GAN architecture, etc. Also there exist some unsolved challenges like identifying the steganographic algorithm, the rate of embedding secret message, estimation of length of the hidden information, and the identification of stego bearing pixels which needs further exploration.

From the never-ending competition between steganography and steganalysis, both the techniques should be robustly developed such that the steganography system should be efficient enough to preserve image statistics, whereas the steganalytic system should be robust enough to detect any deviation in the image statistics. This chapter paves the way for future research direction in developing a Universal Image steganalysis model that would be able to detect any kind of steganographic algorithm requiring fewer computational resources and yielding higher detection accuracy.

REFERENCES

Avcibas, I., Memon, N., & Sankur, B. (2003). Steganalysis using image quality metrics. *IEEE Transactions on Image Processing: A Publication of the IEEE Signal Processing Society*, *12*(2), 221–229.

Balu, S., Babu, C. K., & Amudha, K. (2018). Secure and efficient data transmission by video steganography in medical imaging system. *Cluster Computing*, *22*, 4057–4063.

Boroumand, M., Chen, M., & Fridrich, J. (2019). Deep residual network for steganalysis of digital images. *IEEE Transactions on Information Forensics and Security*, *14*, 1181–1193.

Clevert, D., Unterthiner, T., & Hochreiter, S. (2016). Fast and accurate deep network learning by exponential linear units (ELUs). CoRR. abs/1511.07289.

Corley, I.A., Lwowski, J., & Hoffman, J. (2019). Destruction of image steganography using generative adversarial networks. ArXiv, abs/1912.10070.

Dugas, C., Bengio, Y., Bélisle, F., Nadeau, C. & Garcia, R. (2000). Incorporating second-order functional knowledge for better option pricing. In T. K. Leen, T. G. Dietterich & V. Tresp (Eds.), *NIPS* (pp. 472–478). Cambridge, MA: MIT Press.

Goodfellow, I. J., Warde-Farley, D., Mirza, M., Courville, A. C., & Bengio, Y. (2013). Maxout networks. In International Conference on Machine Learning, (pp. 1319–1327). PMLR.

Guo, Linjie & Ni, Jiangqun & Shi, Y.Q. (2014). Uniform embedding for efficient jpeg steganography. information forensics and security. *IEEE Transactions on Information Forensics and Security, 9,* 814–825. DOI: 10.1109/TIFS.2014.2312817.

He, K., Zhang, X., Ren, S., & Sun, J. (2015). Delving deep into rectifiers: Surpassing human-level performance on ImageNet classification. *2015 IEEE International Conference on Computer Vision (ICCV),* 1026–1034. https://arxiv.org/abs/1502.01852

He, K., Zhang, X., Ren, S., & Sun, J. (2015). Spatial pyramid pooling in deep convolutional networks for visual recognition. *IEEE Transactions on Pattern Analysis and Machine Intelligence, 37,* 1904–1916.

Holub, V., & Fridrich, J. (2012). Designing steganographic distortion using directional filters. *2012 IEEE International Workshop on Information Forensics and Security (WIFS),* 234–239. IEEE. http://citeseerx.ist.psu.edu/viewdoc/summary?doi=10.1.1.352.8640

Holub, V., & Fridrich, J. (2013). Digital image steganography using universal distortion. IH&MMSec '13 - Proceedings of the 2013 ACM Information Hiding and Multimedia Security Workshop. 10.1145/2482513.2482514

Holub, V., Fridrich, J., & Denemark, T. (2014). Universal distortion function for steganography in an arbitrary domain. *EURASIP Journal on Information Security, 1*(2014), 1–13.

Hu, D., Wang, L., Jiang, W., Zheng, S., & Li, B. (2018). A novel image steganography method via deep convolutional generative adversarial networks. *IEEE Access, 6,* 38303–38314.

Ioffe, S. & Szegedy, C. (2015). Batch normalization: Accelerating deep network training by reducing internal covariate shift. *Proceedings of the 32nd International Conference on Machine Learning, in PMLR, 37,* 448–456.

Katzenbeisser, S., & Petitcolas, F. A. P. (2000). *Information hiding techniques for steganography and digital watermarking.* Boston: Artech House.

Kolade, O., Olayinka, A.A., Sunday, F., Adesoji, O.A., & Olubusola, I.F. (2015). Detection of stego-images in communication among the terrorist Boko-Haram sect in Nigeria. *Journal of Data Analysis and Information Processing, 3*(4), 168–174.

Li, B., Wang, M., Huang, J., & Li, X. (2014). A new cost function for spatial image steganography. *2014 IEEE International Conference on Image Processing (ICIP),* 4206–4210.

Maas, A.L. (2013). Rectifier nonlinearities improve neural network acoustic models.

Madasu, A., & Rao, V.A. (2019). Effectiveness of self normalizing neural networks for text classification. ArXiv, abs/1905.01338.

Nair, V., & Hinton, G.E. (2010). Rectified linear units improve restricted Boltzmann machines. *Proceedings of the 27th International Conference on International Conference on Machine Learning,* June 807–814. ICML.

Ozcan, S., &Mustacoglu, A.F. (2018). Transfer learning effects on image steganalysis with pre-trained deep residual neural network model. *2018 IEEE International Conference on Big Data (Big Data),* 2280–2287. DOI: 10.1109/BigData.2018.8622437

Pevný T., Filler T., & Bas P. (2010) Using high-dimensional image models to perform highly undetectable steganography. In R. Böhme, P.W.L. Fong, & R. Safavi-Naini (Eds.) *Information Hiding. IH 2010. Lecture Notes in Computer Science*, Vol. 6387. Springer, Berlin. https://doi.org/10.1007/978-3-642-16435-4_13

Pibre, L., Pasquet, J., Ienco, D., & Chaumont, M. (2016). Deep learning is a good steganalysis tool when embedding key is reused for different images, even if there is a cover sourcemismatch. *Media Watermarking, Security, and Forensics.* arXiv:1511.04855

Provos, N. (2001). Defending against statistical steganalysis. *Proceedings of 10th USENIX Security Symposium.* Washington, DC. The USENIX Association.

Qian, Y., Dong, J., Wang, W., & Tan, T. (2015). Deep learning for steganalysis via convolutional neural networks. *Electronic Imaging.* DOI: 10.1117/12.2083479

Qian, Y., Dong, J., Wang, W., & Tan, T. (2016). Learning and transferring representations for image steganalysis using convolutional neural network. *2016 IEEE International Conference on Image Processing (ICIP)* (pp. 2752–2756). IEEE.

Radford, A., Metz, L., & Chintala, S. (2016). Unsupervised representation learning with deep convolutional generative adversarial networks. CoRR, abs/1511.06434.

Radford, A., Luke, M., & Soumith, Chintala. (2016). Unsupervised representation learning with deep convolutional generative adversarial networks. CoRR, abs/1511.06434

Ramachandran, P., Zoph, B., & Le, Q.V. (2018). Searching for activation functions. ArXiv, abs/1710.05941.

Srivastava, N., Hinton, G.E., Krizhevsky, A., Sutskever, I., & Salakhutdinov, R. (2014). Dropout: A simple way to prevent neural networks from overfitting. *Journal of Machine Learning Research, 15*, 1929–1958.

Sedighi, V., Cogranne, R., & Fridrich, J. (2015). Content-adaptive steganography by minimizing statistical detectability. *IEEE Transactions on Information Forensics and Security, 11*(2), 221–234.

Tan, S., & Li, B. (2014). Stacked convolutional auto-encoders for steganalysis of digital images. *Signal and Information Processing Association Annual Summit and Conference (APSIPA), 2014 Asia-Pacific,*Chiang Mai, Thailand, 1–4.

Wu, S., Zhong, S., & Liu, Y. (2017). Deep residual learning for image steganalysis. *Multimedia Tools and Applications, 77*, 10437–10453.

Wu, S., Zhong, S., & Liu, Y. (2020). A novel convolutional neural network for image steganalysis with shared normalization. *IEEE Transactions on Multimedia, 22*, 256–270.

Westfeld, A. (2001). High capacity despite better steganalysis (F5: A steganographic algorithm). In Moskowitz, I.S. (Ed.) *Information Hiding. 4th International Workshop. Lecture Notes in Computer Science*, Vol. 2137 (pp. 289–302).Springer-Verlag, Berlin.

Xu, G., Wu, H., & Shi, Y. (2016). Structural design of convolutional neural networks for steganalysis. *IEEE Signal Processing Letters, 23*, 708–712.

Ye, J., Ni, J., & Yi, Y. (2017). Deep learning hierarchical representations for image steganalysis. *IEEE Transactions on Information Forensics and Security, 12*, 2545–2557.

Yedroudj, M., Comby, F., & Chaumont, M. (2018). Yedroudj-Net: An efficient CNN for spatial steganalysis. *2018 IEEE International Conference on Acoustics, Speech and Signal Processing (ICASSP)* (pp. 2092–2096). IEEE

Yedroudj, M., Chaumont, M., Comby, F., Amara, A.O., & Bas, P. (2020). Pixels-off: Data-augmentation complementary solution for deep-learning steganalysis. *Proceedings of the 2020 ACM Workshop on Information Hiding and Multimedia Security.* https://dl.acm.org/doi/abs/10.1145/3369412.3395061

Zhang, R., Zhu, F., Liu, J., & Liu, G. (2018). Efficient feature learning and multi-size image steganalysis based on CNN. ArXiv, abs/1807.11428.

Zhang, H., Ping, X., Xu, M., & Wang, R. (2014). Steganalysis by subtractive pixel adjacency matrix and dimensionality reduction. *Science China Information Sciences*, 57(4), 1–7.
Zheng, H., Yang, Z., Liu, W., Liang, J., & Li, Y. (2015). Improving deep neural networks using softplus units. *2015 International Joint Conference on Neural Networks (IJCNN)*, 1–4. http://ir.ia.ac.cn/bitstream/173211/11777/1/IJCNN-2015-1.pdf
Zhou Z., Sun H., Harit R., Chen X., Sun X. (2015) Coverless image steganography without embedding. In Z. Huang, X. Sun, J. Luo, & J. Wang. (Eds.) *Cloud Computing and Security. ICCCS 2015. Lecture Notes in Computer Science*, Vol. 9483. Springer, Cham, Switzerland. https://doi.org/10.1007/978-3-319-27051-7_11

10 Feature Engineering for Presentation Attack Detection in Face Recognition: A Paradigm Shift from Conventional to Contemporary Data-Driven Approaches

Deepika Sharma and Arvind Selwal
Department of Computer Science & Information
Technology, Central University of Jammu, J&K, India

CONTENTS

DOI: 10.1201/9781003132080-10

10.1   INTRODUCTION

The biometric technology is swiftly gaining popularity and has been replacing the conventional authentication techniques of human recognition (Selwal & Gupta, 2017). Among the various human biometric traits, the face is the second most widely deployed after the fingerprint modality (Jain, Ross, & Prabhakar, 2004). The facial biometric recognition has played prominent role in border control, surveillance and forensics, access control, e-government, and e-commerce contexts (Jain & Ross, 2008; Wheeler & Liu, 2014; Sharma & Selwal, 2020). The widespread applications of facial biometric systems have raised the new concerns related to the security of the overall system. Although the existing face recognition systems cover the security aspects efficiently they are vulnerable to a variety of attacks. The most-attempted attacks i.e Presentation Attacks, are the kind of attacks where an intruder presents the fake biometric data in order to bypass the biometric subsystem. PAs do not require the internal information related to the operations of the biometric system, so an adversary can easily use the face artifacts to bypass the sensor module (Menotti et al., 2015; Sharma & Selwal, 2021). There exists a sufficient amount of literature that provides an imminent severity of these attacks in the face biometric system. In 2010, a man boarded a plane in Hong Kong and lands as an Asian man in Canada ("Man boards plane disguised as old man then arrested on arrival in Canada | Daily Mail Online," n.d.). Similarly, in New York, the robbers, while looting a cash checking store, were caught where they imitated white cops by using face masks (Epstein, 2012). The facial systems can be easily spoofed by intruders by making use of various counterfeits, as shown in Figure 10.1.

In photo attacks, the digital photograph of a genuine person is given as an input to the biometric to bypass the sensor. This is the simplest and most frequently attempted attack as the photographs can be easily downloaded from social media applications. Likewise, in video and mask attacks, a video clip and artificially generated face mask of a genuine person is used to attack the system, respectively (Evans et al., 2014). The face PAD is a critical issue in one of the most widely deployed human authentication system in the emerging human authentication tool for Society 5.0. In order to protect face biometric systems from the adverse effect of PAs, a variety of Presentation Attack Detection (PAD) methodologies have been designed in the last decades. These are categorized into two main categories, known as hardware- and software-based face PAD techniques. In the hardware approach, an extra hardware device is integrated within the facial recognition system in order to measure the additional sensor characteristics, vitality features such as eye blink or head movement, or a challenge response approach. On the contrary, in a software-based approach, a software module is used to extract the features from the face images which are then used to discriminate live face images from the fake ones. The software-based approaches prominently rely on texture feature extraction and image quality analysis (Ramachandra, Busch, & Biometric, 2017). A variety of features such texture, image quality, frequency, and motion-based features are extracted from face images by making use of various image feature descriptors such as LBP, BSIF, ULBP, etc. In Society 5.0, the paradigm has shifted to automatic feature extraction for face PAD mechanisms that replace the traditional computational methods based on hand-crafted

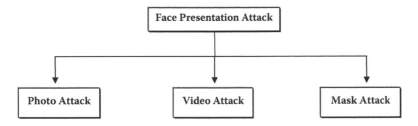

FIGURE 10.1 A taxonomy of face presentation attacks.

image feature extraction. The DCNN models have been deployed widely in face PAD designing. Although the modern deep learning–based techniques provide greater classification accuracy, their training data size and time complexities still remain a major challenge. Abundant techniques have been designed to countermeasure the facial presentation attacks already presented in various surveys (Evans et al., 2014; Ramachandra et al., 2017). In this chapter we restrict our study to feature engineering in software-based techniques and present a review of various image feature descriptors and DCNN-based models that have been prominently deployed in face PAD mechanisms during the era of 2008 to 2020. The major contributions of the present work are summarized as follows:

i. We present a comprehensive study of hand-crafted features that have been extensively used for face PAD methods.
ii. We provide a review of modern deep feature-based face PAD mechanisms by using variants of DCNN.
iii. We suggest various research challenges for feature engineering in face PAD as well as open research issues, which may help prospective researchers to carry out further work in this active field.

The remainder of the chapter is structured as follows: Section 10.2 provides feature engineering for face biometric systems with an insight on facial features. The various PAD methods with a special focus on hand-crafted and deep features approaches are presented in Section 10.3. Section 10.4 expounds an overview of face anti-spoofing data sets. The outcome of this study is summarized with various open research issues and opportunities in Section 10.5. Section 10.6 provides concluding remarks and future scope of the presented work.

10.2 FEATURE ENGINEERING FOR SECURED AND INTELLIGENT FACE BIOMETRIC SYSTEMS

Society 5.0 predominantly utilizes the facial features of the individual for secured human authentication. To countermeasure direct spoofing attacks by the imposters, the efficient feature extraction or selection helps to design better PAD mechanisms. In this section, we explore various techniques that are widely explored by the state-of-the-art face PAD mechanisms.

(a)

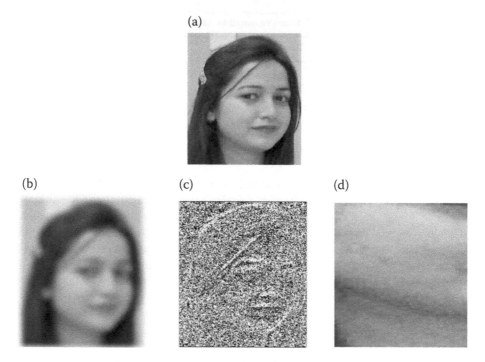

(b)   (c)   (d)

FIGURE 10.2   Features of face at three different levels: (a) Original face image; (b) Level 1 features; (c) Level 2 features; (d) Level 3 features.

10.2.1 FACIAL FEATURES

The characteristics of the face, which has been explored for face recognition and facial PAD mechanisms, are categorized into three levels: **Level 1** details consist of facial feature that are easily observable. It includes the skin color and general face geometry. These features are used for easily discriminating a small round face and an extended thin face or faces demonstrating predominantly female and male characteristics. **Level 2** consists of information related to structure of the facial components such as eyes, nose, lips, etc. and the relationship between these components. Unlike Level 1, these features require a face image with high resolution. Similarly, **Level 3** contain micro-level features of the face that are unstructured and it include moles, marks, and skin discoloration (Jain, Ross, & Nandakumar, 2011). The features of face images are shown in Figure 10.2.

10.3 COMPUTATIONAL IMAGE FEATURES FOR FACE PAD MECHANISMS

The software-based PAD approaches are further divided into static and dynamic ones where static algorithms extract image features from a single face image and dynamic techniques explore multiple frames of a given face image. These techniques are extended further by the complimentary properties such as head movement, motion analysis (Killioglu, Taskiran, & Kahraman, 2017; Kim, Yoo, & Choi, 2011; Komulainen, Hadid, & Pietik, 2013; Xia et al., 2020; Yan, Zhang, Dong, Stan, & Li, 2012), and frequency-based (Li, Wang, Tan, & Jain, 2004; LIU, 2014; Peng & Chan, 2014; Teja, 2011; Zhang et al., 2012) information and most likely the texture features extracted from face images. In texture-based techniques, the texture-based feature information exhibed by the face image is extracted and then used in live and fake face detection mechanisms. The micro-textural features are extracted from the sample of face images and then these extracted patterns are used for discriminating real or bona-fide face images. The texture features can be extracted by a hand-crafted engineering process where a large number of image feature descriptors such as LBP, BSIF, HOG, LPQ, Gabor Wavelet, etc. are used to extract Level 2 details from a given face image. Another method of automatic texture feature extraction is also proposed by various researchers in recent works where bio-inspired models of DCNN are employed.

10.3.1 HANDCRAFTED FEATURES

In this subsection, we briefly describe the various image feature descriptors that are used in the face PAD mechanism. Notably, the variants of Local Binary Pattern (LBP) features are widely deployed by the research community in face PAD mechanism. The various image feature descriptors used to extract the discriminating features are discussed below.

10.3.1.1 Local Binary Patterns

The original LBP descriptor introduced by Ojala et al. (Ojala, Pietikäinen, & Mäenpää, 2000) is a gray-scale texture measure that is derived from the relationship of pixels with its local neighborhood. It is computationally efficient and possesses greater tolerance against the gray-level changes. The operator figured image pixel labels by thresholding the neighborhood of each pixel with the central value and considers binary number as its result. The computation of an LBP descriptor is shown in Figure 10.3. A histogram created from these different labels is then used for texture description.

The LBP code of a pixel is calculated using equations 10.1 and 10.2. The feature vector is constructed by consolidating all the patterns for a given image.

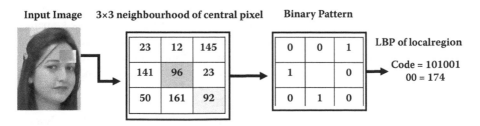

FIGURE 10.3 An illustration of LBP code computation.

$$\text{LBP} = \sum_{j=1}^{8} v(i - c) \times 2^{j} \tag{10.1}$$

$$f(d) = \begin{cases} 1, & \text{if } d \geq 0 \\ 0, & \text{otherwise if } d < 0 \end{cases} \tag{10.2}$$

Where v indicates the quantized values and the central pixel is represented by c and i denotes the considered pixel. The function f computes the difference between the intensities of two given pixels.

The original LBP descriptor has been productively extended, where different-sized neighbors can be considered instead of only eight. By making use of bilinear interpolating values at real valued coordinates allow any radius with a large number of neighborhoods. Equations 10.3 and 10.4 are used for calculating LBP code with a varying neighborhood size.

$$\text{LBP}_{N, R} = \sum_{h=0}^{h-1} f(p_c - p_h) 2^h \tag{10.3}$$

$$f(d) = \begin{cases} 1, & \text{if } d \geq 0 \\ 0, & \text{otherwise if } d < 0 \end{cases} \tag{10.4}$$

Where h indicates the number of the neighborhood of a central pixel c. N denotes the total number of pixels in the given region i.e 8, 24, etc. The function f computes the difference between the intensities of two given pixels.

Another variant to an original LBP descriptor named uniform Local Binary Pattern (ULBP) (Mäenpää, Ojala, Pietikäinen, & Soriano, 2000) was explored, which is inspired due to the reason that few patterns occur frequently in texture images as compared to others. When there exists at most two transitions from 1 to 0 or vice versa while traversing in circular fashion, then the pattern is known as uniform. This descriptor is also computationally efficient like the original version with an additional property of rotation invariance.

The features extracted by using LBP operator on two face images, as depicted in Figure 10.4.

The study of dynamic texture was initiated by Pereira et al. (2012); Pereira, Anjos, Martino, & Marcel (2012); Pereira et al. (2014); Pereira et al. (2014); Komulainen et al. (2012); and Komulainen et al. (2012). The LBP operator from three orthogonal planes (TOP-LBP) was investigated in their work. This extended LBP variant has yielded volume local binary pattern (VLBP), which is a spatio-temporal LBP extension of original one (Ojala et al., 2000). It combines the space and time information into a single descriptor with a multi-solution approach. VLBP combines motion and appearance into the dynamic texture description.

10.3.1.2 Binarized Statistical Image Features

The BSIF descriptor was introduced by Juho Kannala et al. (Kannala & Rahtu, 2012) where the binary string for all the image pixels is generated by computing

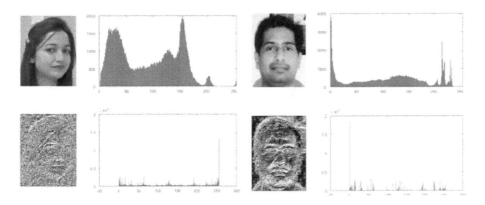

FIGURE 10.4   LBP features extracted from two face images with their respective histogram.

response to a kernel bank which is trained by making use of statistical properties of natural images. The computed code of a pixel is taken as a local descriptor of the intensity pattern of image in pixel's surroundings. For an image segment B(u, v) of size m×n pixels and a linear filter X_i of the same size, the response of the filter r_i is computed as per equation 10.5.

$$r_i = \sum_{u,v} Xi(u, v)B(u, v) = x_i^T p \qquad (10.5)$$

Where the pixels of image B and X_i are denoted by the vector B and x, respectively. The binarized feature b_i is obtained by using equation 10.6.

$$bi = \begin{cases} 1, & ri > 0 \\ 0, & otherwise \end{cases} \qquad (10.6)$$

The feature vectors generated can be used to build histograms to represent the image feature descriptors. Let $I_i(x, y)$, with i = 1, 2, 3 - - 8 represents the n different natural images and X_i be the filter with 11 × 11 size pre-learned from ith image. The ith kernel is combined with the input image B(u, v) for computing the response value $r_i(u, v)$. Additionally, b_i (u, v) is the binary response of the ith kernel at pixel (u, v). Hence, the binary response of all the kernels is computed for the pixels of the whole image. These responses are used to generate a BSIF code for each pixel of the image to construct respective BSIF images. In the last, the normalized histogram of the BSIF image provides the feature descriptor. The BSIF extracted from a face image are shown in Figure 10.5.

10.3.1.3   Local Phase Quantization

The LPQ descriptor was designed by Ojansivu and Heikkil (2008), where Short Term Fourier Transform (STFT), is employed for computing pixel codes. Generally, the responses are computed by making use of STFT in the local neighborhood n and local window (w) by using equation 10.7:

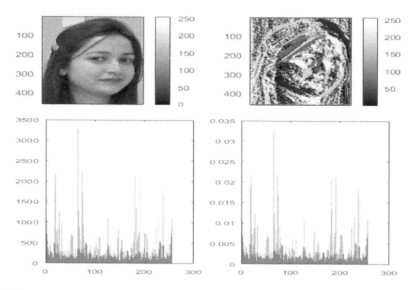

FIGURE 10.5 BSIF extracted from a face image.

$$F1(m, x) = Im(x, y) \; w_R(y - x) \; exp(-j \; 2 \; U^T y) \qquad (10.7)$$

Where the coefficients of Fourier are computed for m1, m2, m3, and m4 frequency points corresponding to $[a, a]T$, $[0, a]T$, $[a, 0]T$, and $[a, -a]T$ so the value of H(mi) spectral response should be greater than zero. The spectral information is decomposed further into real and imaginary parts denoted by Re[F] and Im[F], respectively, the combination of which formed the vector as R = [Re[F], Im[F]]. The response in the final vector is a binarized bit (B_i for ith) and a zero value is assigned for components having response value less than 1, otherwise assigned 1.

At last the response values are coded to generate the compact value of pixel between 0–255 range by employing conversion of binary to decimal as per equation 10.8:

$$f = \sum_{j=1}^{8} B_j \times 2^{j-1} \qquad (10.8)$$

10.3.1.4 Speed Up Robust Features

The SURF (Bay, Tuytelaars, & Gool, 2006) is an efficient rotation and scale-invariant descriptor as well as detector. The detector positioned the key features in an image whereas the descriptor provides the description of these features. Initially it was introduced for reducing the complexity of Scale Independent Feature Transform (SIFT) image descriptor. The SURF makes use of the Haar wavelet, instead of DoG filters for approximating the Gaussian Laplacian. The Haar wavelet computes the vertical (x) and horizontal (y) directional responses for describing the intensity distribution of key points. The region surrounded each concern point is partitioned into 4 × 4 sub-regions. Then for each of these sub-regions i, the x and y directional responses are computed which forms the feature vector FV$_i$ as per the equation 10.9:

$$FV_i = [\textstyle\sum dx, \sum dy, \sum |dx|, \sum |dy|]] \tag{10.9}$$

Where dx and dy denotes the value of Haar wavelet in x and y direction.

These FVs extracted from each sub-region i are combined to form the FV of the SURF descriptor.

$$SURF = [FV_1 - - - - - - - - - - - FV_{16}]$$

The Principle Component Analysis (PCA) was employed for dimensionality reduction of extracted SURF features. Since the SURF descriptor presents images by using local features, it performs effectively for fingerprint and face images.

These micro-textural image features are widely employed in the face PAD mechanisms. Hence, Table 10.1 gives the summary of the various hand-crafted image feature descriptors used in the face PAD mechanisms with their performance.

10.3.2 Deep Features Engineering for Face PAD Mechanisms

Besides exploring the hand-crafted features in face PAD mechanism, we investigate the applicability of naturally learned deep features which makes use of Deep Convolution Neural Networks (DCNNs). The schematic representation of a typical DCNN is shown in Figure 10.6, which clearly shows it consists of a series on interconnected layers that performs a specialized task. The first n-1 layers of DCCN perform the activities of automatic deep feature extraction while the last layer is a conventional fully connected layer that does the job of classification. The research community proposed new Convolution Neural Network (CNN)–based models or modifies the existing models and incorporates them in a face PAD mechanism.

In the past decades, several variants of CNN models have been proposed and these have been used in various biometrical applications. A summary of development in DCNN is highlighted in Table 10.2. Out of these deep leaning models, only a few have been explored for the purpose of face anti-spoofing mechanisms.

The first variant of CNN model which is deployed in face PAD designing is known as the LBPnet. The model is modified by incorporating the LBP layer as the first layer in the basic CNN model which is based on well-referenced Lenet-5 (LeCun, Botton, Bengio, & Haffner, 1998) network. The LBP net consists of the following layers as inherited from the basic Lenet-5 model: (a) Two convolution layers followed by pooling layer, the first one is modified by integrating LBP feature descriptor. (b) a ReLU (Rectified Linear Unit) layer, which performs inner product operation. (c) A fully connected layer which performs the classification operation using Softmax function. The convolution operation can be defined as:

$$C_i(a) = \textstyle\sum_{\forall b \in N(a)} LBP(I(b)). \, K_i j \tag{10.10}$$

TABLE 10.1

An Illustration of Hand-Crafted Features for Face PAD Mechanism

Year	Author	Feature Descriptor	Classifier	Database	PAD Attack	Performance
2011	(Maatta Maatta, Hadid, & Pietikainen, 2011)	Multi-Scale LBP	SVM	NUAA	Photo	98.0% Accuracy
2012	(Chingovska, Rabello, & Marcel, 2014)	LBP	LDA and SVM	REPLAY-ATTACK	Photo and Video	HTER = 15%
2012	(Maatta & Pietika¨inen, 2012)	Multi-scale LBP, Gabor Wavelet, and HOG	SVM	NUAA, Yale Recaptured and Print Attack	Photo	EER = 0.5%
2013	(Waris, Zhang, Ahmad, Kiranyaz, & Gabbouj, 2013)	ULBP, Gabor Wavelet, and GLCM	SVM and PLS	REPLAY-ATTACK	Photo and Video	Gabor feature and ULBP are more reliable
2013	(Yang, Lei, Liao, & Li, 2013)	LBP, LPQ, HOG, etc.	SVM	NUAA, Print-Attack, & CASIA	Photo	0.04% average EER
2014	(Raghavendra & Busch, 2014)	LBP, BSIF	SVM	CASIA, 3DMAD	Mask	ACER of 5.74% for CASIA and 4.78% for 3 DMAD
2014	(Erdogmus & Marcel, 2014)	LBP variants	SVM	Morpho Database & 3DMAD	Mask	7.0% EER for Morpho Database
2016	(Boulkenafet, Komulainen, & Hadid, 2016b)	LBP, CoALBP, LPQBSIF, and SID	SVM	CASIA, MSU, and REPLAY-ATTACK	Photo and video	2.46% average EER
2016	(Boulkenafet et al., 2016a)	SURF	FVE	MSU, CASIA, and Replay Attack	Photo and video	Outperforms the state-of-the-art
2017	(Wang, Nian, Li, Meng, & Wang, 2017a)	Multi-scale LBP	SVM	NUAA, CASIA,MSU, and REPLAY ATTACK	Photo and replay	Promising results are achieved

TABLE 10.1 (Continued)
An Illustration of Hand-Crafted Features for Face PAD Mechanism

Year	Author	Feature Descriptor	Classifier	Database	PAD Attack	Performance
2017	(Zhang, Peng, Qin, & Long, 2018)	GS-LBP and LGBP		MSU, CASIA, Replay Attack, and Replay-Mobile antispoofing databases	Photo and video	8.54% EER for MSU MFSD, 2.53% for CASIA
2019	(Hasan, Mahmud, & Li, 2019)	LBPV and DoG Filtering	SVM	NUAA	Photo and video	99.22% Accuracy

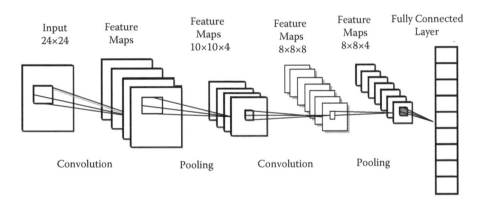

FIGURE 10.6 Schematic representation of convolutional neural network.

Where LBP (I(b)), with $b = (x_b, y_b)$, gives the LBP value of pixel b which belong to the neighbor of pixel $a = (x_a, y_a)$. $Ci(a)$ denotes the value corresponding to position a in the output feature vector Ci, where $i = 1, 2, 3, ..., 20$; and $K_i(j)$, where $i = 1, 2, 3, ..., 20$ corresponds to a value in the ith convolution filter in the position respective to q.

Another modified variant of CNN model is the end-to-end LBP-based CNN network. This network consists of four components: convolution, LBP layer, loss function, and classification layer. The first component is comprised of two convolution layers and one pooling layer with one ReLU layer. The LBP layer extracts the LBP texture features, whereas the loss function is employed for training network parameters.

Another multi-channel CNN is proposed that uses the concept of transfer learning. The LightCNN model is utilized in this network that is pretrained on a large number of images. The number of parameters in this network is much smaller as compared to other CNN models which are used in face PAD mechanisms. The

TABLE 10.2

A brief summary of development of CNN models with architectural specifications

Year	Author(s)	Deep Learning Model	Architecture Specifications
2012	Alex Krizhevsky	AlexNet	• Combination of 8 layers • First 5 convolutional layers • Last three fully connected layers • ReLu layer followed every convolutional Layer
2015	Simonyan and Zisserman	VGG-16	• 16 convolutional layers • 3 fully connected layers • 5 max-pooling layers • A SoftMax layer
2015	Szegedy et al.	GoogLeNet	• 22 convolutional layers • Convolution size is not fixed
2015	Yang and Ramanan	DAG-CNN	• All the layers are combined together termed as a chain-structured backbone
2016	He et al.	ResNet	• Constructed with identity mapping for creating a path between the input and output layer and some layers which are in between are skipped
2016	Iandola et al.	SqueezeNet	• 5MB parameters as compared to 240 MB parameters in original AlexNet

Max-Feature Map function is used instead of ReLU layer. The Binary Cross Entropy is utilized as a loss function, which trains the model by using ground information. The BCE is computed as per equation 10.11:

$$L = -(y \ \log \ (a) + (1 - y) \ \log \ (1 - a)) \qquad (10.11)$$

Where y represents the ground truth and p gave the probability value.

Table 10.3 presents the summary of various CNN models which have been widely deployed in face PAD mechanism. It may be seen that some of the authors such as Li et al. (2018), Chen et al. (2019), and Abbas et al. (2020) have recently exploited hand-crafted features from face images and integrate these with deep features based neural networks.

10.4   SECURITY EVALUATION USING FACE ANTI-SPOOFING DATA SETS

To evaluate the effectiveness of intelligent face PAD mechanism, there is an utmost need of training and testing data sets. The training and testing data set must contain the images of both the genuine and fake face images of several subjects acquired under various conditions. In the literature, researchers have put in tremendous

TABLE 10.3

A summary of deep features based face PAD methods

Year	Author	DCNN Model	Datasets	Attack	Performance
2017	(Souza et al., 2017)	LBPnet and n-LBPnet	NUAA	Photo and Video	EER = 0.019%
2017	(Wang, Nian, Li, Meng, & Wang, 2017b)	Deep texture features & depth cues	REPLAY-ATTACK & CASIA	Photo and Video	HTER = 10.2%
2018	(Li, Feng, Xia, Jiang, & Hadid, 2018)	LBP & CNN	REPLAY-ATTACK & CASIA	Photo and Video	Average EER = 1.55%
2018	Abbas, Rehman, Po, & Liu, 2018)	CNN	CASIA & REPLAY-ATTACK	Photo and Video	HTER = 8.28% and 14.14%, respectively
2019	Chen et al. (Chen et al., 2019)	RI-LBP & CNN	NUAA, REPLAY-ATTACK, CASIA, & MSU	Photo & Video	Average EER = 2.57%
2019	George et al. (George, Mostaani, Geissenbuhler, & Nikisins, 2019)	Multi channel-CNN	WMCA	Photo & Video	ACER = 0.3%
2019	(Abbas, Rehman, Po, & Liu, 2020)	Deep CNN	CASIA, OULU & REPLAY-ATTACK	Photo & Video	HTER = 28.2% for CASIA & HTER = 11.3% for REPLAY-ATTACK
2020	(Abbas, Rehman, Po, & Komulainen, 2020)	LBP & CNN	CASIA, REPLAY-ATTACK, & OULU	Photo & Video	HTER = 1.31% for REPLAY-ATTACK ACER = 22.8% for CASIA, 9.24% for OULU

efforts and have designed various face anti-spoofing data sets that openly available for research community. A brief summary of these face anti-spoofing data sets is depicted in Table 10.4.

It may be clearly observed from the summary in Table 10.4 that the most of the face anti-spoofing data sets contains samples of various subjects in the range of 15 to 80. The size of these data sets may be sufficient to evaluate a hand-crafted feature based face PAD methods, whereas the modern data-driven deep features based PAD methods require a large number of samples in the range of millions. Therefore, to effectively implement the deep-feature based methods, the need is to either design new face-anti-spoofing data sets or use

TABLE 10.4
Publically available face anti-spoofing databases

Authors	Dataset	Subjects	Sensor	Presentation Attack Instruments	Resolution
(Tan, Li, Liu, & Jiang, 2010)	NUAAIMPOSTERDATASET	15	Generic webcam	Paper	640 x 480 pixels
(Anjos & Marcel, 2011)	PRINT ATTACK	50	Apple 13-inch MacBook laptop	Paper, tablet	320 x 240 pixels
(Zhang, Yi, Lei, & Li, 2011)	CASIA-FAS	50	Low and normal-quality camera, Sony NEX-5camera	Paper, tablet, mobile	640 x 480, 480 x 640, 1280 x 720 pixels
(Chingovska, Anjos, & Marcel, 2012)	REPLAY-ATTACK	50	Apple 13-inch MacBook laptop	Paper, tablet	320 x 240 pixels
(Erdogmus & Marcel, 2013)	3D FACE MASK DATABASE	17	Microsoft Kinect sensor	Tablet, mask	640 x 480
(Wen, Han, & Jain, 2015)	MSU-MFSD	55	MacBook Air 13 and Google Nexus 5 camera	Wrap paper, mask, tablet	640 x 480720 x 480
(Raghavendra, Raja, & Busch, 2015)	GUC-LiFFAD	80	Lytro Light field camera	Wrap paper, mask, tablet	1,080 x 1,080
(Costa-Pazo, Bhattacharjee, Vazquez-Fernandez, & Marcel, 2016)	REPLAY-MOBILE	40	1 Smartphone LG-G47 & 1 tablet iPad Mini 26	Printer, displays	720 x 1,280
(Boulkenafet, Komulainen, Li, Feng, & Hadid, 2017)	OULU-NPU	55	6 Smartphones	Printer, displays	6 different resolutions

specialized methods such a data augmentation to increase the number of samples in the existing data sets.

10.5 OPEN RESEARCH ISSUES AND OPPORTUNITIES

In the previous sections we explored the various hand-crafted and deep features that have been used in face PAD mechanism. Presently, the face recognition systems constitute the critical infrastructure for secured human authentication in smart systems in Society 5.0. As these systems are secured mechanisms in contemporary world of computing, therefore their security and privacy is a major concern for the researchers. This comprehensive study leads to many open research issues that will provide a further direction to feature engineering in face PAD mechanism. The various issues are listed below:

    i. **Vulnerabilities due to variations**: Most of the existing feature descriptors suffer from the drawback of variation in different acquisition environments. The variations mainly introduced because of head rotation, lighting effects, illumination conditions, etc. Therefore, there is a need to design more efficient feature descriptor with more description power and which is resistant to above-mentioned variations.

    ii. **Lack of discrimination power**: Our study clearly reveals that the hand-crafted feature engineering for face PAD mechanisms that utilize a single feature descriptor (such as LBP, BSIF, LPQ, etc.) lack significant discrimination capability for classifying real and fake face images. One of the alternative solutions is to develop more robust image feature descriptors to improve the overall classification accuracy. Moreover, discrimination power for face PAD mechanisms may be further enhanced by combining two or more existing feature descriptors.

    iii. **Hybrid approach for face PAD**: The presented review in this work illustrates that both hand-crafted as well as deep feature based face PAD methods have their respective pros and cons. Therefore, one of the opportunities, that are rarely explored, is to make use of hand-crafted features as a perturbation layer in the DCNN models. The hybrid approach will effectively improve the overall robustness of the model for the classification task.

    iv. **Lightweight deep learning models for face PAD**: Most of the existing DCNN models are summarized in Table 10.2 have a very complex structure comprising of several layers (i.e. 152 layers in AlexNet). Therefore, to apply these networks to solve a simple binary classification problem like face PAD adds additional training overhead. Hence, one of the viable solutions is to use the concept of transfer learning where pre-learned models may be integrated with a simple deep neural network comprising of only few layers. Furthermore, face PAD techniques may be designed by exploiting some of the newly introduced lightweight DCNN model such as EfficientNet, SRNet, etc.

v. **Requirement of specialized hardware infrastructure**: For training a deep learning based face PAD model using millions of face images, a simple machine may take several days for complete learning of the model. Therefore, the implementation of DCNN model requires specialized hardware infrastructure such as GPU-based servers having multiple cores. Moreover, these GPU-based servers require large amount of RAM with some special facilities such as maintaining low temperature.

vi. **Insufficient data sets for deep feature based face PAD model**: The existing face anti-spoofing data sets contain sufficient amount of real and fake face images, which works well in case of hand-crafted feature descriptor's training. Whereas the DCNN models require large number of training images, therefore, the existing number of images in face anti-spoofing data sets is inadequate for DCNN model training.

10.6   CONCLUSION

In the contemporary Society 5.0, the computation involving the feature engineering plays an important role for image classification. The face recognition infrastructure has emerged a key tool that offer secured authentication towards computational excellence. In this work, we explored the aspects of feature engineering for face PAD mechanisms during the era of 2008–2020. We discussed prominent hand-crafted image descriptors such as LBP, BSIF, LPQ, and SURF that are used to extract unique features to train the classifier for face PAD methods. We illustrated the development of deep features based engineering for automatic feature extraction to offer efficient face PAD mechanisms. Our study has revealed that both the hand-crafted as well as deep feature based methods have their respective merits and demerits. Therefore, to address the problem of spoof attacks, the future research must focus on finding the viable solutions that integrate the merits of both the hand-crafted and deep level features for accurate face PAD. Moreover, novel data augmentation schemes need to be developed for training DCCN. In the future, the lightweight DCCN models such as Efficient Net, ResNet, or SRNet may be exploited for improving the training overhead and increased classification accuracy in the face PAD models.

REFERENCES

Abbas, Y., Rehman, U., Po, L., & Komulainen, J. (2020). Enhancing deep discriminative feature maps via perturbation for face presentation attack detection. *Image and Vision Computing*, *94*, 103858. https://doi.org/10.1016/j.imavis.2019.103858

Abbas, Y., Rehman, U., Po, L., & Liu, M. (2020). SLNet: Stereo face liveness detection via dynamic disparity-maps and convolutional neural network. *Expert Systems with Applications*, *142*, 113002. https://doi.org/10.1016/j.eswa.2019.113002

Abbas, Y., Rehman, U., Po, L. M., & Liu, M. (2018). LiveNet: Improving features generalization for face liveness detection using convolution neural networks. *Expert Systems with Applications*, *108*, 159–169. https://doi.org/10.1016/j.eswa.2018.05.004

Anjos, A., & Marcel, S. (2011). Counter-measures to photo attacks in face recognition: A public database and a baseline. *2011 International Joint Conference on Biometrics, IJCB 2011.* https://doi.org/10.1109/IJCB.2011.6117503

Bay, H., Tuytelaars, T., & Van Gool, L. (2006). *SURF: Speeded Up Robust Features.* In Leonardis A., Bischof H., Pinz A. (eds) Computer Vision – ECCV 2006. ECCV 2006. Lecture Notes in Computer Science, vol. 3951. Springer, Berlin, Heidelberg, 404–417. doi.org/10.1007/11744023_32

Boulkenafet, Z., Komulainen, J., & Hadid, A. (2016a). Face anti-spoofing using speeded-up robust features and fisher vector encoding. *Signal Processing Letters, 9908*(c), 1–5. https://doi.org/10.1109/LSP.2016.2630740

Boulkenafet, Z., Komulainen, J., & Hadid, A. (2016b). Face spoofing detection using colour texture analysis. *IEEE Transactions on Information Forensics and Security, 6013*(c), 1–13. https://doi.org/10.1109/TIFS.2016.2555286

Boulkenafet, Z., Komulainen, J., Li, L., Feng, X., & Hadid, A. (2017). OULU-NPU: A Mobile Face Presentation Attack Database with Real-World Variations. *Proceedings—12th IEEE International Conference on Automatic Face and Gesture Recognition, FG 2017—1st International Workshop on Adaptive Shot Learning for Gesture Understanding and Production, ASL4GUP 2017, Biometrics in the Wild, Bwild 2017,* Heteroge, 612–618. https://doi.org/10.1109/FG.2017.77

Chen, F., Wen, C., Xie, K., Wen, F., Sheng, G., & Tang, X. (2019). Face liveness detection: Fusing colour texture feature and deep feature. *IET Biometrics, 8*(6). https://doi.org/1 0.1049/iet-bmt.2018.5235

Chingovska, I., Anjos, A., & Marcel, S. (2012). On the effectiveness of local binary patterns in face anti-spoofing. *Proceedings of the International Conference of the Biometrics Special Interest Group (BIOSIG'12).* https://ieeexplore.ieee.org/document/6313548

Chingovska, I., Rabello, A., & Marcel, S. (2014). Biometrics evaluation under spoofing attacks. *IEEE Transactions on Information Forensics and Security, 9*(12), 2264–2276.

Costa-Pazo, A., Bhattacharjee, S., Vazquez-Fernandez, E., & Marcel, S. (2016). The Replay-Mobile face presentation-attack database. *Lecture Notes in Informatics (LNI), Proceedings—Series of the Gesellschaft Fur Informatik (GI), P-260.* https://doi.org/1 0.1109/BIOSIG.2016.7736936

de Souza, G. B., da Silva Santos, D. F., Pires, R. G., Marana, A. N., & Papa, J. P. (2017). Deep Texture Features for Robust Face Spoofing Detection, IEEE Transactions on Circuits and Systems II: Express Briefs, *64*(12), 1397–1401.

Epstein, E.A. (2012). Robbers who disguised themselves as white cops are caught... after they send polite thank-you letter to company that made their "unbelievable" latex masks. *Daily Mail Online.* Retrieved from https://www.dailymail.co.uk/news/ article-2192115/Robbers-disguised-white-cops-caught--send-polite-thank-letter-com-pany-unbelievable-latex-masks.html

Erdogmus, N., & Marcel, S. (2013). Spoofing in 2D face recognition with 3D masks and anti-spoofing with Kinect. *IEEE 6th International Conference on Biometrics: Theory, Applications and Systems, BTAS 2013.* https://doi.org/10.1109/BTAS.2013.6712688

Erdogmus, N., & Marcel, S. (2014). Spoofing Face Recognition with 3D Masks. *IEEE Transactions on Information Forensics and Security,* (July). https://doi.org/10.1109/ TIFS.2014.2322255

Evans, N., Kinnunen, T., Yamagishi, J., Wu, Z., Alegre, F., & Leon, P. De. (2014). Handbook of biometric anti-spoofing. *Handbook of Biometric Anti-Spoofing.* https:// doi.org/10.1007/978-1-4471-6524-8

George, A., Mostaani, Z., Geissenbuhler, D., & Nikisins, O. (2019). Biometric face pre-sentation attack detection with multi-channel convolutional neural network. *IEEE Transactions on Information Forensics and Security, 6013*(c), 1–16. https://doi.org/1 0.1109/TIFS.2019.2916652

Hasan, R., Mahmud, S. M. H., & Li, X. Y. (2019). Face anti-spoofing using texture-based techniques and filtering methods. *Journal of Physics: Conference Series.* https://doi.org/10.1088/1742-6596/1229/1/012044

Jain, A. K., & Ross, A. (2008). *Handbook of biometrics.* A. K. Jain, P. Flynn, & A. A. Ross, (Eds.), Face Recognition. London: Springer.

Jain, A. K., Ross, A. A., & Nandakumar, K. (2011). *Introduction to biometrics.* London: Springer.

Jain, A. K., Ross, A., & Prabhakar, S. (2004). An introduction to biometric recognition. *IEEE Transactions on Circuits and Systems for Video Technology, 14*(1), 4–20. https://doi.org/10.1109/TCSVT.2003.818349

Kannala, J., & Rahtu, E. (2012). BSIF: Binarized statistical image features. *Proceedings of International Conference on Pattern Recognition,* 1363–1366. https://ieeexplore.ieee.org/document/6460393

Killioglu, M., Taskiran, M., & Kahraman, N. (2017). Anti-spoofing in face recognition with liveness detection using pupil tracking. *SAMI 2017, IEEE 15th International Symposium on Applied Machine Intelligence and Informatics,* 87–92. IEEE.

Kim, Y., Yoo, J., & Choi, K. (2011). A motion and similarity-based fake detection method for biometric face recognition systems. *IEEE Transactions on Consumer Electronics, 57*(2), 756–762.

Komulainen, J., Hadid, A., & Pietik, M. (2012). Face spoofing detection using dynamic texture. *Proceedings of 11th International Conference on Computer Vision,* November, 146–157. https://doi.org/10.1007/978-3-642-37410-4

Komulainen, J., Hadid, A., & Pietik, M. (2013). Complementary countermeasures for detecting scenic face spoofing attacks. *2013 International Conference on Biometrics,* 1–7. 10.1109/ICB.2013.6612968

LeCun, Y., Botton, L., Bengio, Y., & Haffner, P. (1998). Gradient-based learning applied to document recognition. *Proceedings of the IEEE,* 1–46.

Li, J., Wang, Y., Tan, T., & Jain, A. K. (2004). Live face detection based on the analysis of Fourier spectra. *SPIE, Biometric Technology for Human Identification.* China, vol. 5404, 296–303. DOI: 10.1117/12.541955.

Li, L., Feng, X., Xia, Z., Jiang, X., & Hadid, A. (2018). Face spoofing detection with local binary pattern network. *Journal of Visual Communication and Image Representation, 54*(December 2017), 182–192.

Liu, W. (2014). Face iveness detection using analysis of Fourier spectra based on hair. *Proceedings of the 2014 International Conference on Wavelet Analysis and Pattern Recognition,* 13–16. 10.1109/ICWAPR.2014.6961294

Maatta, J. M. A. H., & Pietika¨inen, M. (2012). Face spoofing detection from single images using texture and local shape analysis. *IET Biometrics,* 3–10. https://doi.org/10.1049/iet-bmt.2011.0009

Maatta, Jukka, Hadid, A., & Pietikainen, M. (2011). Face spoofing detection from single images using micro-texture analysis. *2011 International Joint Conference on Biometrics(IJCB),* 1–7. https://doi.org/10.1109/IJCB.2011.6117510

Mäenpää, T., Ojala, T., Pietikäinen, M., & Soriano, M. (2000). Robust texture classification by subsets of local binary patterns. *Proceedings of International Conference on Pattern Recognition, 15*(3), 935–938. https://doi.org/10.1109/icpr.2000.903698

Menotti, D., Chiachia, G., Pinto, A., Schwartz, W. R., Pedrini, H., Falc˜ao, A. X., & Rocha, A. (2015). Deep representations for Iris, face, and fingerprint. *IEEE Transactions on Information Forensics and Security, 10*(4), 1–16. https://doi.org/10.1109/TIFS.2015.2398817

Ojala, T., Pietikäinen, M., & Mäenpää, T. (2000). Multiresolution Gray scale and rotation invariant texture classification with local binary patterns. *Lecture Notes in Computer Science (Including Subseries Lecture Notes in Artificial Intelligence and Lecture Notes in Bioinformatics), 1842,* 404–420. https://doi.org/10.1007/3-540-45054-8_27

Ojansivu, V., & Heikkil, J. (2008). Blur Insensitive Texture Classification Using Local Phase Quantization. In Elmoataz, A., Lezoray, O., Nouboud, F., Mammass, D. (Eds.), Image and Signal Processing. ICISP 2008. Lecture Notes in Computer Science, vol. 5099. Springer, Berlin, Heidelberg, 236–243. https://doi.org/10.1007/978-3-540-69905-7_27

Peng, J., & Chan, P. P. K. (2014). Face liveness detection for combating the spoofing attack in face recognition. *Proceedings Ofthe 2014 International Conference on Wavelet Analysis and Pattern Recognition, Lanzhou*, 13–16.

Pereira, T., Anjos, A., Martino, J. M., & Marcel, S. (2012). LBP: TOP based countermeasure against face spoofing attacks. *Asian Conference on Computer Vision, 2012*, vol 7728. Springer, Berlin, Heidelberg. https://doi.org/10.1007/978-3-642-37410-4_11

Pereira, T. D. F., Komulainen, J., Anjos, A., Martino, J. M. De, Hadid, A., Pietikäinen, M., & Marcel, S. (2014). Face liveness detection using dynamic texture. *EURASIP Journal on Image AndVideo Processing*, 2, 1–15.

Raghavendra, R., & Busch, C. (2014). Robust 2D / 3D face mask presentation attack detection scheme by exploring multiple features and comparison score level fusion. *17th International Conference on Informatics Fusion*, (July 2014), 1–7. Salamanca.

Raghavendra, R., Raja, K. B., & Busch, C. (2015). Presentation attack detection for face recognition using light field camera. *IEEE Transactions on Image Processing*, *7149*(c), 1–16. https://doi.org/10.1109/TIP.2015.2395951

Ramachandra, R., Busch, C., & Biometric, N. (2017). Presentation attack detection methods for face recognition systems : A comprehensive survey. *ACM Computing Surveys*, *50*(1), 1–37.

Selwal, A., & Gupta, S. K. (2017). Low overhead octet indexed template security scheme for multi-modal biometric system. *Journal of Intelligent & Fuzzy Systems*, *32*(5), 3325–3337. DOI: 10.3233/JIFS-169274.

Tan, X., Li, Y., Liu, J., & Jiang, L. (2010). Face liveness detection from a single image with sparse low rank bilinear discriminative model. *Lecture Notes in Computer Science (Including Subseries Lecture Notes in Artificial Intelligence and Lecture Notes in Bioinformatics)*, *6316 LNCS*(PART 6), 504–517. https://doi.org/10.1007/978-3-642-15567-3_37

Teja, M. H. (2011). Real-time live fce detection using face template matching and DCT energy analysis. *2011 International Conference of Soft Computing and Pattern Recognition*, 342–346. Dalian.

Wang, Y., Nian, F., Li, T., Meng, Z., & Wang, K. (2017a). Robust face anti-spoofing with depth information. *Journal of Visual Communication and Image Representation*, *49*, 332–337. https://doi.org/10.1016/j.jvcir.2017.09.002

Wang, Y., Nian, F., Li, T., Meng, Z., & Wang, K. (2017b). Robust face anti-spoofing with depth information. *Journal of Visual Communication and Image Representation*, *49*, 332–337. https://doi.org/10.1016/j.jvcir.2017.09.002

Waris, M., Zhang, H., Ahmad, I., Kiranyaz, S., & Gabbouj, M. (2013). Analysis of textural features for face biometric anti-spoofing. *2013 Proceedings of the 21st European Signal Processing Conference (EUSIPCO'13)*, (February 2018). https://ieeexplore.ieee.org/document/681161

Wen, D., Han, H., & Jain, A. K. (2015). Face spoof detection with image distortion analysis. *IEEE Transactions on Information Forensics and Security*, *XX*(X), 1–16.

Wheeler, F., & Liu, X. (2014). *Handbook of Face Recognition*. Springer- Verlag

Xia, Z., Yuan, C., Lv, R., Sun, X., Xiong, N. N., & Shi, Y. Q. (2020). A novel weber local binary descriptor for fingerprint liveness detection. *IEEE Transactions on Systems, Man, and Cybernetics: Systems*, *50*(4), 1526–1536. https://doi.org/10.1109/TSMC.2018.2874281

Yan, J., Zhang, Z., Dong, Z. L., Stan, D. Y., & Li, Z. (2012). Face Liveness Detection by Exploring Multiple Scenic Clues. In International Conference on Control, Automation, Robotics and Vision, 188–193.

Yang, J., Lei, Z., Liao, S., & Li, S. Z. (2013). Face liveness detection with component dependent descriptor. *Internation Conference on Biometrics (ICB)*, 1–6.

Zhang, L. B., Peng, F., Qin, L., & Long, M. (2018). Face spoofing detection based on color texture Markov feature and support vector machine recursive feature elimination. *Journal of Visual Communication and Image Representation, 51*, 56–69. https://doi.org/10.1016/j.jvcir.2018.01.001

Zhang, Z., Yan, J., Liu, S., Lei, Z., Yi, D., & Li, S. Z. (2012). A face antispoofing database with diverse attacks. *5th IAPR International Conference on Biometrics (ICB)*, 26–31. New Delhi: IEEE.

Zhang, Z., Yi, D., Lei, Z., & Li, S. Z. (2011). Face liveness detection by learning multi-spectral reflectance distributions. *2011 IEEE International Conference on Automatic Face and Gesture Recognition and Workshops, FG 2011*, 436–441. https://doi.org/10.1109/FG.2011.5771438

11 Reconfigurable Binary Neural Networks Hardware Accelerator for Accurate Data Analysis in Intelligent Systems

A. Kamaraj and J. Senthil Kumar
Department of Electronics and Communication
EngineeringMepco Schlenk Engineering College, Sivakasi,
Tamil Nadu, India

CONTENTS

11.1 INTRODUCTION

Acceleration is a term used to describe tasks being offloaded to devices and hardware, which are specialized to handle them effectively. They are suitable for any repetitive intensive key algorithm and also can vary from a small functional unit to a larger functional block like motion estimation and video processing. It helps to

DOI: 10.1201/9781003132080-11

perform crucial functions more efficiently than possible in running on a general purpose computer. It offloads certain processes onto the hardware that can be best equipped for boosting the performance of a system. Hardware acceleration utilizes the power of graphical or sound processing units of a computing system to increase performance for certain applications. In most computers, by default, the CPUs may not be powerful to handle complex tasks, and this is where hardware acceleration comes into play. Sound cards in computers are meant for processing and recording of high quality sounds. Graphics cards can similarly be utilized by hardware acceleration to allow quicker higher quality playback of videos and mostly also used for gaming applications. They can also perform complex mathematical computations when compared to CPUs. Generally, processes or sequential instructions are executed one by one and are designed to run general-purpose algorithms controlled by fetching mechanisms. Deploying file hardware acceleration in these kinds of tasks improves the execution of a specific algorithm by allowing greater concurrency, by providing specific data paths for its data, and possibly reducing the overhead of instruction control. Modern processors are multi-core and they often feature parallel computation units.

When we have a powerful, stable GPU, enabling hardware acceleration will allow the users to utilize them to their full extent in all supported applications. Even this can enhance the browsing experience and media consumption tasks such as video editing or rendering programs for the end users. The Intel Quick Sync framework uses, in addition to their modern CPUs designed for fast video rendering and encoding applications. It is also widely used for spam control in server industry. Hardware accelerators are mostly referred to as 3D accelerators, cryptographic accelerators depending on the applications they are being employed.

Artificial intelligence is spreading its wings towards the enhancement of this computation platform to a larger extent. Artificial intelligence–based hardware accelerators support collective streaming, which is used to distribute data to processing elements. Partial results are collected by collective elements to generate the desired output. These categories of accelerators are good in transforming the tensor computation into matrix operations, so that they can quite flexibly accelerate deep learning directly. Peak performance of such computations can be achieved in terms of multiple Gigaflops per second when running the neural networks with the support of FPGA devices. Better efficiency in terms of hardware utilization and energy management are effectively supported with the aid of neural networks utilizing the flexibility of FPGAs for hardware acceleration. Using artificial intelligence, domain-specific architectures could be created and they could be fine-tuned for a class of networks. This enables incorporating custom data flow, custom memory hierarchy, and custom precision. All of these could be customized to improve the latency and reduce the power consumption, cost factors, and improve the efficiency. Fixed function chips are being replaced with FPGAs for configuring them through programming, thereby supporting reconfigurability of the computing elements.

Deep convolutional neural networks (CNNs) are most widely applied in machine learning–based applications such as image classification, speech processing, Internet of Things (IoT), and robotics. The cluster of CPUs and graphics processing units (GPUs) provides Giga or Tera FLOPs per Second (GFLOP or TFLOP) on single-instruction-multiple-data (SIMD) with high clock frequency. One major challenge in deployment of CNN is their larger demands in computation complexity, memory, and resource. For example, AlexNet has a 60M parameter, which needs 240 MB storage of 32-bit data. Also, the VCG-16 network contains 15 GFLOPS of MAC operations, while GoogleNet has 1.43 GFLOPS of MAC operations (Sze, Chen, Yang, & Emer., 2017). In order to improve the hardware utilization, many compressing models are available; some of them are lacking in network connections and limited data-width (Han, Mao, & Dally, 2016; Iandola et al., 2016; Qiu et al., 2016). One of the efficient methods to compress the neural network model is the Binarization Neural Network (BNN) model, in which the 32-bit (Single Precision Floating Point) data into a 1-bit data is either +1 or -1. Also, the multiplication operation is replaced with bitwise operations, which further reduces the computational demand. It has been demonstrated in (Courbariaux, Hubara, Soudry, El-Yaniv, & Bengio, 2016a) on the MNIST, CIFAR, and SVHN networks (Netzer et al., 2011) [6]. Also, recently, ImageNet models are trained in BNN-based XOR-Net with high speed and compression (Rastegari, Ordonez, Redmon, & Farhadi, 2016). Binarization reduces the memory consumption, memory bandwidth, and computation resources.

Hardware accelerations are also widely used for the Internet of Things (IoT) and machine learning applications. Moreover, for processing of unstructured data such as images or acoustic data, deep neural networks are playing a significant role. Running billions of computations of deep neural networks on edge devices for performing IoT based data analytics, we need acceleration of the hardware resources. Certain algorithms need to be accelerated in the cloud as well by utilizing the spectrum of hardware resources available. On the flexibility of implementation side, we can use the conventional CPUs. For the next stage of computing enhancements, GPUs play a significant role for parallel processing on the data. More efficient deployment of acceleration can also be found using Application Specific Integrated Circuits (ASIC). In addition to that, usage of FPGAs enables reconfigurable architectures using the logic blocks and reconfigure their routing characteristics on a chip (Han et al., 2016). Beyond the hardware acceleration tasks, DNN are using streams of data for performing streaming analytics, to support applications that require time-senstive actions.

FPGAs consume much lesser power compared to GPUs. Currently, FPGAs are being adopted in IoT applications, Cloud servers, and leading data centers (Intel Xeon, Microsoft Catapult) (Nurvitadhi et al., 2016). Xilinx-Vivado HLS (Zamacola, Martínez, Mora, Otero, & de La Torre, 2018) and LegUp introduce system-level programming to the designers. Although FPGAs have excellent potential to drive hardware acceleration, they are highlighly challenging to program at the hardware level. Nowadays, diversified categories of frameworks are available to allow the developers to program FPGAs with ease. Further, automation in VLSI tools opened the door for hardware-software co-design and on-chip memory usage.

But, the FPGA-based neural network computation has two more challenges (Guo et al., 2018a):

- The present-day FPGAs are 100 to 300 MHz, which is much slower than GPUs and CPUs.
- Neural Network Implementation on FPGA is much harder than CPU and GPU implementation.

There are few cases where the end users also opt for disabling the hardware acceleration features, where they may not support smooth acceleration. For stable and highly commendable CPU-based applications, the support of acceleration may not be in demand. This may consume additional resources in terms of power, frequency of operation, and additionally the components are prone to overheating or may get damaged when they are continuously operating with high accelerations. If the software utilization for the hardware is not performing well or not stable and can only use CPU, this can also be a reason for disabling hardware acceleration. It is also recommended to disable them, when the end users have poor stability issues and bugs in their system. However, the overwhelming benefits of hardware acceleration in modern computing platforms makes it a strong contender for its deployment in a diversified range of applications.

The remaining sections are organized as follows:

Section 11.2: Presents the research works performed in the BNN and discusses the improvements and performance factors.
Section 11.3: Explains the fundamental details and terminologies of the BNN.
Section 11.4: Illustrates the research considerations details.
Section 11.5: Expresses the proposed BNN and elaborates the internal architectural components with its operation.
Section 11.6: The simulation and implementation results obtained using the synthesis tools are elaborated in comparison with the existing results.
Finally, the proposed research work is summarized and concluded in the conclusion section.

11.2 RELATED WORKS

In this section, we compile the recent works on BNN, carried out by researchers working in this field. In recent days, adopting the FPGA in artificial intelligence (AI) is growing in significant manner. A decade back, CNN was popularly implemented in the FPGAs. But, the problem with CNN is that it needs a lot of computation, which limits the throughput and it is hungry for power. Hence, an alternative methodology to improve the throughput, reduce the computation hardware, and energy is Binarization Neural Network (BNN), which is slowly emerging and moving hand in hand with FPGAs.

Most commonly, the CNN computations are carried out on GPUs and dedicated ASICs. But later, the emergence and the incremental capacity of the FPGAs make the implementation of the CNNs over them. In Nurvitadhi et al. (2016), it has been proved that the FPGAs are performing better over GPUs and CPUs in view of throughput, energy, and customized hardware for the BNN computation. Also, the FPGAs are superior in throughput and power dissipation for trained weights and activation functions (Li et al., 2018). Moreover, FPGA-based accelerators are insensitive to data batch size for SIMD case; that is irrespective of the size of the batch of data, the speed of the FPGAs are nearly constant (Li et al., 2018), which surpasses the capacity of the GPUs. Significantly, the FPGAs have on-chip BRAMs, which speeds up BNN acceleration for AI applications (Li, Zhang, Wang, & Lai, 2020). These research identifications paved a pathway to BNN implementation in FPGA for greater performances.

The major improvements of BNNs are as follows (Liang et al., 2018):

1. Single-bit multiply and accumulate operation; it is performed by XNOR and pop count tasks.
2. Parameter size and intermediate results are miniaturized, which reduces the memory requirement.
3. Wider bandwidth in on-chip memory reduces the data dependency bottleneck.
4. The accuracy of the BNN can be improved by 1-bit padding (Fraser et al., 2017).

Table 11.1 summarizes the performance of the FPGA-based BNN research works. The BNN performances are most widely analyzed using CIFAR-10 alone (Fraser et al., 2017; Li et al., 2018; Yang, 2018); in some research works, MNIST and CIFAR-10 and SVHN databases are considered (Hubara et al., 2016; Umuroglu et al., 2017), and rarely, ConvNet and AlexNet (Liang et al., 2018) is used for the BNN analysis. The BNN has a wide usage in image-processing applications such as face detection, image classification (Fraser et al., 2017; Li et al., 2018; Umuroglu et al., 2017), and edge processing (Murovic & Trost, 2019). Certain researchers are focusing on reducing multipliers (Liang et al., 2018), reducing memory size and accesses (Hubara et al., 2016), parallel-convolution (Murovic & Trost, 2019; Yang, 2018), and scaling BNN (Fraser et al., 2017) towards improving the performance of the BNN architecture.

The performance of the BNN in FPGA largely depends on the type of layers actively incorporated and the type of FPGA adopted. From the analysis, it is observed that the accuracy of the result is dependent on the number of computation layers in the BNN hardware architecture; the performance of the BNN such as speed of operation, TOPS (Tera Operations Per Second), number of frames, and power dissipation are directly proportional to the FPGA resources such as on-chip memory, clock frequency, number of LUTs, etc. Hence, all the research works are being carried on with the latest, higher-end FPGAs such as PYNQ (Yang, 2018), Kintex (Fraser et al., 2017), and Virtex 7 (Li et al., 2018). Almost all FPGA works are designed using Vivado HLS (Murovic & Trost, 2019); few studies used Python (Yang, 2018) as an additional tool for this purpose.

The growing interest in the DNN and hardware acceleration need high-speed computation platforms such as GPUs, FPGAs to clearly carry out their full-fledged implementation, and overcoming the potential challenges in hardware acceleration. So, the previous discussions inspired us to move on with the BNN application to be run on the FPGA with padding. Hence, in the proposed method we have chosen

TABLE 11.1
Performance Analysis of Existing BNN Works

Ref. No.	Contribution	Network	Number of Layers	Application	Target Device	Performance
Farabet, Poulet, Han, & LeCun, 2009	Low power systems	CNN	3 Conv. + 5 FC + 2 Pool	Face Detection	Virtex-4SX35	10 frames per second (512 ×384)
Liang, Yin, Liu, Luk, & Wei, 2018	Resource-Aware Model Analysis (RAMA)	Cifar-10 ConvNet, and AlexNet	6 Conv + 3 FC	Reducing Multipliers	Stratix-V	TOPs speed with only 26 Watt at 150 MHz frequency
Yang, 2018	Parallel-Convolution BNN	CIFAR-10	5 Conv. + 2 Pool + 1 FC.	Image Classification	Xilinx PYNQ Z1	930 frames per second and 387.5 FPS/Watt, 143 MHz frequency 4.9 Mb on-chip Block RAM
Umuroglu et al., 2017	BNN roofline model	MNIST, CIFAR-10, SVHN	K*K Pool	Image Classification	ZC706	25 W Power, 12.3 million image classifications per second
Murovic & Trost, 2019	Parallel BNNs	MNIST	Layers are increased for accuracy	Edge Processing	Zynq7000 Zedboard	2 W power, 60 k FPGA slices, 30 ns inference delays
Li, Liu, Xu, Yu, & Ren, 2018	GPU Vs FPGAs	CIFAR-10	6 Conv. + 2 Pool	Image Processing	Virtex-7 XC7VX690	90 MHz frequency, 7663 GOPS, 8.2W
Fraser et al., 2017	Scaling of BNNs	CIFAR-10	Same as FINN	Image Classification	Kintex - XCKU115-2-FLVA1517E	14.8 TOPS, 671 µs latency, 41 W, 88.7% accuracy
Hubara, Courbariaux, Soudry, El-Yaniv, & Bengio, 2016	BNNs – reduce memory size and accesses	MNIST, CIFAR-10 and SVHN	Torch7 and Theano	-	-	7 times faster than GPU Kernel

image classification using the CIFAR-10 model in Zynq 7000 SoC FPGA with the help of Python code.

11.3   BINARY NEURAL NETWORK FUNDAMENTALS

In this section, we deal with the primaries of BNN in relation with CNN and the main integrated components of its architecture. Today neural networks are doing amazing things, starting from real-time object detection, image recognition, and natural language processing. But in this task, we need high-computation power-intensive devices like GPU, TPU, because all these tasks require millions of addition andmultiplication operations on floating-point numbers. If we have limited resources and need to carry out deep learning operations on hand-held devices, or run machine lerning models in a minimum specification computation devices, in those places BNN could actually come to the rescue. The CNNs are in need of huge resources for computation, so the BNNs are becoming popular in machine learning. In a BNN all the weights and bases are either +1 or -1. So, we can store them in 1 bit, which reduces the storage cost on the devices and binary operations like multiplication and addition can be done by using bit shift operations, which further reduces the computation cost on devices, which makes the BNN advantageous over using DNNs. The main computational bottleneck is the multiplication operation and accumulation operation, which happens many times in the forward pass and backward pass. By binarizing the weights and activation to +1 or -1, the expensive multiplication and accumulation (MAC) operations could be converted to cheap logical operations. The binary values in BNNs can help to execute computations using bitwise operations, so instead of multiplication, using the activation of the pervious layers and all the weights this operation can be carried out. The main idea is to treat binarization to also handle noises and harness the network tolerance to noise, to relax the hardware demands. By binarizing the weights and establishing binary connects between successive layers, we replace the non-linearity function with binarized non-linearity. By incorporating the back propagation, we could differentiate through the binarization function. Some of the commonly used terminologies are as follows:

- Weights: In BNN the real, learned values are used in the multiplication with binary activations, which is derived from preceding layers.
- Activations: The result of the current layer activation function is forwarded to the next layer for dot product. This dot product is performed using multiply accumulate unit.
- Parameters: Weights, bias, and gain are some of the learned values from the back propagation.
- Bias: A scalar value is learned during the computation.
- Gain: Either learned value or statically computed value for the computing process.
- Topology: The arrangement of various layers of the BNN model.
- Fully Connected Layer: For better learning process, all the neurons are interconnected to each other.

11.3.1 BNN REPRESENTATION

The binary weights of the BNN are trained using the back-propagation with the gradient descent based method (Courbariaux, Bengio, & David, 2015). This binarized training leads to loss instead of binarzing the network after the training process. The computation of gradient loss with respect to binary weights through back propagation solves the purpose a little better, even though weight update using the gradient descent methods is a more complex process in the binary weights. Gradient decent methods do minor changes in the weights. This modification could not be performed on binary values.

So, the binarization has been done by keeping the real values (W_r) within the network and can be updated using back propagation and gradient decent. During the interference, real values are not needed, instead binary values (W_b) are stored and utilized during the process. The binarization has been done using the sign function, as shown in equation (11.1):

$$W_b = \text{sign}(W_r) \tag{11.1}$$

The sign function received the real values of the weights in the forward direction of computation and the gradient of the binary weights is traversing in the backward direction, as illustrated in Figure 11.1. Here, in the forward direction, the STE (Straight Through Estimator) is used. The binarization of the activation function is presented in the (Courbariaux et al., 2016b). This process is carried out as similar to the weights binarization shown in Figure 11.1, where the activation function is forwarded using STE through the sign function.

The bitwise operations can be incorporated for the dot product between the weights and activation functions. The binary values are +1 or −1. These values can be encoded as 1 and 0, respectively. The encoded binary values based multiplication has been performed using the EXNOR logical operations, as shown Table 11.2.

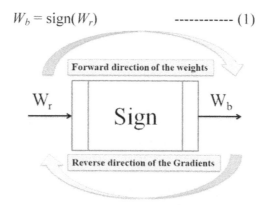

FIGURE 11.1   Representation of sign layer for weights and gradients.

TABLE 11.2
XNOR Operation for the Multiplication Operation

Encoded		Binary Value		XNOR Operation	Multiplication Operation
A	B	A	B		
0	0	−1	−1	1	+1
0	1	−1	+1	0	−1
1	0	+1	−1	0	−1
1	1	+1	+1	1	+1

The multiplication of the dot product in the binarized domain simplified the computation compared to the floating and fixed-point multiplication. In turn, the computation becomes faster and requires less hardware resources. However, the BNN is suffering in accuracy (Lin et al., 2017) and attacks (Szegedy et al., 2014).

There are various improvements that can be done on the BNN network for scaling with a gain term, usage of multiple bases, partial binarization, learning rate convergence, padding with zeros, and sequencing the layer order. These were incorporated in order to increase the efficiency and accuracy of the BNN architecture (Simons & Lee, 2019).

11.4   RESEARCH CONSIDERATIONS

11.4.1   DATA SETS

There are various types of data sets available for BNN analysis like CIFAR-10, MNIST, ImageNet and SVHN. Many number of classes and images are available in the ImageNet compared to other data sets. It originated in 2012 and consists of 1.2×10^6 million images considered for training and 150×10^3 images are taken into account for testing purposes. Moreover, it has more than 1,000 different classes (Simons & Lee, 2019).

11.4.2   TOPOLOGIES

There are various types of topologies currently available for the BNN computation, such as AlexNet or ResNet, DoReFa-Net, and VGG-11. These topologies have different layers and their number of channels is also varying in the topology of the BNN. They are Max Pooling Layer (MP), Fully Connected Layer (FC), and Convolutional Layer (C). For example, CIFAR-10 uses 2C128-MP-2C256-MP-2C512-MP-2FC1024-FC10 topology for its BNN computation, where MP is Max Pooling layer, C represents the Convolutional layer, and FC is Fully Connected layer; also the numbers present in the left side of the layer representation indicate the quantity of layers and quantity of channels are in the order of 128, 256, 512, etc... (Simons & Lee, 2019).

11.4.3 ACCURACY

The accuracy is a significant measure in the BNN network computation. The network architecture or topology, type of data sets, and number of layers have a direct impact on the accuracy of the BNN. The maximum achievable accuracy for the MNIST is 98.77% for the FC784-3FC512-FC10 topology (Chi & Jiang, 2018), SVHN attains 97.47% accuracy for 1/2 BNN (Courbariaux et al., 2016b) and CIAFR-10 achieves 89.85% without padding (Courbariaux et al., 2016b); whereas ImageNet could able to attain 75.6% for the AlexNet topology (Tang, Hua, & Wang, 2017). Hence, the results confirm that the accuracy has deep dependence on topology, number of layers, and data sets.

11.4.4 HARDWARE IMPLEMENTATION

Nowadays, the BNN architectures are implemented on FPGA devices. These FPGA devices have many features like in-built DSP processors and multipliers, block memory (BRAM), semi-customized programmable layers and data paths, and ways and means for optimization in frequency, power, and area. FPGA-based BNN falls under either streaming architectures or layer accelerators. The comparisons of these are shown in Table 11.3.

The effectiveness of the BNN lies on the type of FPGA hardware platform also. The usage of DSPs is playing a significant role in BNN. The FPGAs are chosen based on the LUTs, BRAMs, and clock frequency. The presently available FPGA-based BNN architectures are Zynq and Vertex devices for implementation process. The topology and data set (MNIST, SVHN, CIFAR-10, and ImageNet) perform a key portion in the Power, LUTs, BRAMs, and accuracy of the BNN [24]. From the types of potential applications where BNNs can be used, several of them could

TABLE 11.3
Assessment of Streaming Architectures or Layer Accelerators

S. No.	Parameters	Streaming Architectures	Layer Accelerators
1.	Hardware architecture	Has dedicated hardware for almost all the layers.	Handles only specific types of layers
2.	Pipeline	Possible. Hence, individual layers can process different samples at any instant of time.	Not possible. Every time results are stored in memory.
3.	Throughput	High due to pipeline processing	Low
4.	Hardware resource	High, since all the layers are implemented as a hardware	Minimum.
5.	Application	Suitable for high throughput applications (video processing, speech processing, image processing)	Suitable for constrained resource designs.

benefit from the structure of the BNN: hand-held smart devices, traffic signals, traffic congestion prediction, and generally any applications that benefit from extracting the vital set of features from the existing ones through the BNNs. The following section incorporates the key findings from the literature for implementation of hardware accelerators using FPGAs driven through the explored BNN architecture. The key aspects are highlighted for the image classification application to explore the credibility of the proposed architecture.

11.5 PROPOSED BNN ARCHITECTURE

In this segment, the logical hardware architecture of FPGA-based accelerator systems will be elaborated on, along with internal functional units. In the proposed architecture, it has five convolutional layers with three FC layers, as represented in Figure 11.2. The first layer of the proposed BBN network is provided with a CIFAR-10 RGB with dimensions of $32 \times 32 \times 3$ of each pixel represented with 8-bit resolution. Since the layers are binarized, the multiplication process is carried out on addition and subtraction of the binary values. The convolution equation (11.1) is for the binarized network, where W_b is the binary weight.

In order to retain the weight × height of the output of the volume as 32×32, zero padding has to be applied to the RGB image with an order of one. The first layer architecture of the BNN consists of adder and multiplier blocks, as shown in Figure 11.3. Additionally, it has line buffers to read the image sequentially. The input image has $32 \times 32 \times 3$ size from the CIFAR-10 data set. Each pixel of the image has to be read nine times while the window scans the entire image. The line

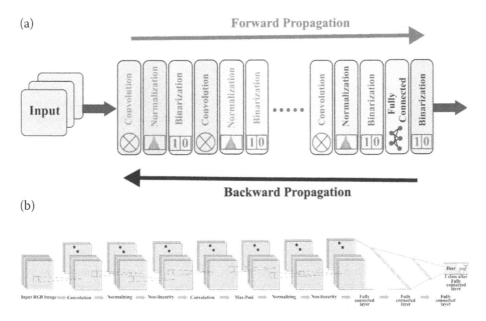

FIGURE 11.2 (a) Block level representation of the BNN layers; (b) architecture of binary neural network.

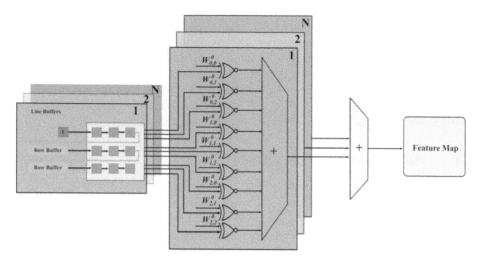

FIGURE 11.3 First convolution layer.

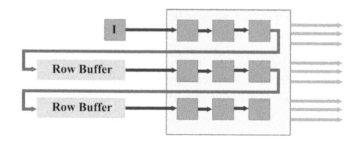

FIGURE 11.4 Line buffer for scanning the 3 × 3 pixels.

buffer is able to read three pixels of the image; for that purpose it has been constructed as 3 × 3 × 3 for each color (RGB). Each buffer delays the input in one row distance. An N-stage shift register is incorporated, in order to implement this delay, in which the N represents the width of the image. The corresponding schematic architecture of the line-buffers are shown in Figure 11.4, which is constructed based on equation (11.2).

$$Y[n, i, j] = \sum_{d=0}^{D-1} \sum_{y=0}^{K-1} \sum_{x=0}^{K-1} \sum_{x=0}^{K-1} W_b[n, d, 2 - X, 2 - y] \times X[d, i + x, j + y]$$

$$(11.2)$$

From Figure 11.3, it has been shown that the N feature maps are computed using N kernels. Each kernel has 3 × 3 × 3, which is convolved with an image size of 128 × 128 × 3. The number of kernels and number of feature maps are made equal. Here, the weights are represented as binary values; hence the MAC (Multiply and Accumulate) operations of the convolution have been performed with chains of additions (for $W_b = +1$) and subtractions (for $W_b = -1$), as shown in Figure 11.5. Also, all the remaining layers of the receiving binary values from the previous layers and their weights are also in terms of binary. Hence, the architecture shown in Figure 11.3 can be updated correspondingly, as in Figure 11.6.

Figure 11.6 consists of N layers of kernels applied to the convolutional layer of the BNN. The binary multiplication process has been converted into a XNOR operation. In the normal CNN, the multiplication consumes almost 90% of the resources, but in the BNN the multiplication operation has been replaced with the XNOR operation. Hence, the amount of hardware resources are greatly reduced in BNN compared to CNN.

The batch normalization is represented as four numbers of fixed point parameters termed μ, α (alpha), γ (gama), and β for every activation layer (Bengio, Léonard, & Courville, 2013). The corresponding output data equation is shown in equation (11.3).

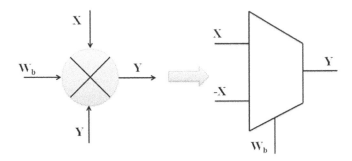

FIGURE 11.5 The initial convolution layer, multiply accumulates realized by addition and subtraction.

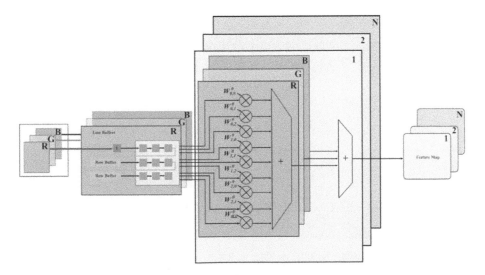

FIGURE 11.6 Second, third, and fourth convolution layer implementation.

$$Y = \frac{X - \mu}{\sqrt{\sigma^2 + \in}}\gamma + \beta \qquad (11.3)$$

In equation (11.2), the x represents the input data, μ is said to be mean, σ^2 is variance, and y is output data. In the presented binary computation, these four parameters are scaled to two parameters based on applying equation (11.4). In this process, shifting and scaling process are carried out.

$$Y = Q_b(X + p) \qquad (11.4)$$

Where $p = \left(\frac{\beta\sigma}{\gamma} - \mu\right)$; $Q = \frac{\gamma}{\gamma}$ and $Q_b = sign(Q)$.

11.6 RESULTS AND DISCUSSIONS

In this work, we have considered performing classification of deer from the CIFAR-10 data set. The same data set is applied towards training and testing phases of the developed BNN. The size of the color images in the CIFAR-10 is 32 × 32 is shown in Figure 11.7. In training, about 5,000 images are used for every class, so that total of 50×10^3 images is there for the 10 classes. In testing, one-fifth of the trained images are considered.

In this BNN, the weights and activation function are binarized. Hence, the amount of hardware resources required for the multiplication process is greatly

FIGURE 11.7 CIFAR-10 data set.

reduced compared to the conventional CNN computation. Also, in BNN the computation speed is also high. The reason is that the multiplication operation is now replaced with a simple XNOR operation. In the proposed architecture, we are implementing five convolutional layers with three fully connected layers. The consolidated list of parameters and total number of parameters are shown in Table 11.4 for this network architecture.

We have used the Xilinx Vivado 2016.4 synthesis tool for hardware realization with a target device as Zynq 7000 (XC7Z020-CLG484) SoC. Table 11.4 shows the BNN implemented with different numbers of layers and the corresponding resource occupation without using the BRAM in the FPGA board. Table 11.5 clearly specifies that growing the amount of layers in the BNN increased the resource occupation.

Then the desired proposed architecture shown in Figure 11.2(b) is implemented with BRAM, and its related other performance parameters such as speed, accuracy, and throughput are estimated and listed in Table 11.6.

Table 11.7 shows the FPGA implementation of BNN research works comparison for the data set CIFAR-10. Here, we have considered only the researches are

TABLE 11.4

Required Parameters for Various Layers of the Proposed Network

Network Layer	Weights Required for the Layers
CL (Convolutional Layer)	3,456 (3 × 3 × 3 × 128)
	1,47,456 (3 × 3 × 3 × 128 × 128)
Pooling Layer	0
CL (Convolutional Layer)	2,94,912 (3 × 3 × 3 × 128 × 256)
	5,89,824 (3 × 3 × 3 × 256 × 256)
Pooling Layer	0
CL (Convolutional Layer)	11,79,648 (3 × 3 × 3 × 256 × 512)
Pooling Layer	0
FL (Fully Connected Layer)	83,88,608 (4 × 4 × 512 × 1,024)
	10,48,576 (1,024 × 1,024)
	10,240 (1,024 × 10)

TABLE 11.5

Resource Utilization for Various Number of Layers (Without BRAM)

Number of Convolution Layers	Number of Fully Connected Layers	Slice Registers	Slice LUTs
1	1	24,188	3,562
2	1	1,78,282	5,07,815
3	2	2,12,935	6,48,296
4	2	9,44,823	8,68,826
5	3	4,88,244	11,12,218

TABLE 11.6
Performance of BNN (Five Convolutional and Three Fully Connected Layers)

S. No.	Parameter	Value
1.	Maximum frequency of operation	360.32 MHz
2.	Accuracy	87.82%
3.	Precision	1–2 Bits
4.	Number of clock cycles	1,670
5.	Total Time	4.9 μs
6.	Throughput	3,42,156 images per second with 32 × 32 resolution

TABLE 11.7
Comparison of FPGA Implementations using CIFAR-10

Ref.	Topology	FPGA	LUTs	BRAMs	Clk (MHz)	Power (W)	Accuracy (%)
Zhou, Redkar, S., & Huang, 2017	2 Conv. 2 F	Zynq7 045	20,264	–	–	–	66.63
Blott et al., 2018	1/2 BNN	Zynq-Ultra 3EG	41,733	283	300	10.7	80.10
	1/2 BNN	Zynq 7020	25,700	242	100	2.25	80.10
Umuroglu et al., 2017	1/2 BNN	Zynq 7045	46,253	186	–	11.7	80.10
Nakahara, Fujii, & Sato, 2017	1/2 BNN	Zynq 7020	14,509	32	143	2.3	81.8
Yang et al., 2018	6 CL.	Zynq 7020	23,426	135	143	2.4	85.9
Ghasemzadeh, Samragh, & Koushanfar, 2018	C64-MP-2C128-MP-2C256-2F-C512-FC10	Zynq 7020	53,200	280	200	–	86.98
Zhao et al., 2017	BNN	Zynq 7020	46,900	140	143	4.7	87.73
Guo et al., 2018b	6 CL 3 FC	Zynq 7020	29,600	103	–	3.3	88.61
Li et al., 2020	5 Conv. 3 FC	Zynq7000 SoC	28,432	106	360.32	2.8	87.82

implemented using the various types of Zynq FPGA devices. The numbers of layers are varying across the research works. The precision of all the works are 1–2 bits, since the network is BNN. The implemented proposed BNN occupies 28,432 LUTs and 106 BRAMs in Zync 7000 Soc FPGA and it is able to operate in 360.32 MHz frequency.

11.7 CONCLUSION

The technological growth is having a direct impact on the revolution on the human being. Nowadays, System on Chip design of neural network in FPGA is gaining its popularity among researchers. In spite of numerous challenges that exist in the use of modern computational devices, use of hardware acceleration still yields better benefits. In the proposed research work, it has been realized with the binary neural network implementation in Zynq FPGA. Here, we have considered five convolutional layers and three fully connected layers for RGB images processing present in the data set CIFAR-10. The entire architecture computation has been binarized in order to reach the resource optimization. All the convolution multiplication operations are converted into simple addition and subtraction operations towards saving the FPGA resources. Upon doing so, we achieve the BNN with frequency of operation as 360.32 MHz, accuracy of 87.82% for deer identification, and its throughput of 3,42,156 images per second with 32 × 32 resolution; for that it occupies 28,432 LUTs, 106 BRAMs in Zync 7000 Soc FPGA. It is observed that the number of layers in the BNN has direct impact on the prediction accuracy, number of LUTs, and BRAMs in FPGA. Hence, SoC design of BNN in FPGA contributes significantly towards the development of intelligent systems in Society 5.0 applications.

As a future work, researchers in this stream could focus on compact FPGA-based accelerators, area and power-efficient BNN, and efficient partitioning of hardware and software for enhancing the acceleration. Runtime reconfigurable processing elements could be incorporated for BNN-based applications by modifying the binary weights on the fly. Real-time stereo estimation through FPGA-based acceleration is also gaining momentum from the research community with the conjugation of BNN. Moreover, for rapid prototyping and millimeter wave optical networks based communications, BNN could be employed as hardware accelerators. Intelligent and smart Society 5.0 applications could be rooted from the impact of BNN for low-power, low-computation-based hand-held and embedded devices used for the society revolution.

REFERENCES

Bengio, Y., Léonard, N., & Courville, A. (2013). Estimating or propagating gradients through stochastic neurons for conditional computation. ArXiv:1308.3432 [Cs].

Blott, M., Preußer, T. B., Fraser, N. J., Gambardella, G., O'brien, K., Umuroglu, Y., … Vissers, K. (2018). FINN-R: An end-to-end deep-learning framework for fast exploration of quantized neural networks. *ACM Transactions on Reconfigurable Technology and Systems (TRETS)*, *11*(3), 1–23.

CIFAR-10 and CIFAR-100 data sets. (2014.) Retrieved from online: https://www.cs.toronto.edu/~kriz/cifar.html

Chi, C. C., & Jiang, J. H. R. (2018, November). Logic synthesis of binarized neural networks for efficient circuit implementation. *2018 IEEE/ACM International Conference on Computer-Aided Design (ICCAD)*, 1–7. IEEE.

Courbariaux, M., Bengio, Y., & David, J. P. (2015). Binaryconnect: Training deep neural networks with binary weights during propagations. *Advances in Neural Information Processing Systems*, 28, 3123–3131.

Courbariaux, M., Hubara, I., Soudry, D., El-Yaniv, R., & Bengio, Y. (2016a). Binarized neural networks: Training deep neural networks with weights and activations constrained to +1 or −1. ArXiv:1602.02830 [Cs].

Courbariaux, M., Hubara, I., Soudry, D., El-Yaniv, R., & Bengio, Y. (2016b). Binarized neural networks: Training deep neural networks with weights and activations constrained to +1 or −1. ArXiv:1602.02830 [Cs].

Farabet, C., Poulet, C., Han, J. Y., & LeCun, Y. (2009, August). Cnp: An fpga-based processor for convolutional networks. *2009 International Conference on Field Programmable Logic and Applications*, 32–37. IEEE.

Fraser, N. J., Umuroglu, Y., Gambardella, G., Blott, M., Leong, P., Jahre, M., & Vissers, K. (2017, January). Scaling binarized neural networks on reconfigurable logic. *Proceedings of the 8th Workshop and 6th Workshop on Parallel Programming and Run-Time Management Techniques for Many-core Architectures and Design Tools and Architectures for Multicore Embedded Computing Platforms*, 25–30.

Ghasemzadeh, M., Samragh, M., & Koushanfar, F. (2018, April). ReBNet: Residual binarized neural network. *2018 IEEE 26th Annual International Symposium on Field-Programmable Custom Computing Machines (FCCM)*, 57–64. IEEE.

Guo, K., Zeng, S., Yu, J., Wang, Y., & Yang, H. (2018). A survey of FPGA-based neural network accelerator. ArXiv:1712.08934 [Cs].

Guo, P., Ma, H., Chen, R., Li, P., Xie, S., & Wang, D. (2018, August). Fbna: A fully binarized neural network accelerator. *2018 28th International Conference on Field Programmable Logic and Applications (FPL)*, 51–513. IEEE.

Han, S., Mao, H., & Dally, W. J. (2016). Deep compression: Compressing deep neural networks with pruning, trained quantization and Huffman coding. ArXiv:1510.00149 [Cs].

Hubara, I., Courbariaux, M., Soudry, D., El-Yaniv, R., & Bengio, Y. (2016). Binarized neural networks. *Advances in Neural Information Processing Systems*, 29, 4107–4115.

Iandola, F. N., Han, S., Moskewicz, M. W., Ashraf, K., Dally, W. J., & Keutzer, K. (2016). SqueezeNet: AlexNet-level accuracy with 50x fewer parameters and <0.5MB model size. ArXiv:1602.07360 [Cs].

Li, Y., Liu, Z., Xu, K., Yu, H., & Ren, F. (2018). A GPU-outperforming FPGA accelerator architecture for binary convolutional neural networks. *ACM Journal on Emerging Technologies in Computing Systems (JETC)*, 14(2), 1–16.

Li, Z., Zhang, Y., Wang, J., & Lai, J. (2020). A survey of FPGA design for AI era. *Journal of Semiconductors*, 41(2), 021402.

Liang, S., Yin, S., Liu, L., Luk, W., & Wei, S. (2018). FP-BNN: Binarized neural network on FPGA. *Neurocomputing*, 275, 1072–1086.

Lin, X., Zhao, C., & Pan, W. (2017). Towards accurate binary convolutional neural network. *Advances in Neural Information Processing Systems*, 345–353.

Murovic, T., & Trost, A. (2019). Massively parallel combinational binary neural networks for edge processing. Elektrotehniski Vestnik, 86(1/2), 47–53.

Nakahara, H., Fujii, T., & Sato, S. (2017, September). A fully connected layer elimination for a binarizec convolutional neural network on an FPGA. *2017 27th International Conference on Field Programmable Logic and Applications (FPL)*, 1–4. IEEE.

Netzer, Y., Wang, T., Coates, A., Bissacco, A., Wu, B., & Ng, A. Y. (2011). Reading digits in natural images with unsupervised feature learning. NIPS Workshop on Deep Learning and Unsupervised Feature Learning. online: https://research.google/pubs/pub37648/

Nurvitadhi, E., Sheffield, D., Sim, J., Mishra, A., Venkatesh, G., & Marr, D. (2016, December). Accelerating binarized neural networks: Comparison of FPGA, CPU, GPU, and ASIC. *2016 International Conference on Field-Programmable Technology (FPT)*, 77–84. IEEE.

Qiu, J., Wang, J., Yao, S., Guo, K., Li, B., Zhou, E., … Yang, H. (2016, February). Going deeper with embedded fpga platform for convolutional neural network. *Proceedings of the 2016 ACM/SIGDA International Symposium on Field-Programmable Gate Arrays*, 26–35. Association of Computing Machinery, New York, USA.

Rastegari, M., Ordonez, V., Redmon, J., & Farhadi, A. (2016, October). Xnor-net: Imagenet classification using binary convolutional neural networks. *European Conference on Computer Vision*, 525–542. Springer, Cham, Swizerland.

Simons, T., & Lee, D. J. (2019). A review of binarized neural networks. *Electronics*, 8(6), 661.

Sze, V., Chen, Y.-H., Yang, T.-J., & Emer, J. (2017). Efficient Processing of deep neural networks: A tutorial and survey. ArXiv:1703.09039 [Cs].

Szegedy, C., Zaremba, W., Sutskever, I., Bruna, J., Erhan, D., Goodfellow, I., & Fergus, R. (2014). Intriguing properties of neural networks. ArXiv:1312.6199 [Cs].

Tang, W., Hua, G., & Wang, L. (2017, February). How to train a compact binary neural network with high accuracy? *Proceedings of the AAAI Conference on Artificial Intelligence* (Vol. 31, No. 1). Online: https://ojs.aaai.org/index.php/AAAI/article/view/10862

Umuroglu, Y., Fraser, N. J., Gambardella, G., Blott, M., Leong, P., Jahre, M., & Vissers, K. (2017, February). Finn: A framework for fast, scalable binarized neural network inference. *Proceedings of the 2017 ACM/SIGDA International Symposium on Field-Programmable Gate Arrays*, 65–74.

Yang, L. (2018). Exploring FPGA implementation for binarized neural network inference. Electronic Theses and Dissertations, 2004-2019. 6205. online: https://stars.library.ucf.edu/etd/6205

Yang, L., He, Z., & Fan, D. (2018, July). A fully onchip binarized convolutional neural network fpga impelmentation with accurate inference. *Proceedings of the International Symposium on Low Power Electronics and Design*, 1–6 online: https://dl.acm.org/doi/abs/10.1145/3218603.3218615.

Zamacola, R., Martínez, A. G., Mora, J., Otero, A., & de La Torre, E. (2018, December). Impress: Automated tool for the implementation of highly flexible partial reconfigurable systems with Xilinx ivado. *2018 International Conference on ReConFigurable Computing and FPGAs (ReConFig)*, 1–8. IEEE.

Zhao, R., Song, W., Zhang, W., Xing, T., Lin, J. H., Srivastava, M., … Zhang, Z. (2017, February). Accelerating binarized convolutional neural networks with software-programmable fpgas. *Proceedings of the 2017 ACM/SIGDA International Symposium on Field-Programmable Gate Arrays*, pp. 15–24. Association of Computing Machinery, New York, USA.

Zhou, Y., Redkar, S., & Huang, X. (2017, August). Deep learning binary neural network on an FPGA. *2017 IEEE 60th International Midwest Symposium on Circuits and Systems (MWSCAS)*, 281–284. IEEE.

12 Recommender System: Techniques for Better Decision Making for Society 5.0

Asha Rani[1], Kavita Taneja[2], and Harmunish Taneja[3]
[1]Gujranwala Guru Nanak Khalsa College, Ludhiana, Punjab
[2]Panjab University, Chandigarh, India
[3]DAV College, Sector 10, Chandigarh, India

CONTENTS

DOI: 10.1201/9781003132080-12

12.1   INTRODUCTION

A recommender system with data science as the backbone is loaded with techniques derived from computer science, mathematics, engineering, statistics, and many more catering to the spectrum of applications emerging in varied domains of Society 5.0. Data science is a unique discipline with many mysteries to unlock that will cater to more specific scientific and societal problems presented by the preponderance of data. Data science helps distributors to affect our buying activities, but the benefit of collecting data goes further. Via wearable trackers that inspire citizens to adopt healthy behaviors and can alert individuals to potentially critical health conditions, data science can enhance public health. Data can also enhance the accuracy of diagnosis, speed up the finding of treatments for particular illnesses, or even avoid the spread of a virus. Scientists were able to monitor the spread of the disease and predict the areas most susceptible to the disease when the Ebola virus epidemic struck West Africa in 2014. This knowledge helped health authorities get ahead of the outbreak and avoid it from becoming a global epidemic. Across most industries, data science has critical applications. Data is used, for example, by farmers for successful food growth and distribution, by food producers to minimize food waste, and by non-profit organizations to support fundraising efforts and forecast financing needs. With the emergence of the Internet, an enormous amount of data is being produced. Big data analytic approaches are being used since long to analyze and create various predictions with these data (Dwivedi & Roshni, 2017). With the growth of data science interest, the need to extract information to identify patterns, correlations, and efficient predictions have emerged (Vartak, Huang, Siddiqui, Madden, & Parameswaran., 2016). A recommendation system can be broadly defined as algorithms aimed at recommending related or similar products to users (depending on the industry, products may be text to read, items to purchase, movies to watch, or something else) (Isinkaye, Folajimi, & Ojokoh, 2015). A RS has been proved to be very useful in ecommerce, service industry, social networking sites, etc. Interest in this area reigns supreme because it is an area of research that is rich in problems and shows abundant realistic trends to support decision making through tailored feedback, for diverse applications adaptive to Society 5.0 (Adomavicius & Tuzhilin, 2005). Figure 12.1 shows the block structure of the recommendation system.

This chapter provides an overview of the recommendation system as an application of data science and tool for efficient decision making for Society 5.0. The generation of recommendation techniques is elaborated in this chapter, which can be classified into five categories: content-based, collaborative, knowledge-based, context awareness-based, and hybrid approaches. The chapter also outlines the trends, merits, and limitations of a recommender system. It also highlights design guidelines to support better decision making for Society 5.0. The presented design guidelines based on issues and trends discussed would help in enhancing comprehension of users and products and integrating contextual knowledge into the recommendation process.

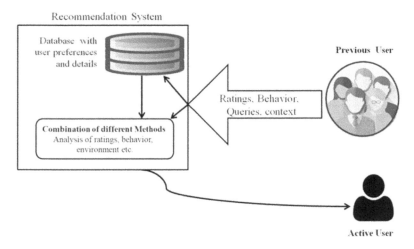

FIGURE 12.1 Block structure of recommendation system.

12.2 LITERATURE SURVEY

The first reference to a recommender system can be traced back to the early 1980s when (Salton, 1985) proposed a word-vector-based algorithm for searching among textual documents. In the mid-1990s, RS found its extensions in industry and academia with the advent of recommendations that rely on explicit ratings (Hill et al., 1995; Resnick et al. 1994; Shardanand & Maes, 1995). Further, research has extended these algorithms from e-mail filtering (Kautz et al., 1997) to document search (Hand, 2007), information retrieval (Huang, Chen, & Zeng, 2004), management science (Li & Kao, 2009), and consumer choice modeling (Scholz et al., 2017).

A recommendation system will predict items that are evaluated based on explicit (rating score) and implicit (logs) data that users have access to, such as movies, food, utilities, songs, clothing, gadgets, etc., as per their preferences. The key methods of recommendation are summarized in Figure 12.2 and elaborated thereafter.

12.2.1 RECOMMENDATION TECHNIQUES: COLLABORATIVE FILTERING BASED

These techniques work by finding other users with similar preferences, and then recommending items; it uses their ratings and preferences to suggest items to the active user. The transaction history, ranking, selection, and purchasing information and actions of other users decide the suggestions. The behavior and expectations of other users for items are exploited to provide suggestions for new users. These techniques do not require the items' characteristics and related documentation. A popular recommender system in this category is based on music recommendations (Shardanand & Maes, 1995), online shopping recommender system: Amazon (Linden et. el., 2003), and others. Table 12.1 shows the matrix of user item ratings on a 1–7 scale where 7 represents the highest rating. Each user gives a rating for each item in the list. When a user with a similar profile arrives, he/she is recommended an item based on the weighted ratings. Collaborative filtering-based recommendation techniques classification is as under:

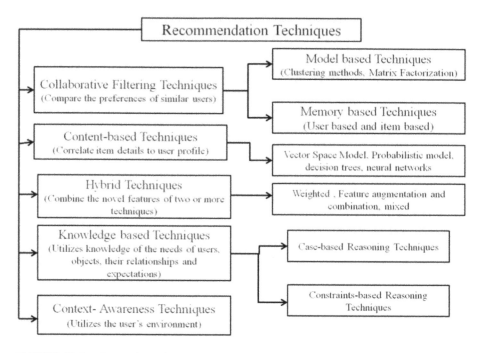

FIGURE 12.2   Generation of recommendation techniques.

TABLE 12.1
User Item Ratings

Client/Item	Item 1	Item 2	Item 3	Item 4	Item 5
Client 1	4	3	7	5	2
Client 2	5	6	3	7	4
Client 3	3	5	6	4	3

Collaborative Filtering - Memory-Based: It advocates exploitation of users' ratings for assessing similarity or weight of client or objects in order to predict between them. It may be dependent on the consumer or the object. The active user receives user-based suggestions based on a common group of clients/users. On the other hand, item-based model depends on the computational similarity across the retrieved objects based on user ratings.

Collaborative Filtering - Model-Based: It advocates prediction by creating a user rating model based on data mining and machine learning. The trends reflect integration of predictive modeling with neural networks and decision trees to support better decision making. In building such networks, the trade off is high computing costs.

12.2.2   RECOMMENDATION TECHNIQUES: CONTENT-BASED

This approach recommends products that are equivalent to products previously desired or rated favorable by the consumer. Typically, a user profile (a set of features derived from the background of evaluated items) and item profile are created using shared attribute space information; for example, NewsDude, (Billsus & Pazzani, 2000) a personal news recommendation system, and LIBRA (Mooney & Roy, 2000), a book recommendation system, etc.

12.2.3   RECOMMENDATION TECHNIQUES: HYBRID

This technique combines two or more recommendation techniques to avoid problems like cold start, sparsity, etc. and improve the performance of recommender systems primarily in terms of the accuracy of decision submitted. Hybrid techniques harness the merits of integrated methods subsiding the limitations of the recommendation techniques used. CinemaScreen (J. Salter and N. Antonoupoulos, 2006) and Fab (M. Balabanovic and Y. Shoham, 1997) are the earliest developed examples of hybrid recommender systems.

12.2.4   RECOMMENDATION TECHNIQUES: KNOWLEDGE-BASED

Knowledge representation, extraction and system design techniques are the foundation of knowledge-based recommendations. Suggestions are made in these methods based on knowledge of the needs of users, objects, their relationships, and expectations of users (Gediminas Adomavicius et al., 2011). Recommendations are not affected by scores or recent preferences. Knowledge-based recommendation techniques can be further classified as:

- Case-Based Systems: Reasoning is case-based, considering the similarity between the active case and its solution present in database.
- Constrained-Based Systems: Provides suggestions based on the user's preferences and also the reason for the particular suggestions. Constraints can be related to product, compatibility, and filter conditions.

12.2.5   RECOMMENDATION TECHNIQUES: CONTEXT AWARENESS-BASED

Context refers to the identified attributes which are known a priori (Gediminas Adomavicius et al., 2011). In this method, the focus is on the additional contextual information such as temperature, season, time, geometrical information, friends, and families etc., which are used to recommend items to users based on circumstances. (Cena et al., 2006; Church, Smyth, Cotter, & Bradley, 2007; Sae-Ueng, Pinyapong, Ogino, & Kato, 2008) used context-aware techniques for recommendation.

12.3   RECOMMENDATION SYSTEMS AND RELATED ISSUES

In formal terms, a recommender system can be defined as (Kumar, 2018):

$$U : CXS \rightarrow R$$

Where

C represents set of all clients/users

S represents the set of all products (movies, songs, books, news, food, etc.) for recommendation to users.

In certain applications, the number of items may range from tens of thousands to millions.

U represents usefulness of a particular product/item, $s \in S$, for a particular client/user, $c \in C$, and R represents the total ordered recommended set.

For efficient recommendations, U is required to be maximized for every user, $c \in C$, providing the recommended items $s` \in S$. Generally, usefulness is represented in terms of ratings (scores), which shows upto what extent a user liked a particular item. However, depending upon the application, U may be any arbitrary function that includes a benefit or cost component. The identified issues which need to be addressed in recommendation systems for applications evident in Society 5.0 are as below.

12.3.1 USER PREFERENCES

A recommender system involves interactions for making different categories of choices, like selecting a video or news or song from a list of predicted items, selecting a particular feature value (such as the cost or mileage of a vehicle) as parameters or selecting the feedback parameters. Users often are not knowing or able to reflect their expectations/preferences in advance, so within a particular recommendation situation, they will need to construct them. In addition, there are numerous popular psychological decision-making criteria, like background, primacy/recency, and framing that potentially affect the preference of referring to recommendation system for decision/advice (Mandl, Felfernig, Teppan, & Schubert, 2011).

People resort to automated to decision-making specifically in situations that require intense options and uncertainty regarding benefits. Recommendation systems come in handy for a spectrum of applications that collaborate towards Society 5.0. Optimization and accuracy are the desired traits of decision making and to achieve this, decision efforts need to be minimized. Collaborative filtering assumes that, which may not always be true, all users are equal. An online shopping website can value the opinions of clients purchasing multiple products over those who only purchase one for recommendation (Shani & Gunawardana, 2011).

12.3.2 SPARSITY IN RATINGS

Collaborative filtering is considered one of the most powerful techniques of recommendation. It is based on recommending content to existing customers primarily on the basis of past transactions and input from the other similar users (Chen, Wu, Xie, & Guo, 2011). A scattered profile matrix results in less reliable recommendations due to the availability of enormous data on items in the catalog and the inadequate amount of transactions and ratings by the customers (Kumar, 2018). This refers to a situation called Sparsity, where there is a lack of transactions and feedback data to derive clients with similar interests and prediction of items. This adversely affect the performance of recommendation system (Anand & Bharadwaj, 2011). When the customer registers for the first time, his/her details and priorities are not known. Consequently, for a newly registered user, the framework would not be efficient enough to provide recommendations. Pearson correlation coefficient (PCC), mean squared difference (MSD), adjusted cosine (AC), and Jaccard similarity are some traditional similarity measurement methods. These methods suffer from various problems like few co-rated item limitation, local information problems, neglecting the proportion of common ratings, etc. The problem of sparsity is found to be handled in many ways: by improving the similarity measurement techniques, finding techniques to predict undeclared ratings;, predicting additional information of user interests under sparse situations, finding out appropriate evaluation metrics for an efficient recommender system, etc. (Idrissi & Zellou, 2020).

To mitigate sparsity problems, it is a cumbersome task to manage the users' identity and prevent malicious behavior. The recommender systems should be able to control the privacy of users with efficient encryption techniques for transmitted and transactional information. There are certain research issues like transparency, cold start, explicit trust, privacy, and dynamics in handling sparsity in recommendation (Chen et al., 2011):

12.3.3 COLD-START PROBLEM (USER AND ITEM POINT OF VIEW)

The missing data problem where preference information is unavailable is also called a cold-start problem; it is another major problem in recommender systems. The cold-start issue arises when only a few ratings have been expressed by a new client using an online platform. Lack of communication between the new client and previous users becomes the hindrance in measuring similarity between them (Kim, El-Saddik, & Jo, 2011). As a consequence, recommendation systems are not in a position to make accurate recommendations. Both collaborative filtering based recommender systems and content based recommender systems have to tackle the cold-start problem (Adomavicius & Tuzhilin, 2005). The problem can be viewed as cold-start products and cold-start clients. Collaborative algorithms desire a firm record of previous product ratings for efficient recommendations. Hence, collaborative approaches do not work properly in certain domains in which new objects are added with zero prior rating records. Therefore, the problem of cold-start items arises in collective filtering methods in certain situations where new items are suggested (Lika, Kolomvatsos, & Hadjiefthymiades, 2014). This problem can be handled effectively in a content-based recommender system predicting product significance without previous ratings (Pazzani & Billsus, 2007). However, the content-based recommender system also suffers from the difficulty of cold-start clients. Suggesting products to new clients is not possible without previous experience of the system. The possible solutions to the cold-start issue can be (Schafer, Frankowski, Herlocker, & Sen, 2007):

- Asking the new users to grade some items
- Explicitly specifying the taste of new users
- Using demographic details to recommend the items to new users

12.3.4 OVERSPECIALIZATION

Overspecialization is the issue faced by a content-based recommender system that limits clients to products similar to those specified in their shared profiles and not exploring new products or services. For example, a person with no experience of mountains would get a recommendation of mountainous tourist places. Several studies have been conducted to solve this problem (Pazzani & Billsus, 2007; Sheth & Maes, 1993; Yi, Callan, & Minka, 2002). Furthermore, the issue with overspecialization is not just that previously viewed things should not be suggested by content-based system but also recommendations of very similar products should be avoided, multiple news articles reporting similar occurrences, or songs of the same movie. Hence, diversity in products to be recommended is often desirable.

12.3.5 NOVELTY AND DIVERSITY OF RECOMMENDATION

Novelty and diversity in spite of being connected notions are distinct. Novelty generally depicts the difference between the items recommended in the past and present by a particular user while diversity caters the variety in recommended items at a particular time (Castells, Vargas, & Wang, 2011). The relation between novelty and diversity can be understood with a simple example as when a user is recommended a book of an author of a different genre, which was not recommended earlier, is termed novelty; while the number of books of different authors and genres indicate diversity. In this research area, major contributions include algorithm to attain diversity. Evaluation of these techniques requires research in methodologies and metrics. Most authors have addressed the problem as a multi-objective optimization problem exhibiting trade-off between accurate recommendation and diverse options. Also, these properties can be seen as opposing objectives to accuracy from the fact that novelty and diversity are fundamentally important to provide dimensions of recommendation utility (Vargas & Castells, 2011).

By extending the variety of possible item types and features at which recommendations are aimed, diversity may increase the likelihood that the user would like at least some items, rather than exhibiting narrow interpretation of user behavior (Hurley & Zhang, 2011). The best way to ensure that suggestions are diverse is to exclude expected ratings and simply rate items randomly. Diversity, however, then comes at the price of accuracy: recommendations are no longer tailored to the preferences of users. Lathia et al. (2010) emphasize to take into account the needs of users while supporting diversity. Two techniques are proposed: temporal hybrid switching, and re-ranking of user's recommendations. Some of the challenges while addressing diversity in a recommender system are (Kunaver & Požrl, 2017) as follows:

- There is limited literature on injecting diversity as a measure in RS. Further studies in this direction could be helpful in developing state-of-the art methods and explanations of diversity.
- Some aspects of diversity are highly subjective, research will benefit from the inclusion of psychological expertise in the creation of new initiatives for diversity.

- RS including diversity initiatives compare things in accordance with their description of meta-data. In this scenario, the challenge is how to manage different types of items in the CF-based systems or with items (with user defined content) that do not have all the meta-data available.
- The problem of overfitting may arise if diversity is not taken into account from the start of the recommendation.

12.3.6 SCALABILITY

Scalability is the property that shows how efficiently a system manages an increasing amount of information. RS faces significant challenges in processing vast volumes of data related to millions of users and items over the Internet. Therefore, estimates to quantify recommendations are exponentially increasing, becoming costly, and often leading to misleading outcomes (Sridevi & Rao, 2016). The computation grows linearly as the data set size grows in proportion to the number of users and artefacts. That is, the algorithm performs well on small data sets but fails to provide adequate results when applied to large data sets. As a result, applying the recommendation technique to large and complex data sets produced by item-user interaction is extremely difficult (Kumar, 2018). Methods suggested for handling the problem of scalability and speeding up the formulation of recommendations are based on approximation mechanisms which however enhance performance, but often result in decline in accuracy (Kantilal & Gheewala, 2018). Bayesian network, clustering, and dimensionality reduction through SVD could be the possible techniques to deal with scalability problems (Takáes, Pilászy, Németh, & Tikk, 2009). In order to deal with information explosion we need to enhance scalability and diminish the time complexity of recommending algorithms. Clustering approaches based on genetic algorithms are finding recognition due to their efficient suitability to the scenario (Georgiou & Tsapatsoulis, 2010). There is again a trade-off between scalability and accuracy. For example, if the algorithm's accuracy is lower than that of other candidates who work on relatively small data sets, the difference in accuracy must be seen on small data sets only. These measurements will provide useful information on both the performance and future research paradigms of a recommendation system (Khusro, Ali, & Ullah, 2016).

12.3.7 CONTEXT-AWARENESS

The recommendation system is said to be context-aware, if it considers the circumstances which aggregates current activity, location, temperature, time, etc. while recommending the products. Schilit and Theimer proposed the term "context-awareness" (Schilit & Theimer, 1994), which explains the ability of a device to sense the environment and act accordingly. The secret to coming up with appropriate recommendation is to define user preferences and contextual information efficiently. In addition, the contextual information identification if done in an unnoticeable manner would enhance the performance (Lee & Lee, 2007). This is because the system makes personal recommendations for different users with distinct circumstances by adding the details. While talking about scalability, accuracy, and time consumption, however, efficiency is still to be improved, so new approaches using hybrid methods are required (Setiowati, Adji, & Ardiyanto, 2018). It is predicted that the future recommendation system may exploit mobile service data as contractual reference like history, location, calendar, and information shared over social network platforms (Felfernig, Ninaus, & Reinfrank, 2014). Society 5.0 promises digitalization and with this beginning of the fifth era in human history, interconnection of data will be unimaginable that hints towards success of such recommender systems in terms of accuracy.

12.3.8 PREDICTION ACCURACY

Prediction accuracy is the common way of measuring the performance of information retrieval systems by correlating the recommendations generated for a user with the predictions a user is actually interested in (Sridevi & Rao, 2016). A fundamental principle in a recommendation system is that the consumer would prefer a system that makes more accurate predictions. Another research area has emerged for discovering algorithms for better predictions. A prediction engine is the heart of any recommendation system, which can measure ratings for items or probability of usage (purchases) or ranking of the items. Interface independency makes prediction accuracy measurement suitable for offline experimentation (Shahbazi, Hazra, Park, & Byun, 2020). In a user analysis, calculating prediction accuracy tests the accuracy in a given recommendation, which further depends on the purpose of the algorithm and the goal of the measurement. Three wide classes of prediction accuracy measures are found in literature:

- Prediction Accuracy by Ratings
- Prediction Accuracy by Usage
- Prediction Accuracy by Ranking Measures

There are various types of measurement metrics available, such as root mean squared error (RMSE), mean absolute error (MAE), precision, recall, f-measure, false positive rate, etc.

12.3.9 PRIVACY

Recommendation systems are dependent on users' personal data of their preferences (ratings, usage history, profiles, etc.) so as to produce reliable and personalized recommendations (Jeckmans et al., 2013). RS are beneficial, but the privacy concerns associated with personal data collection and processing are frequently ignored. It is a matter of great concern that users' personal preferences stay private and safe from intruders or third-party applications (Jeckmans et al., 2013).

Some research questions for constructions of an effective recommendation system arise (Armknecht & Strufe, 2011):

- Does the user need to report his interests to the system?
- Is there a need for the user to report its knowledge regarding locations and preferences?
- How will a correctly calculated recommendation be received by the user?

Work is being done to compare algorithms applied for evaluation of compromised information researchers (McSherry & Ilya Mironov, 2009). In the studies, the presumption is that total privacy is unrealistic and research can be extended to estimate the level of compromises for the violations of privacy.

12.3.10 ADAPTIVITY

Real recommendation systems often work in an environment where the items cluster changes rapidly, or where interest in items can change over time, or users can be navigating web using different devices like desktop PCs, mobiles, etc. Adaptivity can taken in two terms: first, prediction adaptivity of recommendation system in the dynamic setup of changes in items and their preferences and second, the device adaptivity. The recommendation of news items in online newspapers can be an example to understand the first kind of adaptivity (Rosaci, Sarné, & Garruzzo, 2009). The adaptive recommender model proposed by Ricci comprises of two entities: information system and recommender agent (Ricci, 2007). Figure 12.3 shows the structure of an adaptive recommendation system. Information system performs the function of information gathering by generating a query form, query results, showing the top-n results, etc. while the job of RA is completely adaptive. RA deals with the user, IS, and the interactions between them so as to help the user in decision making.

12.4   TRENDS IN RECOMMENDATION SYSTEM TO SUPPORT DECISION MAKING FOR SOCIETY 5.0

The fields like e-commerce, social media, entertainment, education, etc. have seen a significant growth in terms of information size and traditional methodologies of recommendation systems in these areas have had considerable success in delivering item recommendations. But they still suffer from various problems like cold start, data sparsity, scalability, context-awareness, etc. With the recent advances in

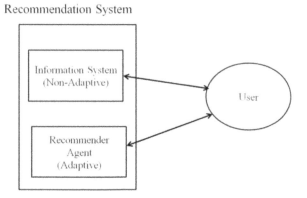

FIGURE 12.3   Block structure of adaptive recommendation system.

machine learning and deep learning, more efforts could be made to exploit these techniques to improve the efficiency of recommendation system to cater the needs of Society 5.0. The literature review in deep learning recommendation systems show that Autoencoder (AE) models, Convolutional Neural Networks (CNNs), and the Recurrent Neural Networks (RNNs) models provide promising results in recommendation systems (Da'u & Salim, 2020).

Natural Language Processing (NLP) in a recommendation system will make them capable of comprehending sentences, idioms, and their forms. In this context, NLP has the ability to analyze different types of data as well as perform syntax, semantics, and pragmatic analyses (Addagarla & Amalanathan, 2019). Such decision making will be the core of autonomous systems that will evolve as integral components of a smart city in Society 5.0.

Another promising trend is applying multi-criteria decision-making methods in recommendation systems. These are the statistical methods which can be applied in recommending products related to health, insurances, finances, etc. These methods are quite effective in scenarios where selection of products or services is crucial for one's life or finances. Similar applications will become exponentially relevant in light of gigantic digitization that will be part and parcel of Society 5.0. For example, recommending an appropriate insurance plan is quite a serious affair as compared to recommending a song or movie for entertainment purposes. These methods provide a rich variety like Analytic Hierarchy Process, Multi Attribute Utility Theory, TOPSIS, Grey Relation Analysis, etc. to enhance the decision-making capability of the developed product or service.

12.5 MERITS AND LIMITATIONS OF RECOMMENDATION SYSTEMS IN LIGHT OF SOCIETY 5.0

Nowadays, a recommender system has become essential for every user on the Internet, whether they are using e-commerce websites like Amazon and Flipkart; online streaming platforms like Hotstar, Netflix, Amazon Prime; or online music like Spotify, etc. (Mohamed et al., 2019). Following are the identified merits of recommendation systems in the current scenerio:

1. **Dealing the problem of information overload:** A recommender system gives efficient and unbiased results when there is huge or complicated data assisting applications of Society 5.0. These are quite fitting methods to deal with the problem of information overload and to filter best suggestions out of enormous options as per users' interests.
2. **Revenue generation:** Researchers have studied and developed a wide variety of algorithms showing the increasing online traffic. Recommender systems have significantly contributed in revenue increase in almost all domains by increasing the sales specifically in digitized platforms that will exponentially expand with Society 5.0.
3. **Client satisfaction:** Customers frequently expect to see close proximity product recommendations based on their previous browsing experience on the web. They feel that by doing so, they would have a greater chance of getting

more serious opportunities for better goods. When they leave the situation and return, it would be helpful if their browsing data from previous shopping or viewing product lists was available.

4. **Personalized suggestions:** A user likes to have personalized recommendations so the data gathered in an indirect manner to enhance the overall services on any website will ensure that they are appropriate for the user's preferences.

A few limitations of such recommendation systems are identified as under:

1. **Human-centered suggestions:** Recommendation systems are presenting suggestions for human beings whose interests and choices are time variant and have variables in multiple dimensions. The mapping of such stochastic information and prediction with acceptable accuracy is a major challenge of a recommender system, particularly as we approach Society 5.0 where major human activities are to be governed with autonomous programmed systems.
2. **Data deficiency:** Recommendation systems need high-quality data for predictions. Well structured and clear data is the prime requirement for accurate recommendations.
3. **Data variability:** In today's world we have plenty of data changing each second. This variability has to be incorporated in recommender system for efficient results and need to do tedious research in the direction.
4. **Unpredictable performance:** We live in the world that is constantly changing and businesses all over the world are chasing production trends and turnover growth. A recommender system can show unpredictable results when the strategies and plans of algorithms are not catering to the requirements of the changing world.

12.6 DESIGN GUIDELINES FOR ATTENUATING THE CHALLENGING ISSUES IN RECOMMENDATION SYSTEM

Recommender systems, most well-known artificial intelligence applications, aim to mathematically model and technically replicate the real-world method of making recommendations. Designing a recommender system is significantly different from designing software for any application (Berka & Plößnig, 2004). Choices and application of various algorithms defines the overall architecture and performance of recommender system. Such design majorly depends on the following:

1. **Domain of the problem:** Each domain has its own implications and requirements. The recommender system in entertainment, shopping, etc. may be taken as casual ones, where the accuracy may be compromised up to a certain level and diversity and novelty is the prime focus while the others may be more serious ones like recommending insurance policies or health advices.
2. **Computational resources:** The efficiency of recommender system not only depends on the algorithms applied but also on the computational

resources. Accuracy in recommender systems is directly proportional to the size of the data set and hence requires quite efficient and advanced computational resources.

3. **The amount of detailed information available to the system:** There is a wide variety of different attributes for entities in RS like user preferences, genre, publishers, year of publication, author, price, etc. in the case of a book recommendation system. These attributes are entity specific and hence depend upon the domain of the recommender system. Other types of information are also required for high-quality recommendations like:

   - Temporal and spatial information: This data is related to the history of user behavior at a particular time and place. The long-term or short-term user preferences data help to design RS with accurate predictions (Al-Hadi, Sharef, Sulaiman, & Mustapha, 2017).
   - Price or any other financial detail: Cost or the financial aspect of the product plays a major role for selection of any product. A high-quality recommender system could not neglect this information.
   - Media files (audio, video, text, or images): Various types of media files are associated with the items to be recommended. The metadata of items or users may be present in any of the formats. An efficient RS should cater to the needs of different types of data and should incorporate the information in providing accurate predictions.
   - Data categorization information: Depending upon the domain of the recommender system, items to be recommended do have some categorization. For example, Genre, lyricist, Singer, movie, etc. in the case of a music recommender system.
   - Item and user context: It is very crucial to make the recommender system context-aware in order to achieve maximum user satisfaction through fine-tuned recommender system.

The analysis of issues discussed in previous sections give us a few guidelines to design an efficient recommender system that can provide the solutions for problems like cold start, sparsity, scalability, context awareness, etc. (Leiva, Budan, & Simari, 2020).

1. A recommender system can focus on the relevant user group rather than the entire data set and may use demographic filtering and clustering to group users with similar interests and demographic characteristics. This will improve consistency, deal with the sparsity issue, and reduce latency.
2. To address the problem of cold start, the registration of the new user should be done and his contextual information like temperature, location, time, etc can be recorded through the Internet protocol address.
3. Diversity and novelty can be addressed by creating multiple recommendation lists for current or previous user preferences.
4. Overspecialization can be dealt with by removing the previous recommended list items and replacing obsolete and old items with somewhat similar items as those are viewed/purchased by similar clients.

12.7 CONCLUSION AND FUTURE SCOPE

There has been considerable progress in the development of a recommender system in the last decade to mitigate the problem of information overload which will see no bounds in Society 5.0. By predicting objects/services to users that are related and relevant, a recommender system plays a major role in day-to-day decision making of a common man. There has been tremendous advancements with a wide variety of approaches from content-based, collaborative, hybrid to industry-fitted recommender system to provide high-quality and fine-tuned recommendations. The recommender systems surveyed in the current scenario requires updating for efficiency of a wide range of applications embedded in smart cities. Recommender system designers face a number of significant issues and challenges in spite of significant research done to address these issues. There is still more work to be done in order to reach the desired target. In this chapter, shortcomings of current methods of recommender system were reviewed and possible research directions were stated to enhance the capability of recommendation technologies for Society 5.0. The properties of recommender systems are confined in the chapter for generic descriptions highlighting design guidelines, trends, and the new research directions in the field of data science for better decision making. The presented design guidelines based on issues and trends discussed would help in enhancing comprehension of users and products and integrating contextual knowledge into the recommendation process for diverse applications pertaining to data science and Society 5.0.

REFERENCES

Addagarla, S. K., & Amalanathan, A. (2019). A survey on comprehensive trends in recommendation systems & applications. *International Journal of Electronic Commerce Studies*, *10*(1), 65–88. https://doi.org/10.7903/ijecs.1705

Adomavicius, G., & Tuzhilin, A. (2005). Toward the next generation of recommender systems: A survey of the state-of-the-art and possible extensions. *IEEE Transactions on Knowledge and Data Engineering*, *17*(6), 734–749. https://doi.org/10.1109/TKDE.2005.99

Al-Hadi, I. A. A. Q., Sharef, N. M., Sulaiman, M. N., & Mustapha, N. (2017). Review of the temporal recommendation system with matrix factorization. *International Journal of Innovative Computing, Information and Control*, *13*(5), 1579–1594.

Anand, D., & Bharadwaj, K. K. (2011). Utilizing various sparsity measures for enhancing accuracy of collaborative recommender systems based on local and global similarities. *Expert Systems with Applications*, *38*(5), 5101–5109. https://doi.org/10.1016/j.eswa.2010.09.141

Armknecht, F., & Strufe, T. (2011). An efficient distributed privacy-preserving recommendation system. *2011 the 10th IFIP Annual Mediterranean Ad Hoc Networking Workshop, Med-Hoc-Net'2011*, 65–70. https://doi.org/10.1109/Med-Hoc-Net.2011.5970495

Berka, T., & Plößnig, M. (2004). Designing Recommender Systems for Tourism. *Proceedings of ENTER 2004*, 11. Retrieved from http://www.salzburgresearch.at/wp-content/uploads/2010/10/enter_ploessnig.pdf

Billsus, D., & Pazzani, M. J. (2000). User modeling for adaptive news access. In *User Modeling and User-Adapted Interaction* (Issue 10, pp. 147–180). Kluwer Academic Publishers. https://doi.org/10.1023/A:1026501525781

Castells, P., Vargas, S., & Wang, J. (2011). Novelty and diversity metrics for recommender systems: Choice, discovery and relevance. *Proceedings of the International Workshop on Diversity in Document Retrieval - DDR'11*, 29–37. http://hdl.handle.net/10486/666094

Cena, F., Console, L., Gena, C., Goy, A., Levi, G., Modeo, S., & Torre, I. (2006). Integrating heterogeneous adaptation techniques to build a flexible and usable mobile tourist guide. *AI Communications, 19*(4), 369–384.

Chen, Y., Wu, C., Xie, M., & Guo, X. (2011). Solving the sparsity problem in recommender systems using association retrieval. *Journal of Computers, 6*(9), 1896–1902. https://doi.org/10.4304/jcp.6.9.1896-1902

Church, K., Smyth, B., Cotter, P., & Bradley, K. (2007). Mobile information access: A study of emerging search behavior on the mobile Internet. *ACM Transactions on the Web, 1*(1). https://doi.org/10.1145/1232722.1232726

Da'u, A., & Salim, N. (2020). Recommendation system based on deep learning methods: a systematic review and new directions. *Artificial Intelligence Review, 53*(4), 2709–2748. https://doi.org/10.1007/s10462-019-09744-1

Dwivedi, S., & Roshni, V. S. K. (2017). Recommender system for big data in education. *Proceedings of 2017 5th National Conference on E-Learning and E-Learning Technologies, ELELTECH 2017*, 2. https://doi.org/10.1109/ELELTECH.2017.8074993

Felfernig, A., Jeran, M., Ninaus, G., Reinfrank, F., & Reiterer, F. (2014). Toward the next generation of recommender systems: applications and research challenges. In Tsihrintzis, G., Virvou, M., & Jain, L. (Eds.), Multimedia Services in Intelligent Environments. Smart Innovation, Systems and Technologies, vol. 24. Springer, Heidelberg. https://doi.org/10.1007/978-3-319-00372-6_5

Georgiou, O., & Tsapatsoulis, N. (2010). Improving the scalability of recommender systems by clustering using genetic algorithms. *Lecture Notes in Computer Science (Including Subseries Lecture Notes in Artificial Intelligence and Lecture Notes in Bioinformatics), 6352 LNCS*(PART 1), 442–449. https://doi.org/10.1007/978-3-642-15819-3_60

Hand, D. J. (2007). Principles of data mining. *Drug Safety, 30*(7), 621–622. https://doi.org/10.2165/00002018-200730070-00010

Hill, W., Stead, L., Rosenstein, M., & Furnas, G. (1995). Recommending and evaluating choices in a virtual community of use. In Proceedings of the SIGCHI Conference on Human Factors in Computing Systems (CHI '95). ACM Press/Addison-Wesley Publishing Co., USA, 194–201. DOI: https://doi.org/10.1145/223904.223929

Huang, Z., Chen, H., & Zeng, D. (2004). Applying associative retrieval techniques to alleviate the sparsity problem in collaborative filtering. *ACM Transactions on Information Systems, 22*(1), 116–142. https://doi.org/10.1145/963770.963775

Hurley, N., & Zhang, M. (2011). Novelty and diversity in top-N recommendation: Analysis and evaluation. *ACM Transactions on Internet Technology, 10*(4), 1–30. https://doi.org/10.1145/1944339.1944341

Idrissi, N., & Zellou, A. (2020). A systematic literature review of sparsity issues in recommender systems. *Social Network Analysis and Mining, 10*(1). https://doi.org/10.1007/s13278-020-0626-2

Isinkaye, F. O., Folajimi, Y. O., & Ojokoh, B. A. (2015). Recommendation systems: Principles, methods and evaluation. *Egyptian Informatics Journal, 16*(3), 261–273. https://doi.org/10.1016/j.eij.2015.06.005

Jeckmans, A., Beye, M., Erkin, Z., Hartel, P., Lagendijk, R., & Qiang, T. (2013). Privacy in Recommender System. *Social Media Retrieval*, 263–281. Retrieved from http://www.csrquest.net/uploadfiles/1D.pdf

Kantilal, S. A., & Gheewala, J. (2018). A survey of recommendation system. *Proceedings of the International Conference on Inventive Research in Computing Applications, ICIRCA 2018, 4*(May), 398–401. https://doi.org/10.1109/ICIRCA.2018.8597427

Kautz, H. A., Allen, E., Bank, R., & Selman, B. (1997). *Patent No. 561968.* Washington, DC: United States Patent office.

Khusro, S., Ali, Z., & Ullah, I. (2016). Recommender systems: Issues, challenges, and research opportunities. *Lecture Notes in Electrical Engineering, 376,* 1179–1189. https://doi.org/10.1007/978-981-10-0557-2_112

Kim, H. N., El-Saddik, A., & Jo, G. S. (2011). Collaborative error-reflected models for cold-start recommender systems. *Decision Support Systems, 51*(3), 519–531. https://doi.org/10.1016/j.dss.2011.02.015

Kumar, P. (2018). Recommendation system techniques and related issues: A survey. *International Journal of Information Technology.* https://doi.org/10.1007/s41870-018-0138-8

Kunaver, M., & Požrl, T. (2017). Diversity in recommender systems – A survey. *Knowledge-Based Systems, 123,* 154–162. https://doi.org/10.1016/j.knosys.2017.02.009

Lathia, N., Hailes, S., Capra, L., & Amatriain, X. (2010). Temporal diversity in recommender systems. *SIGIR 2010 Proceedings of 33rd Annual International ACM SIGIR Conference on Research and Development in Information Retrieval,* 210–217. https://doi.org/10.1145/1835449.1835486

Lee, J. S., & Lee, J. C. (2007). Context awareness by case-based reasoning in a music recommendation system. *Lecture Notes in Computer Science (Including Subseries Lecture Notes in Artificial Intelligence and Lecture Notes in Bioinformatics), 4836 LNCS,* 45–58. https://doi.org/10.1007/978-3-540-76772-5_4

Leiva, M., Budan, M. C. D., & Simari, G. I. (2020). Guidelines for the analysis and design of argumentation-based recommendation systems. *IEEE Intelligent Systems, 35*(5), 28–37. https://doi.org/10.1109/MIS.2020.2999569

Li, Y. M., & Kao, C. P. (2009). TREPPS: A Trust-based recommender system for peer production services. *Expert Systems with Applications, 36*(2 PART 2), 3263–3277. https://doi.org/10.1016/j.eswa.2008.01.078

Lika, B., Kolomvatsos, K., & Hadjiefthymiades, S. (2014). Facing the cold start problem in recommender systems. *Expert Systems with Applications, 41*(4 PART 2), 2065–2073. https://doi.org/10.1016/j.eswa.2013.09.005

Mandl, M., Felfernig, A., Teppan, E., & Schubert, M. (2011). Consumer decision making in knowledge-based recommendation. *Journal of Intelligent Information Systems, 37*(1), 1–22. https://doi.org/10.1007/s10844-010-0134-3

McSherry, F., & Ilya Mironov. (2009). Differentially private recommender system. *15th ACM SIGKDD International Conference on Knowledge Discovery and Data Mining,* 627–636. https://doi.org/10.1007/978-3-319-62004-6_10

Mooney, R. J., & Roy, L. (2000). Content-based book recommending using learning for text categorization. *DL '00: Proceedings of the Fifth ACM Conference on Digital Libraries,* 195–204. https://doi.org/https://doi.org/10.1145/336597.336662

Pazzani, M. J., & Billsus, D. (2007). Content-based recommendation systems. *The Adaptive Web,* 325–341. https://doi.org/10.1007/978-3-540-72079-9_10

Ricci, F. (2007). Learning and adaptivity in interactive recommender systems. In Proceedings of the ninth International Conference on Electronic Commerce (ICEC '07). Association for Computing Machinery, New York, NY, USA, 75–84. DOI: https://doi.org/10.1145/1282100.1282114

Resnick, P., Iacovou, N., Suchak, M., Bergstrom, P., Riedl, J. (1994). GroupLens: an open architecture for collaborative filtering of netnews. In Proceedings of the 1994 ACM Conference on Computer Supported Cooperative Work (CSCW '94). Association for Computing Machinery, New York, NY, USA, 175–186. DOI: https://doi.org/10.1145/192844.192905

Rosaci, D., Sarné, G. M. L., & Garruzzo, S. (2009). MUADDIB: A distributed recommender system supporting device adaptivity. *ACM Transactions on Information Systems, 27*(4). https://doi.org/10.1145/1629096.1629102

Sae-Ueng, S., Pinyapong, S., Ogino, A., & Kato, T. (2008). Personalized shopping assistance service at ubiquitous shop space. *Proceedings of the International Conference on Advanced Information Networking and Applications, AINA*, 838–843. https://doi.org/1 0.1109/WAINA.2008.287

Schafer, B. J., Frankowski, D., Herlocker, J., & Sen, S. (2007). Collaborative Filtering Recommender Systems. *Lncs, 4321*(January 2007), 291–324. http://www.eui.upm.es/ ~jbobi/jbobi/PapersRS/CollaborativeFilteringRecommenderSystems.pdf

Schilit, B. N., & Theimer, M. M. (1994). Disseminating active map information to mobile hosts. *IEEE Network, 8*, 22–32.

Scholz, M., Dorner, V., Schryen, G., & Benlian, A. (2017). A configuration-based recommender system for supporting e-commerce decisions. *European Journal of Operational Research, 259*(1), 205–215. https://doi.org/10.1016/j.ejor.2016.09.057

Setiowati, S., Adji, T. B., & Ardiyanto, I. (2018). Context-based awareness in location recommendation system to enhance recommendation quality: A review. *2018 International Conference on Information and Communications Technology, ICOIACT 2018, 2018-Janua*, 90–95. https://doi.org/10.1109/ICOIACT.2018.8350671

Shahbazi, Z., Hazra, D., Park, S., & Byun, Y. C. (2020). Toward improving the prediction accuracy of product recommendation system using extreme gradient boosting and encoding approaches. *Symmetry, 12*(9). https://doi.org/10.3390/SYM12091566

Shani, G., & Gunawardana, A. (2011). Evaluating recommendation systems. *Recommender systems handbook*, 257–297. https://doi.org/10.1007/978-0-387-85820-3_8

Shardanand, U., & Maes, P. (1995). Social information filtering: Algorithms for automating "word of mouth". *Proceedings of the SIGCHI Conference on Human Factors in Computing Systems*, 210–217. https://doi.org/10.1145/223904.223931

Sheth, B., & Maes, P. (1993). Evolving agents for personalized information filtering. *Proceedings of the Conference on Artificial Intelligence Applications*, 345–352. https:// doi.org/10.1109/caia.1993.366590

Sridevi, M., & Rao, R. R. (2016). A survey on recommender systems. *International Journal of Computer Science and Information Security, 14*(5), 265–272. https://doi.org/10.2 9322/ijsrp.9.09.2019.p9356

Takáes, G., Pilászy, I., Németh, B., & Tikk, D. (2009). Scalable collaborative filtering approaches for large reeommender systems. *Journal of Machine Learning Research, 10*, 623–656.

Vargas, S., & Castells, P. (2011). Rank and relevance in novelty and diversity metrics for recommender systems. *RecSys'11: Proceedings of the 5th ACM Conference on Recommender Systems*, 109–116. https://doi.org/10.1145/2043932.2043955

Vartak, M., Huang, S., Siddiqui, T., Madden, S., & Parameswaran, A. (2016). Towards visualization recommendation systems. *SIGMOD Record, 45*(4), 34–39. https:// doi.org/10.1145/3092931.3092937

Yi, Z., Callan, J., & Minka, T. (2002). Novelty and redundancy detection in adaptive filtering. *SIGIR Forum (ACM Special Interest Group on Information Retrieval)*, 81–88. https://doi.org/10.1145/564392.564393

13 Implementation of Smart Irrigation System Using Intelligent Systems and Machine Learning Approaches

Raghuraj Singh[1], Ashutosh Deshwal[2], and Kuldeep Kumar[1]

[1]Department of Computer Science & Engineering, Dr. B.R. Ambedkar National Institute of Technology, Jalandhar, Punjab, India

[2]Department of Computer Science & Engineering, Thapar Institute of Engineering and Technology, Patiala, Punjab, India

CONTENTS

DOI: 10.1201/9781003132080-13

13.1 INTRODUCTION

Today, India grades second worldwide in farm output. Allied sectors (such as forestry and fisheries) and agriculture accounted for 15.96% of the GDP in 2019 and about half of the total workforce. The pecuniary share of agriculture to India's Gross Domestic Product (GDP) is steadfastly reducing, with the country's broad-based economic growth (Statista, 2020).

At present, the farmers irrigate their fields through manual control at regular intervals, which consumes more water. Using this method, more water is wasted, which is a serious problem. As most of the people in the country are aware, water is deficient in umpteenth regions, and the government is also trying to save the water. The level of groundwater in 2015 was 34.84 mbgl (meters below ground Level), and was 18.77 mbgl in 2005. Hence, a need was felt to design a smart system to irrigate the fields and save the water.

The humidity or moisture level available in the soil determines the amount of water that must be observed or checked frequently or regularly to prevent the crops and plants from wilting; otherwise, plants or crops might die. In a traditional irrigation tract, the agriculturists have to sustain surveillance on the quantum of water in the field for different crops, which differs from crop to crop. Besides, each species of the plant or harvest have their own characteristics. Diverse kinds of crops require a varied quantity of water, as outlined in Table 13.1.

In this work, a smart irrigation system has been implemented using intelligent systems and machine learning approaches. In the proposed system, a moisture sensor is used to check different crops' moisture levels, and accordingly, the pump will be controlled (ON/OFF). The sensor gives the analog values to the microcontroller used in the system. After comparing those values to the threshold value, the microcontroller interrupts the motor, i.e., the pump used to irrigate the fields.

The proposed system saves the farmers' time, and there is no exigency for the farmers to go late at night in their farms to ON/OFF the motor. After the implementation of this system, the farmer will use less amounts of water. It will also convert the farming into a smart way according to the present technology from the traditional farming's irrigation system. In the first instance, this system is competent to lessen water wastage for India's irrigation. Secondly, this system would help make the lives of farmers easy and advance them along with the modern and smart world.

Farmers are the ones who produce crops for the entire world. In the current scenario, where every sector is getting important and evolving with the latest technologies, farmers are in the same position as before. The proposed system would help the farmers to get in contact with the latest technologies to make their work and life slightly easier.

The rest of the chapter is organized as follows: Section 13.2 describes the related work in the concerned area. Section 13.3 discusses data analysis, steps for data analysis, and data visualization. Section 13.4 presents the proposed system architecture. This section further discusses the hardware and software requirements for setting up the proposed system. Implementation details and workings of the proposed system are presented in Section 13.5. Section 13.6 discusses the simulation and implementation results. Finally, Section 13.7 concludes the chapter.

TABLE 13.1
Water Requirements in mm for Various Crops

Crop	Water Requirements (in mm)
Rice	900–2,500
Wheat	450–650
Sorghum	450–650
Maize	500–800
Sugarcane	1,500–2,500
Groundnut	500–700
Cotton	700–1,300
Soybean	450–700
Tobacco	400–600
Tomato	600–800
Potato	500–700

13.2   RELATED WORK

Primary investigations about the work are carried out under the steps, such as understanding, describing, and evaluating the existing approaches used for irrigating the field, plants, etc. In recent years, many researchers have carried out detailed work on various activities in automating the existing irrigation system.

Meena et al. proposed an approach for irrigating the field that uses various sensors to add multi-features in their modules like temperature sensors, moisture sensors, humidity sensors, and many more (Meena 2020). Prakash and Kulkarni used an IoT platform and a GSM (Global System for Mobile Communication) module to make a smart irrigation system. The GSM module is used to provide information to the farmers about their farm (Prakash & Kulkarni, 2020). The authors tried to develop a model that is more helpful over the surface and drip irrigation techniques. They did not explain the proper positing of sensors over a large field, and also mentioned the same as the limitation of their work. Use of the GSM module in the area may result in poor reception due to which farmers may not receive messages on time. The integration of multiple sensors may increase the module's complexity and raise the cost of the module, which will not be affordable for many farmers.

Kumar and Ravi used soil moisture sensors and an IoT platform for home gardening and proposed the same methodology for irrigating the crops. In their study, the authors placed the sensors in the root of the irrigating plants; this may lead to improper functioning of the sensors or damage of sensors and circuits (Kumar & Ravi, 2020). To make a more appropriate methodology comparatively to the IoT-related study's, Agarkhed presented a wireless sensor network (WSN) based approach where sensors shared the data wirelessly to the controller using an IoT platform (Agarkhed, 2017). These studies helped a lot as it solved the more significant challenge in the communication process between sensors and controllers.

Rawal used an IoT platform over the sprinklers to make rotary irrigation techniques better than the traditional ones. In this study, sprinklers are updated on a webpage (Rawal, 2017). Khan et al. have proposed a drip irrigation system technique to automate the irrigation process for optimizing water usage using a distributed wireless network system (Khan et al., 2017). Jain et al. proposed a drip irrigation technique and an IoT-enabled platform to make the system more scalable and efficient. In the drip irrigation technique, the sensors have to place near the root of the plant, due to which there are chances that sensors may get damaged easily. Also, more sensors and valves are required in the drip irrigation technique, which raises the system's cost (Jain et al., 2020). This drip irrigation technique is not well suited to farmers as it consumes more soil in equipment placement and is high in price.

Nandhini et al. focused on wastage of water and proposed methodologies in achieving an optimal irrigation technique for effective monitoring of the water supply (Nandhini et al., 2017). Upadhya and Mathew attempted a method based on fuzzy logic in the estimation of vegetable crops. In this method, humidity, temperature, and moisture of the land are taken in the form of input parameters. In the output parameter, the crisp value of yield is taken out using the fuzzy arithmetic method (Upadhya & Mathew, 2019).

Implementation of emerging technologies like image processing in the agriculture fields for checking and detection of plant diseases was widely adopted by many of the researchers. Sabrol and Satish presented that many researchers have supported collecting the pictures of a plant or its leaves, and then after doing the processing on the collected data set, tried to find out the plant's disease. If there is some disease to a plant, it is reflected when there is a change in the leaf's texture near its surface because it is unique to each plant's disease (Sabrol & Satish, 2016).

János and Matijevics pointed out that many researchers have developed a robotic system for the acquisition of the data and a procedure for controlling the farming limited within the greenhouses. Some of the researchers have used aerial and ground robots for the same. In some works, it has to be noted that mobile-controlled-based robots could be used instead of aerial robots with greenhouse automation infrastructure (János & Matijevics, 2010). Salikhov and Zainitdinova proposed a system that has been developed for the better control of greenhouse micro-climate, which mainly depends on the information provided by ambient climate sensors (Salikhov & Zainitdinova, 2018).

The traditional system that people like to use to irrigate their fields is the pumping system. Many researchers are working to make the pumping autonomous because it is the main part used in irrigation. To make it more eco-friendly, initiatives have been taken by the government and innovators. Kalezhi et al. proposed a sustainable and smart DC microgrid irrigation system with the help of multi-agent systems and IoT (Kalezhi et al., 2019).

Yella et al. developed an embedded component-based off-grid irrigation system that is capable of automatically irrigating the farm immediately when the farmers send a special SMS (consisting of commands to operate motor) to the designed device. The SMS will be received by the device itself and to do so it will take the help of a GSM shield attached to the device. This process will be only for a limited time and is extended as it always needs a farmer's involvement in guiding the whole system. The work done in this paper mainly focused on how the number of resources or energy can be consumed in an irrigation system (Yalla et al., 2013).

Zinnat and Abdullah created a model in which they used fuzzy logic to make decisions for irrigating the farm by performing different tasks. The authors proposed how a fuzzy logic model is used in controlling and managing the irrigation process. The model was very helpful in protecting crops from over-watering, but for managing the whole process of irrigation, the proposed model was not too efficient (Zinnat & Abdullah, 2014).

A wireless sensor network-based model for automatic irrigation of farmers' fields and automation of all the tasks commonly used in the agriculture sector was presented by Sirohi et al., the main problem observed, however, was that the model does not use any single model capable of performing the whole assigned task with the help of any forecasting model in real time (Sirohi et al., 2017). Shabadi et al. put forward a methodology of using GSM in an irrigation control system for efficient and proper use of both water and the power. The authors used a GSM module to administer the system, which may cost more (Shabadi et al, 2014). To overcome this, one can use an Arduino Yun board or any similar microcontroller board, which already has a built-in Wi-Fi module or GSM module. Dursun and Ozden proposed a wireless application approach for drip irrigation automation that is supported by smart sensors like soil moisture sensors. Banumathi et al. proposed an automated irrigation system to supply water to the plantation based on the water-level conditions using an android application, wireless sensor network (WSN), and general packet radio service (GPRS) modules (Banumathi et al., 2017; Dursun & Ozden, 2011).

There are several researchers who carried their researches on three factors of precision agriculture (PA) which are potentials, principles, and applications. The work done by Zhang et al. has highlighted precision agriculture to be a new, latest, unique, and advanced methodology of doing some agriculture-related work, and practice with the help of embedded system components and computers. The authors discussed how precision agriculture is much more efficient and effective compared to traditional agricultural practices (Zhang et al., 2017).

Cassman found that the quantity of soil and the requirements of fertilizers can be easily measured with good accuracy by using precision agriculture. The harvest is going to remain accurate if all the plants' parameters can be monitored properly or assessed accurately with precise and accurate timing of all required processes (Cassman, 1999). Pierce and Nowak presented some of the common aspects of precision agriculture with appropriate examples. Out of those, some of the significant aspects are soil quality, fertilizer and water requirements, and conservation from insects like flies, pests, and more. In all discussed facets, precision agriculture can play a vital part by providing essential or exact requirements for a plant, and finally yielding the optimum harvesting (Pierce & Nowak, 1999).

To provide the food supply for the entire globe is the responsibility of farmers, but a tremendous increase in the world population is leading to food scarcity. Gebbers and Adamchuk addressed food scarcity issues and proposed precision agriculture as a solution for the same. They highlighted that if precision agriculture is implemented properly, it can enhance the harvest, which will resolve food security (Gebbers & Adamchuk, 2010).

Srbinovska et al. used WSN for monitoring several key parameters of precision agriculture. This entire study work was implemented in the greenhouse on peppers. Sensors were used to send the information to a central hub or server remotely (Srbinovska et al., 2015). Kamienski et al. presented an IoT-based precision agriculture system. It aimed to provide adequate water management for agriculture. They demonstrated that a properly regulated water supply helps in saving a lot of water and enhances the harvesting. This methodology is about smartly managing water management, and few countries have implemented similar types of systems for precision agriculture (Kamienski et al., 2018).

The use of nanotechnology is a new technological advancement in the area of precision agriculture (Duhan et al. and Tokekar et al.). In farming, we are well aware that chemicals like fertilizers and pesticides are used. These affect the quality of crops and soil, which results in affecting the quality of most of the foods, which further results in many diseases in human beings. To resolve this issue, biosensors came into existence. Biosensors are used to protect the crops and plants from pests and more. Along with those, the use of nanotechnology can help the plants in getting the required ingredients (Duhan et al. 2017; Tokekar et al., 2016). Table 13.2 concludes all the research work in tabular format, along with the proposed work.

13.3　DATA ANALYSIS

Farming in India mostly depends on the monsoon. But it is difficult for the farmers to predict when it is going to rain. As technology emerges with machine learning, artificial intelligence, etc., this prediction task is now a little bit easier. A data set on the rainfall from 1901 to 2015 is available on the Kaggle (Ilangovan, 2017). This data set is sufficient to understand the records about the rainfall in different states of India easily. After analyzing this data set using machine learning algorithms and approaches, we determined which month maximum and minimum rainfall occur. After that, it is easy to develop a smart and innovative method for irrigating fields using an autonomous embedded system.

For analyzing the data, we followed a five-step process, as shown in Figure 13.1. First, we imported the data set, and then we pre-processed the data set. If there were null values, we filled those null values with the mean. Once the data set was pre-processed, we were ready for visualization to find some linear relationships between variables. The histogram showing the data from January to December for the years 1901 to 2015 is shown in Figure 13.2.

From the visualization of the given data, our motive was to find out the months with the maximum and minimum rainfalls. Initially, we calculated the total amount of recorded rainfall in all months. With the help of probability, we found out the maximum and the minimum rainfall recorded. The results are shown in Figure 13.3. From Figure 13.3, we found that the maximum rainfall occurs in the month of June, and similarly, in the month of March, the minimum rainfall occurs.

Further, we applied a multiple linear regressions model between annual rainfall and periodic rainfall. We calculated mean squared error, root mean squared error,

TABLE 13.2
Comparisons of Research Work

References	Methodology Used	Effective for a Practical and Larger Area	Cost-Effectiveness	Mentioned How to Cover a Large and Irregular Area
(Meena et al.; Prakash and Kulkarni; Yalla et al.; Shabadi et al.)	Sensors and GSM module with WiFi	No	No	No
(Sirohi et al.)	IoT and wireless sensor networks with real-time monitoring	Yes	No	No
(Khan et al.; Jain et al.; Dursum and Ozden)	Drip irrigation	No	No	No
(nandini et al.; Upadhyay et al.; Zinnat and Abdullah)	Other additional techniques like machine learning and fuzzy logic	Yes	Yes	No
(Sabrol and Satish)	Image processing for detection of plant diseases	No	Yes	No
(János and Matijevics; Salikhov and Zainitdinova)	Robotics within greenhouses	Yes	No	Yes
(Kalezhi et al.)	Autonomous pumping system	Yes	Yes	No
(Banumathi et al.)	Autonomous irrigation system with WSN and GPRS	No	No	No
(Zhang et al.; Cassman; Pierce and Nowak; Gebbers and Adamchuk)	Precision agriculture	Yes	Yes	No
(Duhan et al. and Tokekar et al.)	Nanotechnology	Yes	No	Yes
Proposed System	Autonomous irrigation with a combination of both ditch and sprinkle irrigation techniques	Yes	Yes	Yes

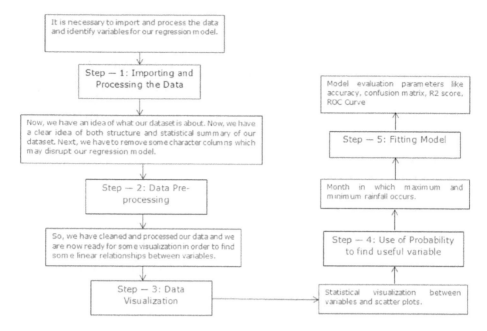

FIGURE 13.1   Steps for data analysis.

Histograms showing the data from attributes (JANUARY to DECEMBER) of the year
s 1901-2015:

<matplotlib.axes._subplots.AxesSubplot at 0x1301c9fa0c8>

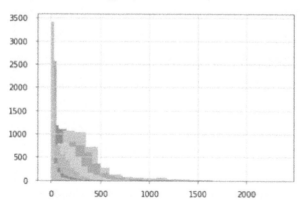

FIGURE 13.2   Histogram visualization of data.

mean absolute error, R2 score, accuracy score, confusion matrix, and classification
report of the model. Figure 13.4 shows the calculated metrics.

The accuracy of the model after successful implementation is found to be
99.83%. The result for this, along with the confusion matrix, classification, and
ROC curve, is shown in Figure 13.5.

```
Months are:  ['JAN', 'FEB', 'MAR', 'APR', 'MAY', 'JUN', 'JUL', 'AUG', 'SEP',
'OCT', 'NOV', 'DEC']
JAN       72.6
FEB      405.6
MAR       44.7
APR      232.6
MAY     1794.0
JUN     2658.3
JUL     2193.1
AUG     2052.1
SEP     2455.2
OCT     1249.8
NOV     1535.7
DEC      803.9
dtype: float64
Total recorded rainfall in these 12 months 5829542.827152476
[1.2453806782557342e-05, 6.957663954552696e-05, 7.667839713227454e-06, 3.9900
212914915124e-05, 0.00030774282875906155, 0.00045600488388529176, 0.000376204
4580554615, 0.00035201731265132127, 0.00042116510210103014, 0.000214390739901
3797, 0.0002634340368591365, 0.00013790103681126512]
Maximum Rainfall will be in the month of JUN
Minimum Rainfall will be in the month of MAR
```

FIGURE 13.3   Minimum and maximum fall of rainfall.

```
___Multiple Linear regression model between annual rainfall and the periodic
rainfall___
Train x shape (2881, 4) ; Test_x (1235, 4)
Train y shape (2881,) ; Test_y (1235,)

Mean Squared Error = 3326.4157535418844
Root Mean Squared Error = 57.675087806971604
Mean Absolute Error = 10.953757241509065
r2_score = 0.9958637383726687
```

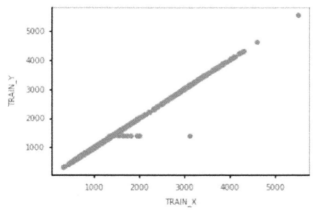

FIGURE 13.4   Fitting of linear regression and output of all calculated errors.

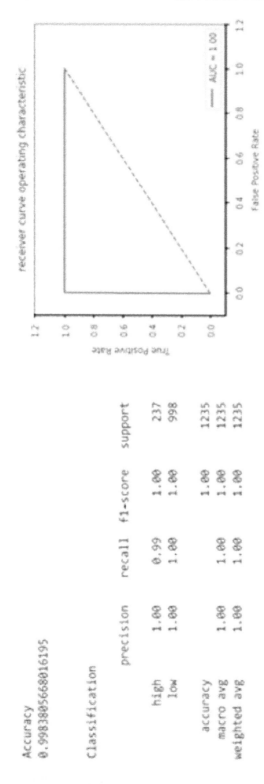

FIGURE 13.5   Accuracy, confusion matrix, classification table, and ROC curve.

After analyzing the data, as there was not too much rainfall in all 12 months of the year, some additional and autonomous irrigation systems were required.

13.4 PROPOSED SYSTEM

The proposed system is based on the concepts of ditch irrigation and sprinkle irrigation. Here, the pipes are used as ditches for storing water and holes in them are used to distribute the water equally in all the parts of the field. In the proposed system, a moisture sensor is placed in the center of the farm's area so that it can detect moisture all over around it. The sensor passes the detected information to the Arduino Uno, or a microcontroller of the system. Further, to cover the enormous area of a farm, the complete farm area is first split into subparts, and then, accordingly, soil moisture sensors are placed in the farm.

An algorithm is then used to measure the threshold value of the soil moisture according to different types of crops, which is programmed into a microcontroller to monitor the soil's humidity content. The algorithm takes the moisture readings from all the moisture sensors and a threshold value is chosen, which is nothing but the average of all readings provided by all sensors. These sensors monitor the soil moisture readings for a particular type of plantation or crop, and the Arduino Uno collates that average value to the threshold value of that specific type of crop. After comparing those values, the microcontroller performs the required action of irrigation. When the moisture content is equal to or more than the threshold value, the pump automatically stops. Crops can now be irrigated automatically even without the presence of a farmer on the farm. The architectural view and the process flow of the proposed system are shown in Figures 13.6 and 13.7, respectively.

13.4.1 SETTING UP SOIL MOISTURE THRESHOLD VALUE

As shown in Table 13.1, the water requirement capacity differs from crop to crop. Since a farmer yields crops according to the seasons and water requirements for a crop varies from season season, the problem arises on how to set up the soil moisture threshold value for a particular type of crop. For that, a programmer takes the readings of the soil moisture for different types of crops at a time interval of 10 minutes, or according to their sketch design, as shown in Figures 13.8 and 13.9.

Figure 13.8 shows that for winter crops, the soil moisture at various time-stamps lies between the ranges of 940–960. Based on that, programmers can set a relevant threshold value range for their program by setting one lower value and the other

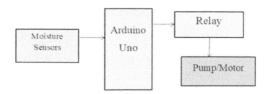

FIGURE 13.6   Architectural view of the proposed system.

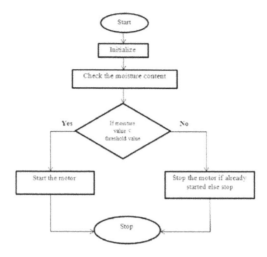

FIGURE 13.7 Process flow of the proposed system.

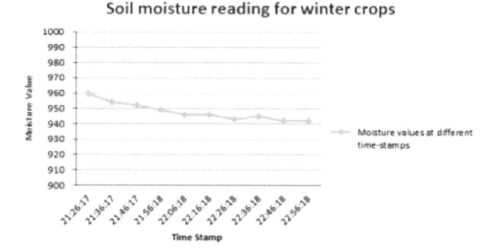

FIGURE 13.8 Soil moisture reading for winter crops.

higher value. Similarly, for different types of crops or plants, they can choose a particular range, as displayed in Figures 13.8 and 13.9. Similarly, some programmers can also calculate moisture reading for other crops like winter crops and rice crops.

Due to the low expenditure of soil moisture sensors and other required hardware, it is convenient to utilize diverse sensors that take less area and easily cover the entire field. The average value obtained from all the sensors' values help design the best-proposed system capable of covering a large area and irrigating it. Further, this system can be implemented where land is not plain or rugged field.

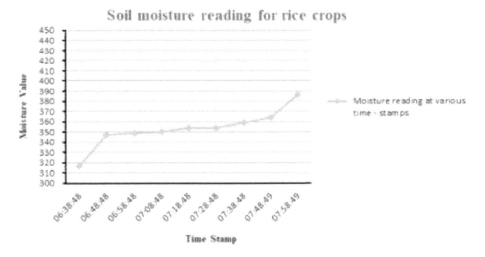

FIGURE 13.9 Soil moisture reading for rice crops.

13.4.2 MATERIALS AND METHODS

This subsection discusses the hardware, software, and other equipments needed to set up the proposed system.

13.4.2.1 Hardware Requirements

Arduino: There are lots of Arduino boards available in the market like Uno, Mega, Lilypad, Nano, etc. In the proposed system, Arduino Uno was utilized to monitor the entire system. It is because, in Uno, we get some extra pins that help to add some extra sensors. All instrumentations or components, for example, relays and moisture sensors, are connected with it. The board features serial communication interfaces, including a USB, on some models used to load Arduino's sketches from an individual's personal computer. The Arduino Uno board used in the introduced system is demonstrated in Figure 13.10(a).

(a)   (b)   (c)   (d)

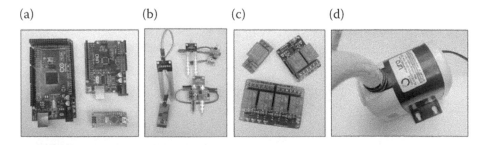

FIGURE 13.10 Hardware components (a) Arduino Uno; (b) FC-28 soil moisture sensor; (c) relay; (d) motor.

Moisture Sensor: A moisture sensor deems the available moisture content in the irrigating soil and shares the data with the Arduino Uno or the micro-controller. It can send the data herein digitally or in the analog configuration. It fetches the required data of the soil's moisture at regular intervals of time. The variation in the obtained soil moisture readings is directly proportional to the quantum of current flowing through the soil. The soil moisture sensor used in the proposed system is demonstrated in Figure 13.10(b). In the proposed system, FC-28 soil moisture sensors are exercised. The sensors are pre-contemplated for both digital as well as analog outputs.

Relay: It is an electrically operated switch that is used to operate higher voltage appliances through a low voltage. Here it is considered to curb the water motor. The relay module utilized for the proposed system is demonstrated in Figure 13.10(c).

Motor: It is used to pump the water. Here, an AC motor is used for the prototype model, shown in Figure 13.10(d).

13.4.2.2 Software Requirements

The software required in developing the model is the Arduino Beta IDE version [29]. The Arduino is open-source software (IDE) that helps users in writing and uploading the code. It is available for almost all operating systems such as macOS, Windows, and Linux. This particular IDE is acquainted with all the types of Arduino boards in existence and can be easily obtained by installing it on personal computers or laptops from the official website of the Arduino. The Arduino Beta IDE and Arduino sketch interface window are demonstrated in Figure 13.11.

13.4.2.3 Supporting Materials

Despite the hardware and software components, some additional supporting equipments are also needed for this proposed system.

Pipes: In this prototype system, PVC pipes are used to transport water from the pump to the field, plants, crops, etc. It works similar to ditch irrigation to store water. Many pipes are available in the market that can be chosen to irrigate fields by farmers like PVCs, Black HDPE coil pipes, etc.

Valve: In this proposed system, valves are used to control the flow of water. As per the landholding, they can be taken into practice to regularize the stream of water in the field. If some want fast flow like high-value crops, they can be opened completely, or else can be controlled according to the need. The valve used for the prototype system

FIGURE 13.11 Arduino IDE interface.

(a)   (b)   (c)

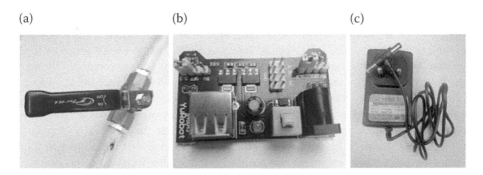

FIGURE 13.12 Supporting Components (a) valve; (b) breadboard power supply module; (c) 12V-2A AC adapter.

is shown in Figure 13.12(a). For the larger area at the time of implementation, the larger size valve according to the size of the pipe can be easily implemented.

Breadboard Power Supply Module: As different sensors operate at different power voltages, a power supply module is used to provide a regular amount of voltage. The main advantage of using this module is that we can easily get 5 V and 3.3 V, which is sufficient to power up the necessary sensors and microcontrollers. It is used to give 5-V power supply to the Arduino board and soil moisture sensors. The breadboard power supply module is displayed in Figure 13.12(b).

12V-2A AC Adapter: To operate the entire circuit, a 12V-2A power adapter is used. Since we cannot provide the necessary power supply to all the actuators or sensors attached with the microcontroller via our personal computers or battery, for the efficient working of the system, we used 12V-2A additional power adapter to provide the appropriate power supply. Figure 13.12(C) shows the 12-V adapter used in the proposed system.

To implement this system practically, only large length wires to connect to the microcontroller or Arduino Uno are required, which is not a major issue. Further, if farmers want to use sprinklers, they can use them. Similarly, for ditch irrigation, they can use pipes at the corner if they wish. But if they want to plough the field again, they may have to remove some of the pipes; otherwise, they may get destroyed during ploughing of the field.

13.5 IMPLEMENTATION AND WORKING

The proposed system is designed to function the irrigation system automatically. The moisture sensors sense the moisture content, and the micro-controller, i.e., Arduino Uno, reads the analog values. We considered analog values because the moisture level changes from time to time and analog values are capable to detect even small changes. Since the Arduino Uno takes only the digital values, it maps all those analog values in a range from 0 to 1023. Considering these values, a programmer analyzes them and sets a threshold value. The changes in analog values for various crops with respect to the time given by the moisture sensor are shown in Figure 13.9. Here, a programmer can set a threshold value from those values, depending on the moisture content of the different types of crops in the program. Using this, farmers can give sufficient and proper amounts of water to the crops as per their requirements. The micro-controller reads all the analog values given by the moisture sensor after a particular time interval and that value is compared with the threshold value. After comparison, the required action will take place, as shown in Figure 13.13. The hardware setup of the prototype model is shown in Figure 13.14.

Controlling the valve can set the flow of water. As now the drift of water is controlled, so finally the water reaches those areas that really have to irrigate uniformly. As shown in Figure 13.13(b), the flow of water is distributed so that the water cannot enter from one particular position, but can enter from various positions to the irrigating field. This results in complete irrigation of the field as water is coming from all sides and in sprinkle form.

When the water reaches to the sensor, the soil's moisture again changes the readings, and the sensor gives some analog value. That value is further compared with the threshold value, and based on that a specific operation takes place. The main benefit of this module is that one can simply change the lower and upper limit of the threshold range in the sketch (program) and set it according to other crop's requirements. It results in proper irrigation by using a fewer amount of water. Further,

(a)   (b)

FIGURE 13.13   Prototype model (a) before; (b) during the irrigation.

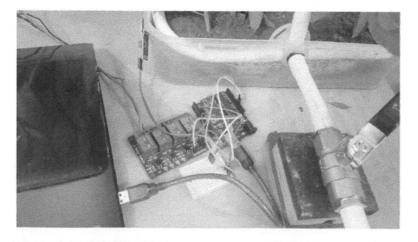

FIGURE 13.14   Hardware setup for prototype testing.

we can assemble all the hardware parts in a box or in some other similar item so that the hardware can remain safe from natural climatic conditions such as rain, fog, etc.

13.6   RESULTS AND DISCUSSION

A smart irrigation system is basically designed to help the farmers to reduce their work pressure. This proposed system is helpful for the persons working in fields or to the farmers. It helps them save time and in between that, they can make use of their time in manifold activities, which will be the main consideration for agriculturists or farmers. Certainly, it would help them in improving their economic conditions by producing good crops.

The self-operating capability of this module solicits the minimal quantum of water for the irrigation purpose-related works. It extends the contribution of itself in reducing the groundwater scarcity issue. Further, due to this model, the energy utility is lower than the conventional method.

The sensors calculate the average of the moisture and the microcontroller performs the required action. However, sometimes it may be possible that hardware or sensors can face some problems. So here the main question arises, how can farmers rectify those problems? Farming cannot become fully automatic, as sometimes farmers have to visit their fields to check the system's performance by observing moisture or water in the fields. If there are problems in the hardware components, there may be an excessive amount of water present in the field or there may be no water or less water. Whether the farmers are implementing this system on their own or with the help of some other person, they can easily track such problems. Once tracked, the system can be tested by observing the output values using personal computers. Based on that, the requisite changes can be easily made in the program. After performing testing again, the system can be implemented again at its place.

13.6.1   SIMULATION RESULTS

We used Arduino software for better functioning of the introduced hardware system. Figure 13.15(a) shows the simulation results of the moisture content of the winter crops after 10 minutes. This data helps us in choosing the lower and higher limits for the threshold range. Similarly, Figure 13.15(b) represents the simulation results of the rice crops. All the values are analog values whose range lies from 0 to 1023. If we want to see a scale of 1 to 100%, we can map these values.

As all the required data of soil moisture for different types of plantations, crops, and others are available, a working prototype model is also ready, which demonstrates the smart irrigation system. Now the main concern is implementing this model in daily life like a flower bed, field, garden, etc. Figure 13.16 shows the implementation of this proposed system in a flower bed (or a small garden). Figure 13.16(a) represents the smart irrigation system setup and it also represents the flower bed before irrigation. Initially, the soil is dry, which means it requires water. The Arduino compares the moisture value to the flower bed's threshold value and an action takes place. It turns on the motor. Figure 13.16(b) demonstrates the outlet of water from the pipes.

(a)   (b)

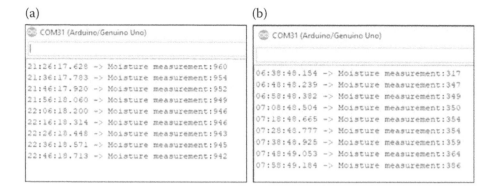

FIGURE 13.15 Simulation results of the moisture content of the winter and rice crops.

(a)   (b)

FIGURE 13.16 Flower bed (a) before irrigation; (b) after irrigation.

13.7   CONCLUSION

In this work, a smart irrigation system has been implemented using intelligent systems and machine learning approaches. A moisture sensor is used to check different crops' moisture levels, and accordingly, the pump will be controlled. The sensor gives the analog values to the microcontroller used in the system. After comparing those values to the threshold value, the microcontroller interrupts the motor, i.e., the pump used to irrigate the fields. The mean of implementing the work is successfully achieved and fulfills the expected objectives. Using this system, farmers do not need to irrigate manually in extremely odd weather conditions, nights, etc. The implemented prototype module does not require the farmers' availability at the time of irrigation. The proposed system automatically monitors and controls the pump status by turning it on and off according to the soil's moisture in fields. The software and hardware utilized in the proposed system performed their functioning exactly as per the specifications. Usage of time concept in a micro-controller makes the model more flexible with time stamp settings for efficiently running a water pipeline for irrigating the farm.

13.8 FUTURE RESEARCH DIRECTIONS

The proposed system in the future can gain ground by making it more intelligent. A tract may be prepared that will utilize less quantities of hardware and perform well. It should be made simpler or more advanced by using the next upcoming technologies or sensors. Despite using sensors at different locations, one can simply use an artificial intelligence system that will pay attention to the soil moisture content after a particular time interval at various places. On the basis of this, the water motor will be curbed.

REFERENCES

Agarkhed, J. (2017). IoT Based WSN for Irrigation System- A Review. *International Journal of Research in Advent Technology*, 5, 26–29.

Arduino IDE. (2020). Retrieved from https://www.arduino.cc/en/software

Banumathi, P., Saravanan, D., Sathiyapriya, M., & Saranya, S. (2017). An Android Based Automatic Irrigation System Using Bayesian Network with SMS and Voice Alert. *International Journal of Scientific Research in Computer Science, Engineering and Information Technology*, 2(2), 573–578.

Cassman, K. G. (1999). Ecological intensification of cereal production systems: Yield potential, soil quality, and precision agriculture. *Proceedings of the National Academy of Sciences*, 96(11), 5952–5959.

Duhan, J. S., Kumar, R., Kumar, N., Kaur, P., Nehra, K., & Duhan, S. (2017). Nanotechnology: The new perspective in precision agriculture. *Biotechnology Reports*, 15, 11–23.

Dursun, M., & Ozden, S. (2011). A wireless application of drip irrigation automation supported by soil moisture sensors. *Scientific Research and Essays*, 6(7), 1573–1582. https://doi.org/10.5897/SRE10.949.

Gebbers, R., & Adamchuk, V. I. (2010). Precision agriculture and food security. *Science*, 327(5967), 828–831.

Ilangovan, R. (2017, August 5). Rainfall in India. *Kaggle*. Retrieved from https://www.kaggle.com/rajanand/rainfall-in-india.

Jain, R. K., Gupta, B., Ansari, M., & Ray, P. P. (2020). IOT enabled smart drip irrigation system using web/Android applications. *11th International Conference on Computing, Communication and Networking Technologies (ICCCNT)*, 1–6. 10.1109/ICCCNT49239.2020.9225345.

János, S., & Matijevics, I. (2010). Implementation of potential field method for mobile robot navigation in greenhouse environment with WSN support. *IEEE 8th International Symposium on Intelligent Systems and Informatics*, Subotica, 319–323. 10.1109/SISY.2010.5647434.

Kalezhi, J., Rwegasira, D., Dhaou, I. B., & Tenhunen, H. (2019). A DC microgrid smart-irrigation system using internet of things technology. *2019 IEEE PES/IAS PowerAfrica*, 318–322. 10.1109/PowerAfrica.2019.8928795.

Kamienski, C., Soininen, J. P., Taumberger, M., Fernandes, S., Toscano, A., Cinotti, T. S., ... Neto, A. T. (2018, June). Swamp: an IoT based smart water management platform for precision irrigation in agriculture. 2018 Global Internet of Things Summit (GIoTS), 1–6.

Khan, A., Singh, S., Shukla, S., & Pandey, A. (2017). Automatic irrigation system using internet of things. *International Journal of Advance Research, Ideas and Innovations in Technology*, 3(2), 526–529.

Kumar, M. K., & Ravi, K. S. (2020). Automation of irrigation system based on Wi-Fi technology and IOT. *Indian Journal of Science and Technology, 9*(17), 1–5. 10.17485/ijst/2016/v9i17/93048.

Meena, H., Nandanwar, H., Pahl, D., & Chauhan, A. (2020). IoT based perceptive monitoring and controlling an automated irrigation system. *11th International Conference on Computing, Communication and Networking Technologies (ICCCNT)*, 1–6. 10.1109/ICCCNT49239.2020.9225455.

Nandhini, R., Poovizhi, S., Jose, P., Ranjitha, R., & Anila, S. (2017). Arduino based smart irrigation system using IoT. *3rd National Conference on Intelligent Information and Computing Technologies IICT*, Coimbatore, India, 1–5.

Pierce, F. J., & Nowak, P. (1999). Aspects of precision agriculture. *Advances in Agronomy, 67*, 1–85. Academic Press, , Newark, DE, USA.

Prakash, B. R., & Kulkarni, S. S. (2020). Super smart irrigation system using internet of things. *2020 7th International Conference on Smart Structures and Systems (ICSSS)*. DOI:10.1109/icsss49621.2020.9202361.

Rawal, S. (2017). IOT based Smart Irrigation System. *International Journal of Computer Applications, 159*(8), 7–11.

Sabrol, H., & Satish, K. (2016). Tomato plant disease classification in digital images using classification tree. *2016 International Conference on Communication and Signal Processing (ICCSP)*, 1242–1246. https://doi.org/10.1109/ICCSP.2016.7754351.

Salikhov, R. B., & Zainitdinova, A. A. (2018). System of monitoring and remote control of microclimate in greenhouses. *2018 XIV International Scientific-Technical Conference on Actual Problems of Electronics Instrument Engineering (APEIE)*, 265–267. https://doi.org/10.1109/APEIE.2018.8545889.

Statista. (2020). *India: Distribution of gross domestic product (GDP) across economic sectors from 2009 to 2019*. Retrieved from https://www.statista.com/statistics/271329/distribution-of-gross-domestic-product-gdp-across-economic-sectors-in-india/

Sirohi, K., Tanwar, A. H., & Jindal, P. (2017). Automated irrigation and fire alert system based on hargreaves equation using weather forecast and ZigBee protocol. *2016 2nd International Conference on Communication Control and Intelligent Systems (CCIS)*, 13–17. https://doi.org/10.1109/CCIntelS.2016.7878191.

Shabadi, L., Patil, N., Nikita, M., Shruti, J., Smitha, P., & Swati, C. (2014). Irrigation control system using android and GSM for efficient use of water and power. *International Journal of Advanced Research in Computer Science and Software Engineering, 4*(7), 607–611.

Srbinovska, M., Gavrovski, C., Dimcev, V., Krkoleva, A., & Borozan, V. (2015). Environmental parameters monitoring in precision agriculture using wireless sensor networks. *Journal of Cleaner Production, 88*, 297–307.

Tokekar, P., Vander Hook, J., Mulla, D., & Isler, V. (2016). Sensor planning for a symbiotic UAV and UGV system for precision agriculture. *IEEE Transactions on Robotics, 32*(6), 1498–1511.

Upadhya, S. M., & Mathew, S. (2019). Implementation of fuzzy logic in estimating yield of a vegetable crop. *Third National Conference on Computational Intelligence (NCCI)*, Bangalore, India, 1427, 1–11. 10.1088/1742-6596/1427/1/012013.

Yalla, S. P., RajeshKumar, K. V., & Ramesh, B. (2013). Energy management in an automated solar powered irrigation system. *2013 International Conference on Information Communication and Embedded Systems (ICICES)*, 1136–1140. 10.1109/ICICES.2013.6508260.

Zinnat, S. B., & Abdullah, M. (2014). Design of a fuzzy logic based automated shading and irrigation system. *17th International Conference on Computer and Information Technology (ICCIT)*, 170–173. 10.1109/ICCITechn.2014.7073098.

Zhang, N., Wang, M., & Wang, N. (2002). Precision agriculture—A worldwide overview. *Computers and Electronics in Agriculture, 36*(2-3), 113–132.

14 Lightweight Cryptography Using a Trust-Based System for Internet of Things (IoT)

Amit Chadgal and Arvind Selwal
Central University of Jammu, Jammu and Kashmir - 181143, India

CONTENTS

14.1   INTRODUCTION

An innovative technology wherein multiple sensors, smart devices, and items are associated with one another to carry out interaction without utilizing any human endeavors is known as the Internet of Things (IoT). The devices work autonomously, depending on the association among devices. Analyzing collected data for decision making, the provision of lightweight data, the extraction of useful data, and validating cloud-oriented resources are many of the acts IoT devices perform. Users, devices, sensors, and items are extremely closely connected to one another via IoT. The technologies that range from smart grid medical applications to intelligent transportation networks incorporate IoT within them. Due to high opportunities for business presented in the IoT cases (Rose, Eldridge, & Lyman, 2015), the scale of intelligent devices and services delivered by IoT networks has been enormous. This heterogeneity of IoT devices on the cloud infrastructure; IoT networks have been built so that the data can be distributed through various applications.

There are certain applications in which IoT architectures are being deployed lately. Some of these applications are presented below:

i. Medical and healthcare industry. Patients' medical parameters and drug distribution are tracked using the RFID-sensor capabilities of the mobile phone, which is one of several healthcare sector IoT applications (Bandyopadhyay & Sen, 2011). In case of any accidents, prompt medical attention is required and this is easily provided by IoT systems. The diseases of patients can easily be identified, monitored, and prevented by these systems. Also, the records of patients can be stored, which can be useful in future emergency conditions through implantable and addressable wireless devices.

ii. Manufacturing industry: The optimization of production processes and monitoring of complete cycle of objects from generation to their disposal is possible by linking objects with information technology. This connection can be established either by embedding the smart devices or by using the distinctive identifiers and data carriers, which can be helpful in interaction among the information systems and the intelligent supporting network architecture. Very detailed data or status of various objects and production machines can be defined by tagging the objects and containers. For the refined production schedules and improved logistics, the fine grain information is given as input. Around the identifiable objects, self-organizing and intelligent manufacturing solutions are designed.

iii. Smart cities: For improving the smartness of cities an important role is played by the IoT, which helps in monitoring the parking spaces existing in the city, examination of building and bridge conditions, and examining the pedestrian levels and vehicles. Further, the sensitive regions of cities, the vehicles levels, and adaptive lighting in street lights are monitored through these systems. Depending upon the climatic conditions and unexpected events occurring in the surroundings, various warning messages and diversions are provided to the intelligent highways and smart roads.

iv. Smart agriculture and smart water: The soil moisture and trunk diameters of vineyards are monitored using IoT systems so that the agricultural field can be improved and strengthened. The amount of vitamins present in agricultural products are maintained through this monitoring process, which can thus result in increasing the growth and production. The fungus and other microbial contaminants that cause problems in plants are prevented by controlling the humidity and temperature level. Further, the various environmental conditions are also forecasted easily when IoT systems are deployed (Bhuvaneswari & Porkodi, 2014).

v. Security and emergencies: In this field, the use of IoT technologies has increased to a great extent. It is possible to monitor the radiation levels, liquid presence, and perimeter access controls in industries to reduce the possibility of disasters. The entry of unauthorized people to restricted regions is detected and controlled by using the perimeter access control mechanism. Further, various breakdowns and corrosions of huge data centers, sensitive buildings, and warehouses can be prevented using liquid presence. Any kind of gas leakages and levels within industrial applications and the surroundings of mines and chemical factories are monitored by applying IoT. Within nuclear power stations, the radiation levels are measured using radiation levels applications.

14.1.1 CHALLENGES IN INTERNET OF THINGS

Divergent protocols and technologies are utilized to implement the IoT components. Thus, the configurations are complex and designs are poor in case of the components.

There are five important parameters which contribute to arise in technological challenges which are presented below:

i. **Security:** Several public and private sector organizations have been focusing on the major security problems that are occurring in IoT. It will become very easy for the experienced attackers to recognize the security holes and enter the applications due to the presence of such large numbers of new hubs within the systems and web (Surender et al., 2013). The past studies have shown that some basic applications have been intruded and deployed against their own servers by the IoT devices which are under the influence of malware. Depending upon the type of manner, in which IoT has been included in our lives, security advancements can be done in future.

ii. **Connectivity:** Connecting various devices would be another issue which will be commonly found in future IoT systems. The structures and relevant technologies of IoT being used today will be resisted to solve this issue. For the authentication, connection, and authorization of different terminals present within the network, a centralized architecture has been used currently. The models deployed today are appropriate based only on today's conditions. Therefore, for today's current scenarios, it is appropriate to apply this method. However, the future scenarios in which billions of devices are a part of one individual network, the scalability of this model is not enough. The current centralized system is blocked due to the occurrence of such conditions. Since a total system shutdown is possible due to the unavailability of servers, huge investments and expenditures are provided on the cloud clusters of servers such that the humongous amount of information being transmitted can be controlled.

iii. **Compatibility and Longevity:** There is a huge growth in demand for IoT within different technologies. The Internet of Things can be used with several technologies but the additional hardware and software are required. Different technologies have been incorporated already and a convention will be advanced soon. Serious threats will be professed and additional software and hardware requirements will be set up such that communication amongst the devices can be made possible. Few other compatibility problems are also faced within these systems. Since these technologies are becoming out of date soon, the devices working on these technologies have become purposeless in the future.

14.1.2 Destination-Oriented Directed Acyclic Graph

DODAG is a type of Directed Acyclic Graph that uses RPL routing protocol and is rooted at the sink. DODAG Information Object (DIO) messages are periodically originated from the root for initiating DODAG creation. This produced DODAG is advertised using the link-local multicast (Baccelli, Philipp, & Goyal, 2011). The DIO messages contain details related to DODAG's root name, the routing parameters used, and the rank of originating router. The router entering the DODAG decides its own rank according to the details its neighbors advertise within their DIOs.

Figure 14.1 shows a basic example of the DODAG construct process. Node 3 is beyond the root node radio range of all three client nodes mentioned in the figure. The DIO control message is broadcast with its rank and the id is transmitted to the client nodes when the root node initiates a network topology generation (Zhang & Li, 2014). Here, DAO messages are sent from client node 1 and node 2 to the root to enter DODAG as they are within the root's radio range. Client node 3 waits to hear from the root node for a period of time but does not receive any response. So, the neighboring nodes start proactively sending DIS message to the solicit DIO. If client node 2 receives this DIS message in any event, the DIO message which is sent earlier to client node 3 will be forwarded. After receiving the DIO message, a DAO message is sent back by client node 3 to client node 2. This message is then forwarded from client node 2 to the root node. Therefore, the development process should be completed when the client node 3 ends up joining the DODAG.

14.1.3 IPv6 Low Power and Lossy Networks Routing Protocol (RPL)

RPL is a structured protocol for remote routing of vectors. Due to the way the devices are connected, there is no process in the protocol, and therefore no loop. The DODAG is avoiding cycles of internet wired border routers. All DODAG-connected devices are linked via the border router to the Internet. The protocol avoids loops when measuring a node's location regarding the root node (Internet Engineering Task Force, 2017). This location is termed as ranking of the root node, and the ranking improves if one steps away from the edge router. Loops are prevented by avoiding transmissions from a moving child device. A node has a parent

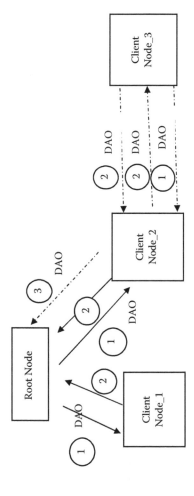

FIGURE 14.1   DODAG construction process.

that transfers data to the border router from the nodes, and may have multiple kids. The node is liable to transmit children's packets to the border router.

The control packets known as DIO and DIS are deployed in the Destination Oriented Directed Acyclic Graph (DODAG) for expressing the information regarding it. The formation of DODAG is described using some phases. Initially, a root is set to forward local multicast DIO messages. These messages are received by the nodes nearby to the root. The processing of message is done as they are taken from the node which is of lower rank in comparison to them. The root is chosen as their parent. After that, these nodes assist in transmitting the link-local multicast DIOs to the other nodes. The rank of DIOs is evaluated on the basis of the OF and the rank of the DIO sender by the node when it received these messages. Meantime, a node sends a request for DIO from the root with the help of DODAG Information Solicitation (DIS). The DIO is sent instantly via DODAG root after receiving the request. When the DODAG is designed, the DIO messages are employed to maintain it.

14.1.4 Version Number Attack

The root node uses the version number to guarantee that the global RPL repair mechanism is managed and that the routing status for all DODAG nodes is changed. The history of the IoT system is one due to the nature of version numbers attacks. This attack can be carried out by the attacker at even less cost and by using a global recovery system, which is known as a resistant protocol, the network can be resolved. The root undertakes a global repair (Mayzaud, Badonnel, & Chrisment, 2017a) with several anomalies in the network. For the final DODAG rebuild, the DODAG version number is raised. The DODAG Information Object (DIO) is the version number control file. There are similarities between the current version number and the one obtained from its parent from each receiving node. In the case of a higher variant, information about the current rank is discarded, the trickle timer reset is reset, and the DODAG entry mechanism is added. A loop-free topology is assured for this global repair given its very high costs. You can presume that the node did not update to the new version of DODAG if the previous version value is shown in DIO post. The other nodes do not pick a node like the one chosen by their parent. The version number must be propagated unchanged through DODAG so as to prevent any potential network inconsistencies. No technique has been provided in RPL for ensuring the validity of version number available in received DIO messages. If a malicious node may change this value in its own DIO messages, the network can be harmed here. When a malicious DIO is received with a new version number, the trickle timer is reset, the version is modified and the new version is advertised to the neighbors using DIO Messages. The fraudulent version number is propagated across the network. The valid network nodes in this scenario immediately spread the incremented version number that is introduced by malicious node.

14.1.5 Trust and Reputation

Trust and reputation is a tool for security in environments where various groups are communicating and interacting.

i. Trust: We need to identify trust in general before we consider at trust in the manner of trust systems. Firstly, there are various meanings of trust, but compared to the confidence and reputation structures, the concept of Gambetta suits better "Trust is a specific degree of subjective likelihood with which a node assesses that another node, will carry out a certain action, both before and may control that action (or regardless of their ability to carry out that action at all times)". Based on this precision, trust can be defined as the probability of an individual behaves as expected.

ii. Reputation: It is an approximation of how a node will act in future, dependent on his previous behavioral experiences. Reputation may be the sum of various agent experiences or interactions based solely on the experience gained in the past by one single agent. It is used to provides an added source on which nodes should dependent when making informed decisions.

14.2 LITERATURE REVIEW

Due to the dynamic nature of Internet of Things, routing, security, and quality of service are the three major issues of this network. The various type of routing protocols are used for the path establishment from source to destination. This research work is related to detection and isolation of malicious nodes from the network. In this chapter, various techniques are presented to increase the security of the Internet of Things (Table 14.1).

14.2.1 RELATED WORK

Ahmet Aris et al. (2017) proposed two simple alleviation approaches for the elimination of RPL VN intrusions. The primary mitigation technique eliminated version number changes utilizing leaf gadgets. The underlying plan gave virtual numbers insights regarding the toughness of the areas. The underlying mitigation procedure was not valuable to limit the attack from the other intrusion locales. The second strategy of mitigation was fit for moderating the effect of the assault without thinking about where the assault was occurring. In the subsequent strategy, the nodes could possibly change their virtual number when the mass number of nearest hubs with better positions announced a virtual number changed. The possibility of the proposed arrangement was tried with the utilization of numerous topologies. Further examinations in the field of various virtual number attacks circumstance can be completed later on. Hybrid mitigation technique examination may likewise be done in the short term.

Dvir et al. (2011) recommended another routing protocol to be utilized to wipe out the issues introduced in the LPLNs. This protocol has been called routing protocol IPv6. The primary goal of this routing protocol was to give the Low Power and Lossy Networks usefulness. TheRPL provided pathway multiplicity by developing and dealing with the DAGs through at least one gateway. Thus, an opponent who had placed an individual node close to the gateway could by itself be able to divert a longer phase of network traffic. An up-to-date version called DODAG has been used to reconstruct the topology of the routing. It was also

TABLE 14.1

Comparative Summary of Lightweight Cryptographic Techniques

Reference No. & Author's Name	Technique	Advantages/ Features	Disadvantages/ Improvements
(Ahmet Arıs et al., 2019)	Two simple alleviation approaches for the elimination of RPL version number attacks are proposed.	The efficiency of the proposed solution was tested with the use of many topologies.	The study and implementation of the multiple Version Number attackers situation was not performed in this study.
(Dvir et al., 2011)	This routing protocol's main objective was to provide the flexibility in the Lossy Networks with Low Power.	Through this process, a greater segment of DODAG was combined with existing DODAG via the intruder, for a greater part of network traffic forwarding.	The proposed technique was only implemented in standard systems for evaluations and the future work can be extended by implementing this technique in RPL and real wireless sensor deployments.
(Mayzaud et al., 2017a)	A new classification method for categorizing the attacks found in addition to the RPL in which there were considered primarily three types of attacks.	The researchers have suggested a number of methods based on different properties for avoidance of these types of attacks.	This proposed technique is only able to secure the networks that include one malicious node and thus, the future work can be extended for making improvements in this technique to ensure that systems with multiple malicious nodes can be secured.
(Khan, Herrmann, Ullrich, & Voyiatzis, 2017)	For managing the status information about the neighbors, the proposed approach used the faith management technique. The proposed approach proved very successful for singling out nastily behaving units.	For the validation of MATLAB simulations, a test bed involving Z1 components will be developed in near future.	The proposed technique is not enough computing intensive due to which several issues are being faced.
(Abdo et al., 2018)	A novel method has been suggested to ensure the	The outcomes tested showed the suggested	The issue of uncertainty still exists after

TABLE 14.1 (Continued)
Comparative Summary of Lightweight Cryptographic Techniques

Reference No. & Author's Name	Technique	Advantages/ Features	Disadvantages/ Improvements
	health and security of industrial hazard investigations. In this reason, the lately developed version of the security analysis was combined with a usually used safety investigation program called bowtie analysis.	solution was working well.	implementing the proposed approach which can be resolved by making improvements in this approach in future.
(Aris et al., 2016)	It also carried out the analysis of the attacks, which was focused on various cases. The research was done on a functional topology of the network that included both mobile and stationary nodes.	The simulation displayed that the performance of the mobile attackers and distant nodes had marginally similar effects on the network.	It is still easy to drain the resources and reduce the lifetime of networks if the fraudulent node is kept on mobile network identity.
(Abdul-Ghani et al., 2018)	A new approach to the internet of things was suggested based on building blocks policy.	An IoT invasion model was developed for the very first time that backed on building block reference model. The tested results clearly depicted the effectiveness of the proposed model.	No standard method has still been defined for securing the IoT systems.
(Mayzaud et al., 2020)	A solution based on distributed monitoring design was implemented for the conversation of node resources. For the identification of the attacker in the proposed approach, the collaboration of observed nodes was exploited.	With the help of strategic monitoring nodes placement, the false positive rate could be reduced.	Only one certain parameter is emphasized here by the proposed technique which makes it possible for the technique to lack in accuracy and privacy aspects.

(Continued)

TABLE 14.1 (Continued)
Comparative Summary of Lightweight Cryptographic Techniques

Reference No. & Author's Name	Technique	Advantages/ Features	Disadvantages/ Improvements
(Chen et al., 2016)	An adaptive trust management technique was proposed through which the social relationships could be evolved dynamically across the authorities of IoT devices.	A table-lookup method was applied here, which demonstrates better performance of proposed technique.	The mutual authentication technique needs to apply to validate the IoT devices.

suggested that the same approach should be used for avoiding of the primary invader, to publish the decreasing rank value. This approach merged a larger segment of the DODAG with the DODAG via the invader for major network traffic to be forwarded. Consequently, the illegitimate increment in version number could be evaded using this new security element.

Mayzaud et al. (2017b) presented a fresh approach to classify the intrusions found besides the RPL. Generally, three classes of were taken for that approach. The network's life had been reduced by exploiting resources. These intrusions generated many bogus communications or built a number of loops. When the attacks against the topologies were carried out, the network aggregated to a sub-optimal arrangement. In case of attacks, the malicious node captured and the examined a large part of the network. The researchers have given various approaches based on different properties for the avoidance of these kind of attacks. The RPL specification technique didn't mention the implementation and the control of the security modes. Thus, it was concluded that a big challenge to the structure of RPL networks was the transaction between different security levels.

Khan and Herrmann (2017) proposed some new approaches for IDS which were very suitable for the tiny devices. For managing the status information about the neighbors, the proposed approach used the faith management technique. The proposed approach proved very successful for singling out nastily behaving units. This process was completed in a power-oriented system. The main aim of the trust management subjective logic was the recognition of the attacker nodes presented in the system.

Abdo et al. (2018) proposed an approach to ensure the safety and security of industrial threat investigations. To accomplish this, the new version of the security analysis was combined with a previously used safety investigation system called bowtie analysis. The updated variant was used for examining the attack tree.

A detailed evidence of the risk scenario was given with regard to security and privacy, by combining the approach of bowtie analysis and attack tree. A fresh mechanism was presented for the risk range evaluation relying on two term similar parts. The first included security while the next one was for safety. The outcomes displayed that proposed scheme was working great. In future, the specialists must set up a more exact and intense method for assessing the likelihood of attack.

Aris et al. (2016) conducted an in-depth analysis of attacks on number version of the RPL. It also carried out the analysis of the attacks, which was focused on various cases. The research was carried out on a functional topology of the network that included both mobile and stationary nodes. There were multiple cardinalities in those nodes. The examination work was centered around the particulars for IETF routing. Likewise, the impact of attacks of version number on devise power usage was resolved. A probabilistic methodology was utilized to quantify the chances for the assaults. The impacts of the achievement were displayed by the different estimations of p. The aftereffect of reenactment demonstrated that exhibition of the portable aggressors and inaccessible hubs had almost comparable impacts on the organization. An examination will be directed later on the coming activities of DIO data to recognize the conceivable part of virtual number assault.

Abdul-Ghani et al. (2018) introduced a new web-based approach to suggestive issues focused on building blocks regulation. This was, essentially, a reference model of four layers. An inclusive model of IT invasion including four main phases was created. First was provided an IoT asset based on invasion plane consisting of four mechanisms. These elements were details, software, protocol muffle the entire objects, and IoT stack of substance. A pattern of the purpose of IoT security was established in the second step. In the third step the IoT invasion classification has been established for each part. In the final step, there was established breach of security goals and the connection within each attack. Also, in the last step a collection of solution was defined for protecting each asset. An IoT invasion model was developed for the very first time that relied on building block reference model. The test results clearly showed the effectiveness of the proposed model.

Aris & Oktug (2020) focused on understanding the impact of multiple VNA in IoT networks based on RPL. Various attackers were considered during the formation of security systems. The analysis of impact occurred due to multiple attackers was carried out from different perceptions. It was observed in the simulations and analysis that the maximization of attackers had affected only PDR and there was not any effect on the network delay and average energy utilization. The outcomes demonstrated that longer delays and superior power consumption was occurred in case the attacking were nearer to the root. At last, the performance of suggested mitigation method was quantified while dealing with multiple attackers.

Chen et al. (2016) suggested an adaptive confidence management strategy that could dynamically develop social relationships through IoT system authorities. Here the design trade-off among confidence linkup and trust variation was seen within the adaptive trust management protocol design. To ensure high accuracy of trust evaluation along with performance increases, the best confidence parameter setting for modifying IoT social conditions was selected via a social IoT framework using the protocol for adaptive confidence management. A table-lookup approach was applied here to evaluate the results and to demonstrate the viability of the proposed adaptive trust management strategy, which proved that the performance of the proposed technique was better.

14.2.2 RESEARCH CHALLENGES

Following are the various research gaps:

- The techniques which are proposed so far to increase security of the network are the encryption schemes. The encryption schemes which are proposed so far are complex which increase security of the network.
- The techniques to increase the security has various type of security attacks are possible like brute force attack, man-in-middle attack, etc. The technique needs to propose which have the least number of attacks are possible in the network.
- The lightweight cryptographic techniques need to propose to increase security in which key should be generated dynamically.

The objective is to design trust-based technique for the detection of multiple version number attack in IoT, implement a designed scheme, and compare it with the already existing scheme for the detection of Version Number Attack in IoT.

14.3 RESEARCH METHODOLOGY

Following are the various steps of the implementation strategy (Figure 14.2):

- **Deployment of the Network:** The network will be set up with the confined number of sensor devices and with the base station. The sensor devices are responsible to sense various types of conditions like pressure, temperature, etc. The sensor devices are of heterogeneous in nature means sensors have different battery and processing power. The DODAG protocol will arrange the network into treelike structure.
- **Trigger of version number attack:** The malicious nodes will be formed in the network, which are responsible for triggering a VN attack. The contagious nodes will update the version number in the DODAG protocol. The DODAG protocol will select the path which has high version number this leads to creation of loop-based paths in the network.
- **Trust Calculation:** The trust-based scheme will be implemented in this research for the mitigation of the VN (Version Number) attack. The trust-based scheme will work, in the three phases which are pre-processing, trust calculation, and trust updation. The sensor devices which have least trust will be marked as malicious from the network.
- **Analyze network performance:** In the last phase, the network performance will be analyzed in terms of certain parameters to check the effectiveness of the proposed technique.

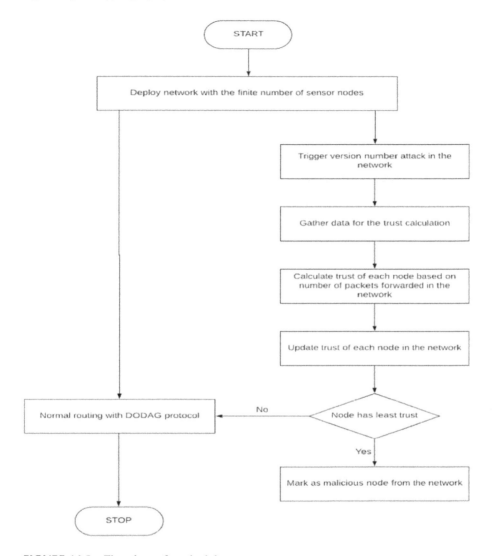

FIGURE 14.2 Flowchart of methodology.

14.3.1 PROPOSED ALGORITHM

1. Deployment network with the finite number of sensor nodes
2. Define source and destination nodes in the network
3. If (Path exists from source to destination)
   3.1. Transmit data on established path

4. Else
   4.1. Source flood route request packets in network
   4.2. Destination reply back with route request packets

    4.3. Check version number of each path

5. Calculate trust of each node
    5.1. if (version number from least trusted node)
        5.1.2. Mark node

Else
    Continue on established
    End

Eventually, the keys are created from the encryption. The information is isolated into blocks and each square is encrypted separately to produce last encode. The chaos-based information cryptosystem generally comprises of two phases. The plain information is given at its info. There are two phases in the chaos-based information cryptosystem (Figure 14.3). The confusion phase is where the situation of the information is mixed over the entire information without angering the estimation of the information and the sensed data winds up doubtlessly non-identifiable. The data change is finished by a chaotic framework. The chaotic conduct is constrained by the underlying conditions and control boundaries which are gotten from the 16-character key. To improve the security, the second phase of the encryption cycle targets changing the estimation of each data to shield sensed data from attackers.

The confusion phase is the data change where the situation of the data is mixed over the sensed data without bothering the estimation of the data and the sensed data winds up discernibly unidentifiable. In this manner, these underlying conditions and control boundaries fill in as the secret key. It isn't particularly secure to have quite recently the change stage since it very well may be broken by any attack. To improve the security, the second phase of the encryption cycle targets changing the estimation of each data in the whole sensed data. The cycle of diffusion is moreover brought out through a chaotic map which is generally reliant on the underlying conditions and control boundaries. In the diffusion stage, the data esteems are adjusted successively by the arrangement created from one of the three chaotic frameworks picked by

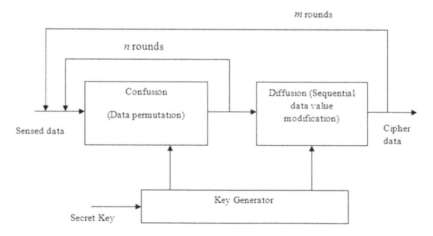

FIGURE 14.3   Architecture of chaos-based data cryptosystem.

outside key. The whole confusion-diffusion round rehashes for different occasions to achieve a pleasing degree of security. The chaotic property trademark in chaotic map makes it more sensible for sensed data encryption.

14.4 RESULT AND DISCUSSION

This research work relates to the isolation of attacks on version number in the IoT (Table 14.2). In the DODAG protocol, the version number attack is triggered. The outcome and discussion section compares the three scenarios, which are DODAG protocol and effect of version number attack on DODAG protocol efficiency in terms of throughput and packet loss. The third scenario relates to the isolation in DODAG protocol of version number attack. It is studied that given scenario has least packet loss and maximum throughput as compared to attack scenario and base paper scenario.

14.4.1 NETWORK DEPLOYMENT

The network is configured in Figure 14.4, with the confined sensor nodes. The network is distributed at random, and the whole network is classified into different clusters. The data will be sent to the base station, divided into clusters, and sense information.

14.4.2 VERSION NUMBER ATTACK

The version number attack trigged in the DODAG protocol is shown in Figure 14.5. The version number attack is triggered in the network in which malicious will create the loop and data will be transmitted in that loop to increase the loop.

14.4.3 ANALYSIS OF PACKET LOSS AND THOUGHPUT

The packet loss of the three scenarios, which are attack scenario, base paper scenario, and new scenario, have been compared for analyzing performance level (Figure 14.6). The results reveal that the proposed scenario has a less amount of packet loss (30.6% approx) as compared to the other two scenarios (Table 14.3).

TABLE 14.2
Simulation parameters

Parameter	Values
Simulator	Ns2-2.35
Number of nodes	32
Area	800 * 800 meter
Antenna type	Omi-directional
Channel	Wireless channel
Propagation Model	Two ray

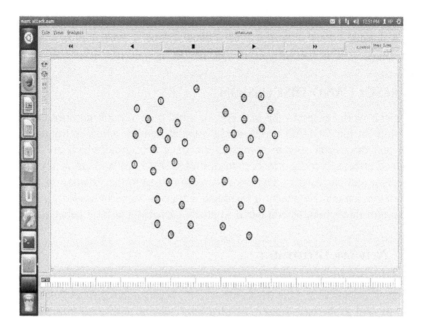

FIGURE 14.4 Network deployment.

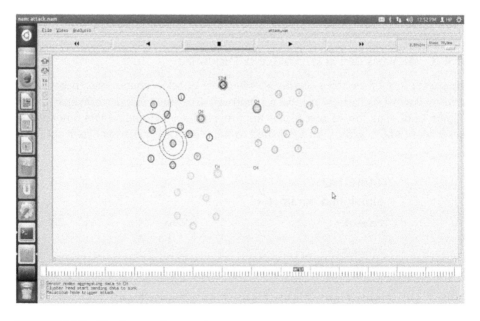

FIGURE 14.5 Version number attack.

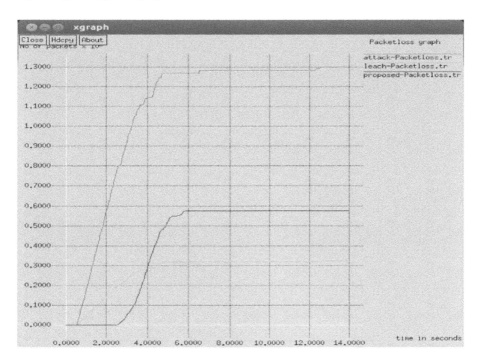

FIGURE 14.6   Analysis of packet loss.

TABLE 14.3
Packet loss analysis

Simulation Time	Existing Algorithm	Proposed Algorithm
4 second	15 packets	12 packets
6 second	21 packets	14 packets
8 second	26 packets	16 packets

The throughput of the three scenarios, which are attack scenario, base paper scenario, and new scenario, have been compared for analyzing performance level (Figure 14.7). The results reveal that the preferred scenario has greater throughput (17.15% approx.) relative to two other scenarios (Table 14.4).

14.5   CONCLUSION

This work relates to minimizing attack by version number in IoT. RPL for IoT devices uses hierarchical topology DODAG, where attacker nodes increment the number of variants, which causes the loop route in the network. In this analysis, the

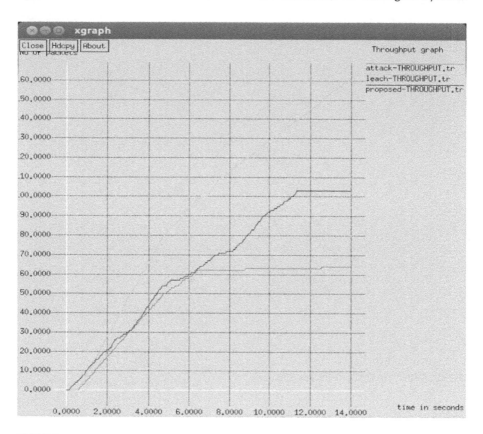

FIGURE 14.7   Analysis of throughput.

TABLE 14.4
Throughput analysis

Simulation Time	Existing Algorithm	Proposed Algorithm
4 second	20 packets	24 packets
6 second	22 packets	26 packets
8 second	25 packets	31 packets

trust-based framework is implemented to help mitigate the attack on the version number of the network. The trust-based system can use less of the network's resources. The suggested scenario is implemented in version 2 of the network simulator and effects are evaluated in terms of both the throughput and the loss of packets. It is analyzed that the suggested scenario's throughput is high relative to two other scenarios. Compared with other scenarios, the packet loss of the proposed scenario is small.

REFERENCES

Abdo, H., Kaouk, M., Flaus, J. M., & Masse, F. (2018). A safety/security risk analysis approach of Industrial Control Systems: A cyber bowtie – combining new version of attack tree with bowtie analysis. *Computers and Security*, 72, 175–195. https://doi.org/10.1016/j.cose.2017.09.004

Abdul-Ghani, H. A., Konstantas, D., & Mahyoub, M. (2018). A comprehensive IoT attacks survey based on a building-blocked reference model. *International Journal of Advanced Computer Science and Applications*, 9(3), 355–373. https://doi.org/10.14569/IJACSA.2018.090349

Aniruddha Chakrabarti. (2015). Emerging open and standard protocol stack for IoT. *LinkedIn*. Retrieved from https://www.linkedin.com/pulse/emerging-open-standard-protocol-stack-iot-aniruddha-chakrabarti

Aris, A., & Oktug, S. F. (2020). Analysis of the RPL version number attack with multiple attackers. *2020 International Conference on Cyber Situational Awareness, Data Analytics and Assessment, Cyber SA 2020*. https://doi.org/10.1109/CyberSA49311.2020.9139695

Aris, A., Oktug, S. F., & Berna Ors Yalcin, S. (2016). RPL version number attacks: In-depth study. *Proceedings of the NOMS 2016: 2016 IEEE/IFIP Network Operations and Management Symposium*, (Noms), 776–779. https://doi.org/10.1109/NOMS.2016.7502897

Arış, A., Örs Yalçın, S. B., & Oktuğ, S. F. (2019). New lightweight mitigation techniques for RPL version number attacks. *Ad Hoc Networks*, 85, 81–91. https://doi.org/https://doi.org/10.1016/j.adhoc.2018.10.022

Baccelli, E., Philipp, M., & Goyal, M. (2011). The P2P-RPL routing protocol for IPv6 sensor networks: Testbed experiments. *2011 International Conference on Software, Telecommunications and Computer Networks, SoftCOM 2011*, 172–177.

Bandyopadhyay, D., & Sen, J. (2011). Internet of things: Applications and challenges in technology and standardization. *Wireless Personal Communications*, 58(1), 49–69. https://doi.org/10.1007/s11277-011-0288-5

Bhuvaneswari, V., & Porkodi, R. (2014). The internet of things (IOT) applications and communication enabling technology standards: An overview. *Proceedings of 2014 International Conference on Intelligent Computing Applications, ICICA 2014*, (October 2017), 324–329. https://doi.org/10.1109/ICICA.2014.73

Chen, I. R., Bao, F., & Guo, J. (2016). Trust-Based Service Management for Social Internet of Things Systems. *IEEE Transactions on Dependable and Secure Computing*, 13(6), 684–696. https://doi.org/10.1109/TDSC.2015.2420552

Chung, B., Kim, J., Jeon, Y., Oss, A. I., & Environments, D. (2016). On-demand security configuration for IoT devices: IEEE Xplore Document, 1082–1084. Retrieved from http://ieeexplore.ieee.org/document/7763373/

Dvir, A., Holczer, T., & Buttyan, L. (2011). VeRA: Version number and rank authentication in RPL. *Proceedings of 8th IEEE International Conference on Mobile Ad-Hoc and Sensor Systems, MASS 2011*, 709–714. https://doi.org/10.1109/MASS.2011.76

Eder, T., Nachtmann, D., & Schreckling, D. (2013). Trust and reputation in the internet of things. *Sec.Uni-Passau.De*, 1, p. 1.

Guo, J., Chen, I., & Tsai, J. J. P. (2016). A survey of trust computation models for service. *Computer Communications*. Retrieved from http://dx.doi.org/10.1016/j.comcom.2016.10.012

Hopalı, E., & Vayvay, Ö. (2018). *Internet of things (IoT) and its challenges for usability in developing countries*, 2(January), 6–9.

Khan, Z. A., & Herrmann, P. (2017). A trust based distributed intrusion detection mechanism for internet of things. *Proceedings of- International Conference on Advanced*

Information Networking and Applications, AINA, 1169–1176. https://doi.org/10.1109/AINA.2017.161

Khan, Z. A., Herrmann, P., Ullrich, J., & Voyiatzis, A. G. (2017). A trust-based resilient routing mechanism for the internet of things. *ACM International Conference Proceeding Series, Part F1305.* https://doi.org/10.1145/3098954.3098963

Mayzaud, A., Badonnel, R., & Chrisment, I. (2016). A taxonomy of attacks in RPL-based internet of things. *International Journal of Network Security, 18*(3), 459–473.

Mayzaud, A., Badonnel, R., & Chrisment, I. (2017a). A distributed monitoring strategy for detecting version number attacks in RPL-based networks. *IEEE Transactions on Network and Service Management, 14*(2), 472–486. https://doi.org/10.1109/TNSM.201 7.2705290

Mayzaud, A., Badonnel, R., & Chrisment, I. (2017b). Detecting version number attacks in RPL-based networks using a distributed monitoring architecture. *2016 12th International Conference on Network and Service Management, CNSM 2016 and Workshops, 3rd International Workshop on Management of SDN and NFV, ManSDN/NFV 2016, and International Workshop on Green ICT and Smart Networking, GISN 2016*, 127–135. https://doi.org/10.1109/CNSM.2016.7818408

Onwuegbuzie, I. U., Razak, S. A., & Isnin, I. F. (2019). Performance evaluation for ContikiMAC, XMAC, CXMAC and NullMAC protocols for energy efficient wireless sensor networks. *2019 IEEE Conference on Wireless Sensors, ICWiSe 2019, 2019-Jan* (November), 12–17. https://doi.org/10.1109/ICWISE47561.2019.8971832

Rose, K., Eldridge, S., & Lyman, C. (2015). The internet of things: An overview. *Internet Society, 5*(October), 53. Retrieved from http://www.internetsociety.org/doc/iot-overview

Savira, F., & Suharsono, Y. (2013). 济无. *Journal of Chemical Information and Modeling, 01*(01), 1689–1699.

Sharma, D., & Selwal, A. (2020). A novel transformation based security scheme for multi-instance fingerprint biometric system. In Badica, C., Liatsis, P., Kharb, L., & Chahal, D. (Eds.), Information, Communication and Computing Technology. ICICCT 2020. Communications in Computer and Information Science, vol. 1170. Springer, Singapore. https://doi.org/10.1007/978-981-15-9671-1_12

Song, Z.; Lazarescu, M. T.; Tomasi, R.; Lavagno, L.; Spirito, M. A. (2014). Internet of Things Applications: Challenges and Opportunties. *Internet of Things*, (May), 75–109.

Suo, H., Wan, J., Zou, C., & Liu, J. (2012). Security in the internet of things: A review. *Proceedings of 2012 International Conference on Computer Science and Electronics Engineering, ICCSEE 2012, 3*(March), 648–651. https://doi.org/10.1109/ICCSEE.2012.373

Surender, J., Arvind, S., Anil, K., & Yashwant, S. (2013). Low overhead time coordinated checkpointing algorithm for mobile distributed systems. In Chaki, N., Meghanathan, N., & Nagamalai, D. (Eds.), Computer Networks & Communications (NetCom). Lecture Notes in Electrical Engineering, vol. 131. Springer, New York, NY. https://doi.org/10.1007/978-1-4614-6154-8_17

Zhang, T., & Li, X. (2014). Evaluating and analyzing the performance of RPL in Contiki. *Proceedings of the International Symposium on Mobile Ad Hoc Networking and Computing (MobiHoc), 2014-Aug* (August), 19–24. https://doi.org/10.1145/2633675.2633678

15 Innovation in Healthcare for Improved Pneumonia Diagnosis with Gradient-Weighted Class Activation Map Visualization

Guramritpal Singh Saggu[1], Keshav Gupta[1], and Palvinder Singh Mann[2]
[1]Indian Institute of Information Technology and Management, Gwalior, India
[2]DAV Institute of Engineering and Technology, Jalandhar, India

CONTENTS

DOI: 10.1201/9781003132080-15

15.1   INTRODUCTION

The danger of pneumonia is gigantic for some, particularly those who belong to the countries which are developing and where most people have to encounter power scarcity and depend on the type of power source that pollutes the environment. The WHO appraises that more than 4 million untimely deaths happen yearly from air-contamination-related infections including pneumonia and 30% of newborns die because of pneumonia (Organization, 2018).

Pneumonia is an inflammation infection that infects the air sacs in either one or both the lungs, which then might get filled with fluid. Pneumonia can happen to anyone irrespective of their age but it is really dangerous for newborns, children, elderly people and individuals with breathing-related disorders. The severity is determined by the root of the infection. For example, while the bacterial pneumonia is far more fatal and could occur suddenly or maybe gradually, viral is less fatal and shows symptoms gradually. However, bacterial infection alongside viral pneumonia can complicate the situation (Bouch & Williams, 2006). Fungal pneumonia also exists, but it only occurs in people with impaired immune systems, which makes it very dangerous and patients require significant time to regain their vitality. Pneumonia usually affects the lobes of the lungs and if multiple lobes are affected an individual needs to be kept under the supervision of medical professionals (Scott, Brooks, Peiris, Holtzman, & Mulholland, 2008).

In pneumonia without early interventions, it advances to a state where even intravenous antibiotics do not show much effect, which in turn leads to high clinic casualty rates for kids (Berkley et al., 2003). In some cases of pneumonia, death can occur within three days of the infection onset. Early detection, followed by early intervention in the case of pneumonia can save people's lives in many cases and in others treat the infection before it can impact the health of an individual in any major way. Hence, there is a pressing requirement to conduct experiments and research and in order to innovate new strategies to provide identification based on computer vision methods in the case of pneumonia.

X-ray works by using a low dose of ionizing radiation and providing photographs of the inside of the desired part. Around the chest area, an X-ray can be used to evaluate the condition of lungs, heart, and chest to help with problems like chest discomfort, chest injury, or breathing issues. Furthermore, it can help detect various diseases in the lungs like pneumonia, emphysema, etc. Since a chest X-ray is quick and simple, it is especially valuable in emergency determination and treatment. In pneumonia, the examination of the chest radiography plays a critical role in its treatment. According to the CDC, more than 50,000 individuals passed away in the United States in the year 2015 and more than 1.7 million grown-ups consistently looked for medical help because of pneumonia. Chronic obstructive pulmonary disease i.e., a group of lung diseases that block the airflow, are the main reason for deaths of kids under the age of 5 worldwide (Heron, 2013). It results in the deaths of about 1.4 million kids, which accounts for 18% of the total deaths. It also has been observed that pneumonia is more widespread in developing countries or the undeveloped countries where clinical assets are either unavailable or difficult to access. Thus, there is a need to develop a low-cost, modest, and precise solution for the diagnosis of pneumonia.

15.1.1 Deep Learning and Convolutional Neural Networks

Deep learning is a subclass of artificial intelligence that imitates the activities of the human mind in preparing information for use in identifying objects, perceiving discourse, translating dialects, and making choices and decisions. Deep learning can learn without human management, drawing from information that is both unstructured and unlabeled. Deep learning is a type of artificial intelligence which can be utilized for detection, recognition, estimation, classification, and various other tasks. The current deep learning and CNN models can arrive at human-level precision in predicting and examining any representation (Liu et al., 2018). Deep learning plays an important role in the medical domain. Deep learning can be utilized in a wide range of problems such as diagnosis and detection of various types of tumors and sores in medical images (Brunetti, Carnimeo, Trotta, & Bevilacqua, 2019; Litjens et al., 2017), technology-based diagnosis (Asiri, Hussain, Al Adel, & Alzaidi, 2019), the examination of electronic medical records (Shickel, Tighe, Bihorac, & Rashidi, 2017), the automated treatment and drug discovery and intake (Boldrini, Bibault, Masciocchi, Shen, & Bittner, 2019), environment analysis (Malukas, Maskeliunas, Damasieviciius, & Wozniak, 2018), and to come up with ways to support and assess the human's health. The major component of the accomplishment of deep learning depends on the capacity of the deep neural networks to take in significant levels of knowledge extraction from input data and information through some of the common purpose algorithms (Bakator & Radosav, 2018). Current, deep learning implementations can't surpass or replace medical practitioners or specialists in diagnosis of diseases, but it can offer help for specialists in the medical field in performing tedious work, for example, inspecting chest radiographs for the indications and detection of pneumonia.

Recently, CNNs have become popular and are used often in the field of computer vision and are attracting interests from various domains like radiology. CNN is developed for processing data with grid type patterns such as images. The inspiration for CNN comes from the architecture of the animal visual cortex (Fukushima & Miyake, 1982). which is intended to naturally and adaptively acquire knowledge of spatial progressive systems of highlights, from low level to significant level of the patterns. Basic CNNs are usually constructed using three layers which are named as convolution layers with different kernel sizes, pooling layers such as average and max and fully connected dense layer. The first two layers perform the task of extracting features whereas the third layer does the mapping between extracted features and the output. The convolution layer is able to extract basic features like lines, color, bends, etc. Stacking multiple layers on top of each other such that output of one layer feeds into the next layer allows it to extract more complex features such as shapes, which are important features in an image.

Currently in radiology handcrafted knowledge extraction or feature extraction techniques like gabor filter to extract features which are then used in more conventional machine learning techniques like support vector machine and boosting approach such as random forest (Lambin et al., 2012). A convolutional neural network will not require a hand-crafted knowledge extraction or feature extraction it will learn the best representation with respect to the task at hand itself.

CNN has hundreds of hyper parameters that need to be fine tuned using various methods in order for the model to get up to the mark accuracy and to tune those parameters huge amounts of data are required. Since it is not always feasible to gather or produce large amounts of data for the task at hand, a technique or a tool is required that can transfer the knowledge acquired in one task to effectively solve the other task. This way the data is gathered for a few tasks only instead of every task.

Transfer learning is a machine learning technique that tries to solve the problem of insufficient training data by attempting to transfer the knowledge obtained from one domain into another by removing the assumption that both test and training data must be independently and identically distributed. It is possible to transfer knowledge for tasks belonging to the same domain as well as for tasks belonging to different domains effectiveness however differ in both. The most common approach with transfer learning in the case of CNN is to use the model trained on large data and fine tuning it for the task at hand. Since the convolutional layers are trained already they are able to extract the shapes and edges by fine tuning we are able to tell the model shapes and edges relevant to the task at hand and we can train the fully connected layers to map appropriately.

15.1.2 Image Preprocessing and Visualization

The term *image pre-processing* is used for operations that are applied on images at the lowest level of abstraction (Sonka, Hlavac, & Boyle, 1993). Rather than increasing the information content they decrease it but their main aim is to reduce the noise, undesired distortions, artifacts, etc. and enhancing the features that might be relevant to the task. It can range from something as simple as an image from RGB to grayscale or a binary image to something bit more complicated like detecting edges in the images or eroding the images. It is not necessary that complex image pre processing would give better results, it is completely subjective and depends on the task at hand. For example, for face detection, converting an image to grayscale works well as for face detection color does not provide much information and converting to grayscale allows the algorithm to run much faster.

Image pre-processing is not necessary, as CNN itself has a feature extractor that tries to learn the best knowledge for the task, but enhancing the certain knowledge which might be relevant or important to the task makes the job of extracting features much easier for CNN and it is also possible that now CNN might be able to extract more detailed features which would be more relevant to the task.

While current progress of artificial intelligence progress is imposing, it isn't without distinctive challenges. For instance, the absence of interpretability and explainability of neural networks (Holzinger, Biemann, Pattichis, & Kell, 2017), from the extracted and learned features to the decision making process, is a significant area of research. While describing neural networks choices is significant, there are various different issues that emerge from deep learning, for example, security and trust because of bias in the deep learning models and data sets, just to name a few. As more and more AI-powered models and systems are deployed in the real world, these issues will be more widespread and likely to occur. Subsequently, an overall feeling of model results isn't just useful; we need to address the mentioned issues.

Data visualizations and visual analytics dominate at conveying data and finding bits of knowledge by utilizing visual encodings to change theoretical information into significant representations. Numerous strategies (Hohman, Kahng, Pienta, & Chau, 2018) produce static visualizations demonstrating which parts of an image are generally critical to a model's classification. These methods and results give understanding into what sorts of features deep neural networks are learning at particular layers in the model, and give a debugging tool to improve and understand a model. As per a survey's takeaways, interpretability is the most popular reason for deep learning visualization. There are various ways such as temporal metrics, instance-based analysis, and interactive experimentation algorithms for attributions and feature visualization and have applications in many domains.

15.1.3 RELEVANCE AND CONTRIBUTIONS

The contribution of our work and its relevance is defined as below:

1. Image Preprocessing: We present a new image pre-processing pipeline that makes it possible to increase the accuracy and efficiency using the same model.
2. CNN-based models: We used multiple CNN-based models, which allows us to extract the hierarchical features from the images without explicitly programing the same.
3. Ensemble: We use ensemble learning in order to improve the reliability and accuracy of our technique.
4. Visualization: To better understand the outputs of our models we have used Grad-CAM visualization so that infected regions can be identified and be a help to the medical practitioners.

5. Results: Our proposed method has achieved remarkable results that are obtained by using an average ensemble of five different models trained for pneumonia diagnosis.

15.2 BACKGROUND AND MOTIVATION

Image classification of the medical images is a necessary task in therapy and treatment. The conventional methods have reached the peak of their performance. However, using the conventional methods require much effort while developing it and they do not perform all that well. Recently deep neural networks have shown the capacity to perform well in the case of classification tasks; in particular CNNs have been producing the best results in the case of classification tasks (Gu & Wang et al., 2018). That's why there is growing enthusiasm around using CNN for purposes of medical image classification. Another reason for using CNN for medical image classification is the fact that CNN are very good at knowledge extraction or feature extraction and learn the relevant features for the tasks by themselves. So by using them it is possible to save the effort that goes into extensive feature engineering; moreover it is possible to help the clinician by showing them the parts in the image of X-ray that prompted the model to produce certain results.

Collecting and labeling of medical images is a difficult job because a lot of time needs to be invested in properly collecting and getting experts to uniformly and accurately annotate the whole data. Moreover, there are privacy concerns with respect to the patients. So there is a need for the method that can effectively transfer the knowledge from one task or domain to another so that it would be possible to get accurate and precise results without the huge number of images and CNN have proved to be very effective in the same. Since they perform the knowledge extraction or feature extraction in hierarchies that is extracting simple features and then extracting complex features from the simple one. The initial part of the hierarchy remains the same for most tasks; only the later part needs to be fine tuned according to the task at hand.

15.3 LITERATURE SURVEY

In Wang et al. (2017), the authors have contributed by putting forward another data set called chest X-ray 8 that contains pictures of more than 30,000 patients, with more than 100,000 pictures of frontal X-rays. These authors have accomplished promising outcomes utilizing CNN with more depth. Ronneberger et al. (2015) leveraged data augmentation and CNN to accomplish extraordinary outcomes, in any event, preparing a few examples of images not only with chest X-ray images. In a different research Roth et al. (2014) demonstrated experiments that helped to distinguish lymph nodes by utilizing deep CNN. They had encouraging outcomes surprisingly with low quality images also. Shin et al. (2016) tried using various CNN models to handle the issue of lymph and lung infection detection.

In Rajpurkar et al. (2018), the authors have proposed CheXNeXt, a CNN-based deep neural network with 121 layers, which helped distinguish 14 distinct pathologies, and also includes pneumonia, while using chest X-ray images. To begin with, neural networks were supposed to estimate the likelihood of 14 variations from normal in X-ray images. Then, the networks are stacked to give results by taking the average estimation of each model. CNN-based networks were developed for all 14 distinct pathologies and their corresponding probabilities were obtained in the X-ray images. These were then ensemble to get the final predictions by using the average of the predictions of various individual networks. In Wozniak et al. (2018), authors proposed a new technique for developing neural networks that are probabilistic in nature, and permit the development of compact networks. Gu & Lu et al. (2018) worked on lung nodule detection by utilizing a three-dimensional deep CNN and a multi-scale predicting methodology.

In Khatri et al. (2020), authors have introduced the use of the distance earth moves to recognize the infected lungs from the non-infected ones for pneumonia. Abiyev & Ma'aitah (2018) and authors of Stephen et al. (2019) performed pneumonia classification using the CNN model. In Cohen, Bertin, & Frappier (2019) and Rajaraman, Candemir, Kim, Thoma, & Antani (2018) authors have achieved promising results by using self-designed CNN for detection and explained their results. In Lakhani & Sundaram (2017) authors have achieved the AUC score of mid-nineties by using AlexNet and GoogLeNet architecture for neural network network with data augmentation techniques without pre-training them. In Sirazitdinov et al. (2019), pneumonia classification is performed using a region-based CNN with image augmentation for segmenting the pulmonary images. Ho & Gwak (2019) fine tuned a DenseNet to classify 14 illnesses using a localization-based approach. In Xiao et al. (2019), authors introduced a computed tomography (CT)-based network known as multi-scale heterogeneous 3D CNN (MSH-CNN). Sin-loss is introduced by Xu, Wu, & Bie (2018) and uses a hierarchical CNN network for pneumonia identification. In Jaiswal et al. (2019), authors used mask regional convolutional neural network (RCNN), which uses global as well as local features to segment images and further uses regularization and dropout for pneumonia classification. In Jung, Kim, Lee, Lee, & Kang (2018), authors used three-dimensional deep CNN, which used resnet architecture i.e. shortcut or skip connections. Authors of Chouhan et al. (2020) used majority voting to aggregate the estimation of various networks to provide final estimation.

15.4   DATA ANALYSIS

Chest X-ray images in the data set were taken from a Children's Medical Center located in China. It contains the chest X-ray images of children from age one to five. The images in the data set were not taken specially and were part of the routine process that the patient went through. Whilst creating the data set, all the images went through a screening process and the low quality images were removed. Labelling of the data set was done by two experienced physicians. A third physician also verified the evaluation set to check for the errors that might have crept in while labelling initially. Since the data set size was small it was only divided into two parts i.e., training set and the test set. Table 15.1 shows the division of the data set according to class and set. In case of pneumonia chest X-ray image of a person consists of abnormal opacification in the lungs. We can see in Figure 15.1 that a normal chest X-ray has clear

TABLE 15.1

Analysis of the Research Findings, Work Performed by Different Authors and Their Limitations

Authors	Work Performed	Limitations in Their Work
Ronneberger et al., 2015	Used data augmentation and CNN to achieve remarkable outcomes.	This work used single CNN models to give the results.
Rajpurkar et al., 2018	CheXNeXt, a deep CNN network with 121 layers	Trained the CNN model without leveraging the power of transfer learning
Abiyev & Ma'aitah, 2018	CNN model is used for pneumonia identification	This work did not use pretrained network architecture.
Lakhani & Sundaram, 2017	Used AlexNet and GoogLeNet with data augmentation	This work didn't try out different ensembling approaches to find the best one.
Ho & Gwak, 2019	Used pre-trained DenseNet-121 to classify	Work doesn't use any kind of data augmentation or pre-processing.

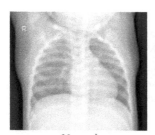 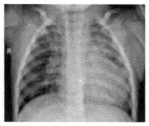 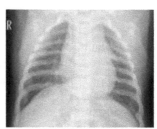

Normal   Bacterial Pneumonia   Virus Pneumonia

FIGURE 15.1 Sample of X-ray images of normal, bacterial, and virus pneumonia taken from the data set used.

lungs, with no abnormalities. On the other hand, an X-ray with pneumonia has an abnormality. How the abnormality is spread through the lungs is predominantly determined by the nature of pneumonia i.e., if it is bacterial or viral (Table 15.2).

15.4.1 Exploratory Data Analysis

Ben Graham Method: First we experimented with the Ben Graham method, which involves first converting the images into gray-scale and then adding Gaussian blur to it; this is a very insightful way to improve lighting conditions. This way we are able to see important details in a much better way, as shown in Figure 15.2.

Image Resizing and Normalization: Image resizing is defined as downscaling or up-scaling images to deal with different-sized images to match the

TABLE 15.2
Data set Partition for Training and Testing

Category	Training Set (No. of Images)	Test Set (No. of Images)
Normal	1,349	234
Pneumonia	3,883	390
Total	5,232	624

dimensions of the smallest image available or match the dimensions of the desired input size. Normalization is defined as rescaling data to be projected into a predefined range such as [−1, 1] or [0, 1]. This is performed so that the same algorithms can be applied to the data set.

Histogram Equalization and Color space Conversion: Image contrast is enhanced using histogram equalization this is one of the most prominent computer image processing techniques and the shape of the histogram changes, whereas in histogram stretching the shape of histogram remains the same. Color spaces are the different types of modes of the color which are used in image processing and signals for various purposes.

Morphological Dilation and Erosion: Dilation helps to increase the size of the objects by adding the pixels to the boundaries whereas erosion results in decreasing the size of the objects by removing pixels on boundaries of the object. Dilation results in filling of the holes and broken areas in an image and increases in the brightness while erosion helps remove a few irregularities as compared to structuring elements. Mathematically, dilation is the XOR of X and Y while erosion is dual to dilation. Some useful insights are drawn by using erosion and dilation and results are shown in Figures 15.3 and 15.4, respectively.

15.5 FRAMEWORK AND METHODOLOGY USED

In this section, we discuss framework and methodology, which involve convolutional neural networks and its layers, data pre-processing, and augmentation and we have used the architecture of various pretrained models.

15.5.1 CONVOLUTIONAL NEURAL NETWORK

Convolutional neural networks (Saha & Saha, 2018) or more commonly known as CNNs, are a type of neural network. It is mostly used for deep neural networks in image processing. The network requires an image as an input and calculates the weights and biases to parameters and helps identify the distinction between images. These types of networks normally do not require a high level of pre-processing. CNNs are a combination of multiple layers. These are discussed next.

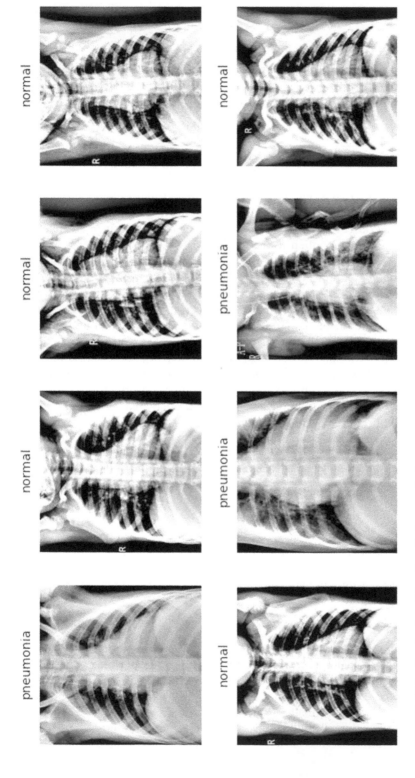

FIGURE 15.2   Result of applying Ben Graham method.

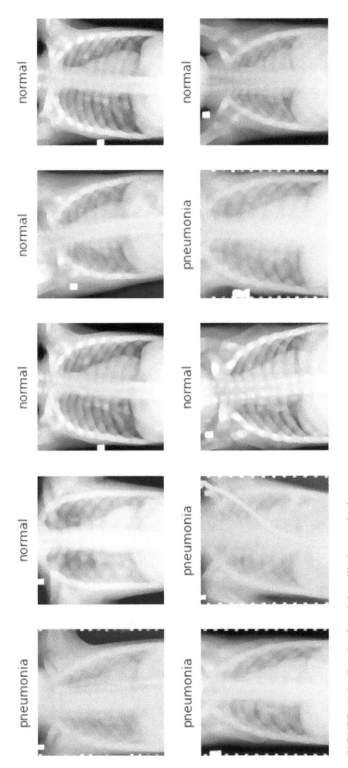

FIGURE 15.3 Result of applying dilation method.

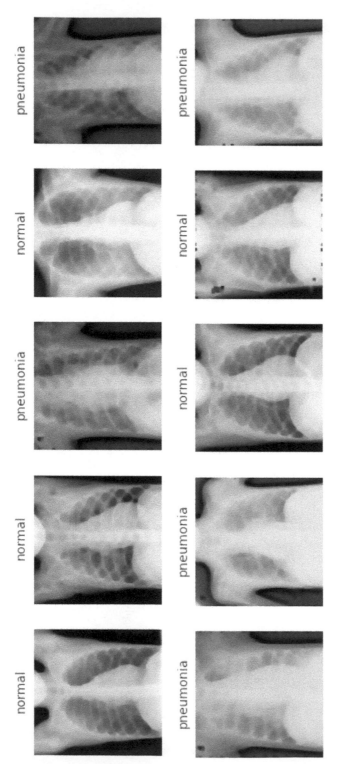

FIGURE 15.4 Result of applying erosion method.

Input Layer - This layer has the dimensions that are the same as that of the input image. It has to be taken care that all images have been re-scaled to the same dimension to avoid data inconsistency.

Hidden Layers - These layers are the primary part of the architecture and are fed by the input layer. These are normally used to produce more complex output features, through a combination of calibrated weights and biases. A convolution filter is used on the previous layer to produce the complex output features.

Output Layer - This layer is used to generate probabilities, which are used for classification of the image into a category. This layer utilizes the result of the last hidden layer as input and generates two outputs, which are probabilities corresponding to each class. The class, which has the higher assigned probability, will be predicted to be the right answer.

Activation Function - Sigmoid Activation Function is applied in the final output layer, which is used to calculate probabilities and thus perform binary classification. The outputs of hidden layers are given to ReLu Activation Function.

Figure 15.5 depicts the primary tasks, which are performed for image processing in the CNN. The image is taken as an input and the network throws out a probability vector as an output. The vector depicts the probability associated with every class. The class having the highest probability is the predicted answer for image classification. Operations applied in the network are explained next.

Convolution - A kernel/filter with dimensions less than the dimensions of the input image is used in this operation. The filter is placed upon the upper-left corner of the picture and the terms in the pixel matrix and the filter which overlaps with it is multiplied and the summation is used to obtain the first value corresponding to the next feature map. The filter is then slid ahead and similar estimation is done again to get the whole feature map. When the filter reaches the rightmost position, we slide the filter below by one row and the estimation/calculation is repeated using the same procedure.

Max Pooling - This layer utilizes a filter (2×2 in the figure), which is slid over the feature map with a stride of 2 and we take the maximum of each 2×2 matrix. In some cases, average pooling is used, which takes the average of the 2×2 matrix. In this case, the dimensions of the image are halved.

Fully Connected - A fully connected layer more commonly known as a dense layer is the most common layer in the artificial neural network. In the fully connected layer output of the previous layer, the neuron is passed through each neuron of the current layer. So if the size of the previous layer is "M" and the output of the current layer is "N" the parameter matrix would have dimensions "MxN."

15.5.2 DATA PREPROCESSING AND AUGMENTATION

Augmentation is a data pre-processing technique that basically serves two purposes: first to produce more training samples from the given training samples and second to reduce the overfitting of the models and make it more generalizable. It does so by taking the existing data and making slightly modified copies of the existing data and adding them to the set. In the cases of images, there are two types of augmentations: geometric augmentation and photometric augmentation. Geometric augmentation

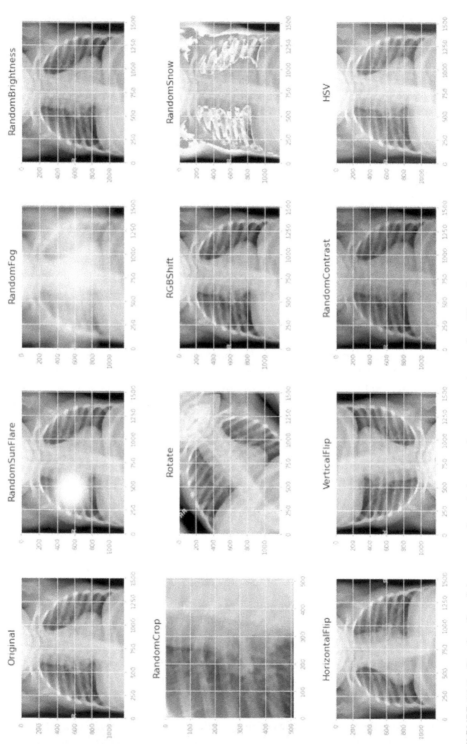

FIGURE 15.5　Results of applying geometric and photometric augmentation to chest X-ray images.

includes making modifications to the geometry of the image by rotating it, shifting it, or flipping it. Photometric augmentations include modifying how the image looks by changing the pixel values by permuting the channels, adding noise, or adding whitening, etc. Since in x-ray images lighting and intensity play an important role, not all augmentations are suitable for them. The augmentation we do is geometric heavy. We do include some photometric augmentation like channel shifting, color space shifting, etc.

15.5.3 TRANSFER LEARNING: PRETRAINED ARCHITECTURE

15.5.3.1 ResNet-50

Residual Networks (He, Zhang, Ren, & Sun, 2016), popularly known as ResNet, are classic convolutional neural networks that are being used in many computer vision tasks. While trying to make the neural network deeper we cannot simply stack the layers on top of each other because that results in a vanishing gradient problem, in which gradient becomes extremely small while back propagating, and makes negligible changes in the weights of the neural network. Hence, rendering it untrainable. ResNet solves this issue of a vanishing gradient by using the skip connections, known as identity shortcut connection. These connections skip one or more layers. Layers are still stacked in the case of ResNet but the original input is added to the output. Due to these advantages, we have used ResNet-50 model architecture to train our model.

15.5.3.2 EfficientNet

EfficientNet (Tan & Le, 2019) is the latest architecture proposed by researchers at Google, which largely focuses on the efficiency of the network while also improving the accuracy. CNNs are defined in terms of three scaling dimensions known as depth of the network, width, and the resolution. Depth of the network is defined in terms of its number of layers. Width is associated with how wide the network is. Resolution of the network is defined as the input size of the image that is passed into our network. Scaling networks based on any one have several disadvantages and the model starts to degrade after some time. In order of the scale the width, depth, and resolution of the network in a principled way, (Tan & Le, 2019) proposed a simple yet effective scaling technique that uses the concept of compound scaling. A base network was obtained using a Neural Architecture search that optimizes both accuracy and FLOPS. This network architecture uses an MBConv block, which is an Inverted Residual Block that was previously used in the MobileNetV2 architecture with an Excite and Squeeze block injected.

15.5.3.3 VGG-16

VGGNet (Simonyan & Zisserman, 2014) was invented by Visual Geometry Group. While the previous convolutional neural networks like AlexNet were more focused on strides and window size, VGG mainly focused on the depth of the convolutional neural network. Increase in depth of convolutional neural networks in VGG allows it to use a smaller window size and strides. This also decreases the number of parameters to make the decision function more nonlinear without affecting the window size and stride; VGG uses 1x1 convolutional layers. A decrease in the size of the window allows it to have a greater number of weight layers, which leads to an improvement in performance. VGG outperformed various baselines for many image recognitions after its introduction.

15.5.3.4 MobileNetV2

MobileNetV2 (Sandler, Howard, Zhu, Zhmoginov, & Chen, 2018) brings in new possibilities to use neural networks in mobile devices by helping deliver high-accuracy results while keeping the operations and number of parameters as low as possible. In order to make the network as efficient as possible, normal convolution operation is replaced by the depth separable convolution operation which is more efficient. Furthermore, the concept of bottlenecks between the layers was introduced, which is used to encode both the output and intermediate input. Desired accuracy or performance can be achieved by fine tuning the hyper-parameters such as image resolution and width multiplier as they are tuneable, thus making the architecture customizable for different performance points. MobileNetV2 is faster in terms of operations and has good enough accuracy, which makes it highly acceptable for use in mobile devices and other applications which are required to use memory efficient inference.

15.5.3.5 DenseNet

The researchers were able to increase the depth of neural networks up to a certain point using ResNet; after which it didn't really improve results. To increase the depth further while still improving the results, DenseNet (Huang, Liu, Van Der Maaten, & Weinberger, 2017) was introduced. The problem arises with traditional CNNs when we increase the depth and this results in the path from the input layer to output becoming large and the gradient vanishes before reaching the other side. DenseNet tries to solve the problem by ensuring maximum gradient flow by connecting every layer directly with each other. DenseNet leveraged the potential of the network through feature reuse instead of representation power from wide architectures. DenseNet requires fewer parameters and does not require learning redundant feature maps as in traditional CNN and has narrow layers, so the set of feature maps that is added is small in size. This architecture allows each layer to have direct access to the gradients. DenseNet concatenates the incoming feature maps rather than summing them.

15.6   DESIGN AND ARCHITECTURE

In this section, we explain the flow of data in our model design and architecture developed for the system.

15.6.1   DATA FLOW ANALYSIS

Chest X-ray images are used as input for our models. They are first pre-processed by a pipeline that is explained in the next section, following which they are passed through five pre-trained convolutional neural networks that are fine tuned for the task of pneumonia diagnosis from the chest X-ray images of the data set. The results of each model are averaged to obtain the final probability, which is then used for the final prediction. The whole data flow is visually represented in the Figure 15.6.

15.6.2 ARCHITECTURE DETAILS

Image Preprocessing: Pneumonia detection from chest X-ray images is a resolute task because the detection can be made on the basis of complex features, which means that detection on the basis of simple features will not give high accuracy. To extract complex features using convolutional layers, we would need to build a deep convolutional neural network which would then tend to overfit in the training data, resulting in poor generalizability. In order to solve this problem in the case of pneumonia detection, we use image pre-processing.

Image pre-processing would reduce the irrelevant information hence enhancing the features which are relevant to the task at hand. This allows us to train either shallow convolutional neural networks or use the pre-trained convolutional neural network to extract features that would give acceptable accuracy for the classification task. The classification technique or pipeline also needs to be efficient, as otherwise, it would take a significant time to process multiple images. Image pre-processing would also help in making the technique more efficient as it would remove the significant amount of information from the image, which means the model would have to process less information, thus allowing it to be more efficient. Image pre-processing used here involves moving through multiple steps.

Image Resizing: Image is resized to 512 * 512 pixels.
Erosion: Image is then eroded using a kernel of size 5 * 5
Histogram equalization: Histogram equalization is performed on each channel of the eroded image

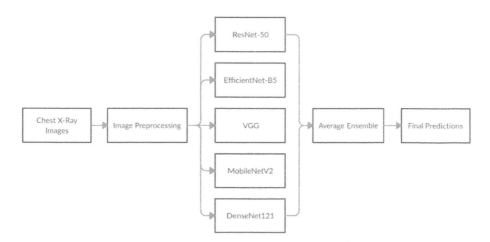

FIGURE 15.6   Data flow diagram.

Color Space Conversion: Equalized image is then converted to a grayscale image
Ben Graham: Gaussian noise is added to the grayscale image
Normalization: Each pixel of the image is divided by 255, to ensure that the value of
each pixel is between 0 and 1.

This image pre-processing allows us to achieve efficiency as the result of image pre-processing has only one channel instead of three and multiple information has been taken out of that as well. Moreover, this image processing allows us to see the X-ray images in a better lighting condition while improving the global contrast, which makes it easier to detect the pneumonia.

Architecture: The final presented model is an ensemble of convolutional neural networks trained on the same data set. Each CNN has a different backbone that is the way convolutional layers are arranged and some CNN has extra layers as well. We trained different CNNs because this way if one CNN architecture is bad at a particular feature detection, the others might cover up for the same. We have used transfer learning while training each individual CNN, more specifically we use the CNN architecture that was trained on ImageNet data set, remove the classification layer and add the new one and train its weights as well as fine tune the weights of the layers that were not removed according to the current task. We trained each model for 15 epochs using Adam as an optimizer. While training we use a variable learning rate which was dependent on the model's performance on the validation set i.e after each epoch the model is tested on the validation set and according to the performance the learning rate is adjusted this is done to avoid the plateaus encountered while training the model loss function used while training is focal loss.

Focal Loss: Class imbalance problem is very common in the medical domain and its related data sets, it is observed when information related to one class is over-represented than the other class. The dominated representation of one class over the other in the data set makes the neural network biased towards learning that dominated class then making the model biased. Therefore it is important to maintain the balance. Focal loss (Zhou, Wang, & Krahenbuhl, 2019) is used to address the issue of the class imbalance problem. A modulation term applied to the cross-entropy loss function makes it efficient and easy to learn the examples which were earlier difficult to learn for the model. In this loss all the false negative or misclassified samples are given more weight when training as compared to those which are correctly classified also known as true negatives. Below is the equation of focal loss. Here, the hyper-parameter handles the class imbalance problem and this form is also known as the Alpha form.

$$FL(\rho) = -\alpha(1-\rho)^\gamma log(\rho) y = 1$$

$$FL(\rho) = -(1-\alpha)\rho^\gamma log(1-\rho) y \neq 1$$

Equation 15.1 Final Focal Loss in Alpha Form

Gradient-weighted class activation map: Grad-CAM (Selvaraju et al., 2017) also known as gradient-weighted class activation map, is the method for making CNN-based neural networks more understandable by visualizing the regions of the image which are important for predictions. Grad-CAM can give us these visualizations without changing the architecture of the base model and without re-training any layer or model unlike CAM which modify the base network and to get the feature maps remove the last fully connected layer, thus requiring the model to be retrained. Visualizations of the Grad-CAM are class-discriminative and this localize well for relevant image regions. As we want to preserve the spatial location information of the object, so in this technique the last convolutional layer is used as that layer identifies parts specific to that class and thus gives effective visualizations. To obtain a Grad-CAM for any class, we compute the gradient of the score for that class with respect to feature maps of a convolutional layer. Importance of each feature map for a specific class is obtained using average pooling technique and then a weighted combination of forward activation maps is performed followed by ReLU.

15.7   EXPERIMENTAL SETUP

In this section, we discuss the evaluation metrics used and implementation details of our method.

15.7.1   EVALUATION

Recall, precision, f1-score, and accuracy are used to measure the performance of the model i.e., multiple metrics instead of a single one. The test set is used for validation of the proposed method. The test set was created earlier alongside the training set on which the models are trained. The metrics mentioned are primarily used for the classification task. We use the stated metrics instead of simple accuracy because accuracy does not deal with true positives, false positives, true negatives, and false negatives, which are important when the data set has imbalanced classes or when the results of classification tasks carry certain significance with them, for example giving wrong results in the case of pneumonia can have severe consequences.

15.7.2   IMPLEMENTATION DETAILS

All the CNN models described are developed using the open source deep learning framework's tensorflow and keras in Python 3 language. Input images were split into three sets training, validation, and testing. The model was trained on the training set and the validation set was used to continuously evaluate the model's performance and change the appropriate parameters for performance enhancement. The final results of the model are calculated using the performance of the model on the test set, which gives the estimate of how well the model will perform on the data it has not seen or the generalizability of the model.

15.8   RESULTS AND DISCUSSION

In this section, we will discuss the experiments we have performed and results of the model and visualization of Grad-CAM.

TABLE 15.3
Results of Each Model and Ensemble Model

Model	Precision (%)	Recall (%)	F1-Score (%)	Test Accuracy (%)
ResNet-50	90.48	93.54	92.01	93.45
EfficientNet B4	94.20	96.50	95.35	94.34
VGG 16	91.27	89.03	90.15	93.23
MobileNet V2	86.76	93.02	89.89	88.76
DenseNet 121	88.78	94.35	91.56	97.50
Ensemble Model	97.47	98.09	97.77	97.69

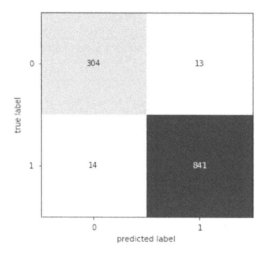

FIGURE 15.7 Confusion matrix.

15.8.1 RESULTS

Here, we visualize the results of different deep neural nets as well their ensemble in the Table 15.3. We notice that while individual neural networks perform fairly well, the ensemble of all these models beats all these models in every metric which is consistent with the expectation from the ensemble technique. We have also included the confusion matrix for the ensemble method. The confusion matrix shown in Figure 15.7 visualizes true positives, false positives, true negatives, and false negatives that are used to calculate all the necessary metrics. The graph depicting how training and validation accuracy changes as model trains is also shown in Figure 15.8.

For explainability of the model and visualization we produce GradCAM visualization from CNN, which is used for the prediction task as well. Examples of the Grad-CAM visualization can be seen in Figures 15.9 and 15.10.

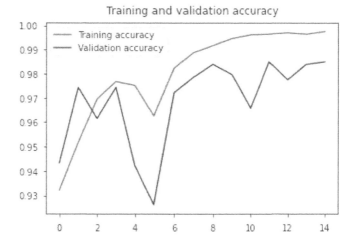

FIGURE 15.8 Graph between accuracy and number of epochs.

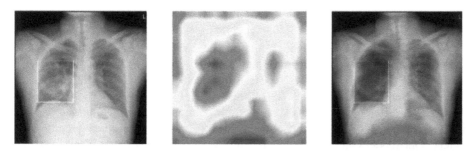

FIGURE 15.9 The first image is input chest X-ray image, second is heat map generated from the CNN and third image is heatmap superimposed on the chest X-ray to show inflamed area.

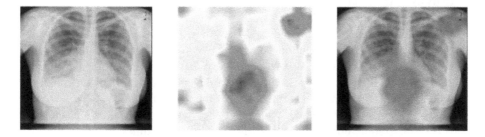

FIGURE 15.10 The first image is input chest X-ray image, second is heat map generated from the CNN and third image is heatmap superimposed on the chest X-ray to show inflamed area.

TABLE 15.4
Comparative Results of Model on tTest Set

Model	Precision (%)	Recall (%)	F1-Score (%)	Test Accuracy (%)
Liang & Zheng, 2020	89.10	96.7	92.7	90.05
Kermany et al., 2018	–	93.20	–	92.80
Chouhan et al., 2020	93.28	99.62	–	96.39
Our Method	97.47	98.09	97.77	97.69

15.8.2 DISCUSSION

In this section, we will analyze and study the proposed methodology and compare it with previous works based on the experimental results and discuss the improvement in pneumonia diagnosis by this method.

Comparison with the previous words:

According to the results of the proposed method shown in the table and figures, the accuracy on the test data set achieved is 97.49% with 98.01% precision, recall, and f1-score for the ensemble model. As compared with all the models trained and tested on the same data set, we have achieved the best score on our average ensemble model as compared to ResNet-50, EfficientNet B4, VGG-16, MobileNetV2, and DenseNet121 (Table 15.4). The models used for these methods have been trained with different algorithms such as their data pre-processing and augmentation, activation function, loss function such as focal loss, and other hyper parameters. Our model compared well with the previous studies and not only shows better performance as compared to existing solutions but also shows improvement in terms of training time, accuracy optimization, efficient scaling, parameter optimization, and better inference time on mobile devices as in the case of MobileNetV2.

Analysis of the proposed method:

We have shown the performance of our models and have further analyzed the results with visualizations using Grad-CAM, as shown in the figure. We have used feature maps of the last convolutional layer of every model and used them to visualize the results. Convolutional layers can learn these edge modes of the images and this abstracts them as they pass through the layers. The heat map generated is imposed on the original image to get the idea of the regions of interest and features learned. In summary, our method achieved better results and has advantages in terms of visualization, efficiency in training, and inference time.

15.9 FUTURE RESEARCH DIRECTIONS

Our proposed models are optimized in terms of inference, accuracy, and other parameters which is very critical for commercial use of any applications and can be deployed to the real world. For actual clinical practice, we need to further optimize the neural networks in terms of inference time and parameters without compromising the accuracy. Further, the data set used in this research is collected from a single organization, so to make our models more robust to changes and accuracy we need to continue the research in this field and collect data from different medical organizations and build a continuous pipeline to train and test our models. We have used Gradient-weighted Class Activation Mapping for visualization and that will help medical practitioners to better visualize the results and make artificial intelligence more explainable; thus this can be improved by using techniques like Deep SHAP, which is a high-speed approximation algorithm to visualize the models. Further, the model can be optimized and used for computed tomography and magnetic resonance imaging analyses.

15.10   CONCLUSION

In this paper, we have presented an improved pneumonia diagnosis using new image pre-processing and visualization to make the model explainable. This automated diagnosis method will classify the chest X-ray images into normal and pneumonia. We have a method using image pre-processing and transfer learning in which we use state-of-the-art pre-trained model architectures using different structures such as residual for ResNet, compound scaling with Inverse residual block in EfficientNet, low latency inference for MobileNet, and reusing the feature map in case of DenseNet and ensemble model leverages the power of all these model to give results which are better than prior results. We have used state-of-the-art visualization algorithms called Gradient-weighted Class Activation Mapping to analyze the results of models. There is a possibility of further optimizing the models and results by hyper-parameter tuning and at the same time, this work can be extended to make medical analysis feasible in underdeveloped areas of the world and can be a boost to the telemedicine sector.

GLOSSARY

True Positives: True positives are defined with respect to a particular class. They indicate that the given class is the predicted and it is correct.

True Negatives: True negatives are defined with respect to a particular class. They indicate that the given class is not the predicted class and it is correct.

False Positives: False positives are also defined with respect to a particular class. They indicate that the given class is the predicted class and it is incorrect.

False Negatives: False negatives are also defined with respect to a particular class. They indicate that the given class is not the predicted class and it is incorrect.

Confusion Matrix: Confusion Matrix divides the whole prediction by model into true positives, true negatives, false positives, and false negatives and visualizes them in a matrix format.

Precision: Precision is the ratio of the predictions that truly belongs to the class out of predictions that model predicts to belong to the class; that is, true positive divided by sum of true positive and false positives.

Recall: Recall is the ratio of the predictions that the model predicted to belong to the class out of the total number of predictions that did belong to the class; that is true positive divided by sum of true positive and false negatives.

f1 Score: Since we usually want one metric to evaluate the performance, the f1 score combines both precision and recall. The f1 score is a harmonic means of precision and recall.

REFERENCES

Abiyev, R. H., & Ma'aitah, M. K. S. (2018). Deep convolutional neural networks for chest diseases detection. *Journal of Healthcare Engineering*.

Asiri, N., Hussain, M., Al Adel, F., & Alzaidi, N. (2019). Deep learning based computer-aided diagnosis systems for diabetic retinopathy: A survey. *Artificial Intelligence in Medicine, 99*, 101701.

Bakator, M., & Radosav, D. (2018). Deep learning and medical diagnosis: A review of literature. *Multimodal Technologies and Interaction, 2*(3), 47.

Berkley, J., Ross, A., Mwangi, I., Osier, F., Mohammed, M., Shebbe, M., ... Newton, C. (2003). Prognostic indicators of early and late death in children admitted to district hospital in kenya: Cohort study. *BMJ, 326*(7385), 361.

Boldrini, L., Bibault, J.-E., Masciocchi, C., Shen, Y., & Bittner, M.-I. (2019). Deep learning: A review for the radiation oncologist. *Frontiers in Oncology, 9*, 977.

Bouch, C. & Williams, G. (2006). Recently published papers: Pneu monia, hypothermia and the elderly. *Critical Care, 10*(5), 167.

Brunetti, A., Carnimeo, L., Trotta, G. F., & Bevilacqua, V. (2019). Computer-assisted frameworks for classification of liver, breast and blood neoplasias via neural networks: A survey based on medical images. *Neurocomputing, 335*, 274–298.

Chouhan, V., Singh, S. K., Khamparia, A., Gupta, D., Tiwari, P., Mor- eira, C., ... De Albuquerque, V. H. C. (2020). A novel transfer learning based approach for pneumonia detection in chest x-ray images. *Applied Sciences, 10*(2), 559.

Cohen, J. P., Bertin, P., & Frappier, V. (2019). Chester: A web delivered locally computed chest x-ray disease prediction system. *arXiv preprint arXiv*:1901.11210.

Fukushima, K., & Miyake, S. (1982). Neocognitron: A self-organizing neural network model for a mechanism of visual pattern recognition. In M. A. Arbib & S. Amari (Eds.), *Competition and cooperation in neural nets* (pp. 267–285). Berlin: Springer.

Gu, J., Wang, Z., Kuen, J., Ma, L., Shahroudy, A., Shuai, B., ... Cai, J. (2018). Recent advances in convolutional neural networks. *Pattern Recognition, 77*,354–377.

Gu, Y., Lu, X., Yang, L., Zhang, B., Yu, D., Zhao, Y., ... Zhou, T. (2018). Automatic lung nodule detection using a 3d deep convolutional neural network combined with a multi-scale prediction strategy in chest cts. *Computers in Biology and Medicine, 103*,220–231.

He, K., Zhang, X., Ren, S., & Sun, J. (2016). Deep residual learning for image recognition. *Proceedings of the IEEE Conference on Computer Vision and Pattern Recognition*, 770–778. IEEE.

Heron, M. P. (2013). Deaths: leading causes for 2010.

Ho, T. K. K., & Gwak, J. (2019). Multiple feature integration for classification of thoracic disease in chest radiography. *Applied Sciences, 9*(19), 4130.

Hohman, F., Kahng, M., Pienta, R., & Chau, D. H. (2018). Visual analytics in deep learning: An interrogative survey for the next fron- tiers. *IEEE Transactions on Visualization and Computer Graphics, 25*(8), 2674–2693.

Holzinger, A., Biemann, C., Pattichis, C. S., & Kell, D. B. (2017). What do we need to build explainable ai systems for the medical domain? *arXiv preprint arXiv*:1712.09923.

Huang, G., Liu, Z., Van Der Maaten, L., & Weinberger, K. Q. (2017). Densely connected convolutional networks. *Proceedings of the IEEE Conference on Computer Vision and Pattern Recognition*, 4700–4708.

Jaiswal, A. K., Tiwari, P., Kumar, S., Gupta, D., Khanna, A., & Rodrigues, J. J. (2019). Identifying pneumonia in chest x-rays: A deep learning approach. *Measurement*, *145*,511–518.

Jung, H., Kim, B., Lee, I., Lee, J., & Kang, J. (2018). Classification of lung nodules in ct scans using three-dimensional deep convolutional neural networks with a checkpoint ensemble method. *BMC Medical Imaging*, *18*(1), 48.

Kermany, D. S., Goldbaum, M., Cai, W., Valentim, C. C., Liang, H.,Baxter, S. L., ... Yan, F. (2018). Identifying medical diagnoses and treatable diseases by image-based deep learning. *Cell*, *172*(5), 1122–1131.

Khatri, A., Jain, R., Vashista, H., Mittal, N., Ranjan, P., & Janard- hanan, R. (2020). Pneumonia identification in chest x-ray images using EMD. In *Trends in Communication, Cloud, and Big Data* (pp. 87–98). Berlin: Springer.

Lakhani, P., & Sundaram, B. (2017). Deep learning at chest radiography: Automated classification of pulmonary tuberculosis by using convolutional neural networks. *Radiology*, *284*(2), 574–582.

Lambin, P., Rios-Velazquez, E., Leijenaar, R., Carvalho, S., Van Stiphout, R. G., Granton, P., ... Dekker, A. (2012). Radiomics: Extracting more information from medical images using advanced feature analysis. *European Journal of Cancer*, *48*(4), 441–446.

Liang, G., & Zheng, L. (2020). A transfer learning method with deep residual network for pediatric pneumonia diagnosis. *Computer Methods and Programs in Biomedicine*, *187*,104964.

Litjens, G., Kooi, T., Bejnordi, B. E., Setio, A. A. A., Ciompi, F., Ghafoorian, M., ... Sanchez, C. I. (2017). A survey on deep learning in medical image analysis. *Medical Image Analysis*, *42*,60–88.

Liu, N., Wan, L., Zhang, Y., Zhou, T., Huo, H., & Fang, T. (2018). Exploiting convolutional neural networks with deeply local description for remote sensing image classification. *IEEE Access*, *6*,11215–11228.

Malukas, U., Maskeliunas, R., Damasıevicıius, R., & Wozniak, M. (2018). Real time path finding for assisted living using deep learning. *Journal of Universal Computer Science24*,475–487.

Organization, W. H. (2018). Household air pollution and health. Retrieved from https://www.who.int/news-room/fact-sheets/detail/household-air-pollution-and-health.

Rajaraman, S., Candemir, S., Kim, I., Thoma, G., & Antani, S. (2018). Visualization and interpretation of convolutional neural net- work predictions in detecting pneumonia in pediatric chest radiographs. *Applied Sciences*, *8*(10), 1715.

Rajpurkar, P., Irvin, J., Ball, R. L., Zhu, K., Yang, B., Mehta, H., ... Langlotz, C. P. (2018). Deep learning for chest radiograph diagnosis: A retrospective comparison of the chexnext algorithm to practicing radiologists. *PLoS medicine*, *15*(11), e1002686.

Ronneberger, O., Fischer, P., & Brox, T. (2015). U-net: Convolu- tional networks for biomedical image segmentation. *International Conference on Medical Image Computing and Computer-Assisted Intervention*, 234–241. Springer.

Roth, H. R., Lu, L., Seff, A., Cherry, K. M., Hoffman, J., Wang, S., ... Summers, R. M. (2014). A new 2.5 d repre- sentation for lymph node detection using random sets of deep convolutional neural network observations. *International Conference on Medical Image Computing and Computer-Assisted Intervention*, 520–527. Springer.

Saha, S. & Saha, S. (2018). *A comprehensive guide to convolutional neural networks—The eli5 way*. Retrieved from https://towardsdatascience.com/a-comprehensive-guide-to-convolutional-neural-networks-the-eli5-way-3bd2b1164a53

Sandler, M., Howard, A., Zhu, M., Zhmoginov, A., & Chen, L.-C. (2018). Mobilenetv2: Inverted residuals and linear bottlenecks. *Proceedings of the IEEE Conference on Computer Vision and Pattern Recognition*, 4510–4520. IEEE.

Scott, J. A. G., Brooks, W. A., Peiris, J. M., Holtzman, D., & Mulholland, E. K. (2008). Pneumonia research to reduce childhood mortality in the developing world. *The Journal of Clinical Investigation*, *118*(4), 1291–1300.

Selvaraju, R. R., Cogswell, M., Das, A., Vedantam, R., Parikh, D., & Batra, D. (2017). Grad-cam: Visual explanations from deep networks via gradient-based localization. *Proceedings of the IEEE International Conference on Computer Vision*, 618–626. IEEE.

Shickel, B., Tighe, P. J., Bihorac, A., & Rashidi, P. (2017). Deep EHR: A survey of recent advances in deep learning techniques for electronic health record (ehr) analysis. *IEEE Journal of Biomedical and Health Informatics*, *22*(5), 1589–1604.

Shin, H.-C., Roth, H. R., Gao, M., Lu, L., Xu, Z., Nogues, I., ... Summers, R. M. (2016). Deep convolutional neural networks for computer-aided detection: CNN architectures, data set characteristics and transfer learning. *IEEE Transactions on Medical Imaging*, 35(5):1285–1298.

Simonyan, K., & Zisserman, A. (2014). Very deep convolutional networks for large-scale image recognition. *arXiv preprint arXiv*:1409.1556.

Sirazitdinov, I., Kholiavchenko, M., Mustafaev, T., Yixuan, Y., Kuleev, R., & Ibragimov, B. (2019). Deep neural network ensemble for pneumonia localization from a large-scale chest x-ray database. *Computers & Electrical Engineering*, *78*, 388–399.

Sonka, M., Hlavac, V., & Boyle, R. (1993). Image pre-processing. *Image Processing, Analysis and Machine Vision*, 56–111. Springer.

Stephen, O., Sain, M., Maduh, U. J., & Jeong, D.-U. (2019). An ef- ficient deep learning approach to pneumonia classification in health- care. *Journal of healthcare engineering*, 2019.

Tan, M. & Le, Q. V. (2019). Efficientnet: Rethinking model scaling for convolutional neural networks. *arXiv preprint arXiv*:1905.11946.

Wang, X., Peng, Y., Lu, L., Lu, Z., Bagheri, M., & Summers, R. M. (2017). Chestx-ray8: Hospital-scale chest x-ray database and benchmarks on weakly-supervised classifica-tion and localization of common thorax diseases. *Proceedings of the IEEE conference on com- puter vision and pattern recognition*, 2097–2106. IEEE.

Wozniak, M., PoYap, D., Capizzi, G., Sciuto, G. L., Kosmider, L., & Frankiewicz, K. (2018). Small lung nodules detection based on local variance analysis and probabilistic neural network. *Computer Methods and Programs in Biomedicine*, *161* ,173–180.

Xiao, Z., Du, N., Geng, L., Zhang, F., Wu, J., & Liu, Y. (2019). Multi-scale heterogeneous 3d cnn for false-positive reduction in pulmonary nodule detection, based on chest ct images. *Applied Sciences*, *9*(16), 3261.

Xu, S., Wu, H., & Bie, R. (2018). Cxnet-m1: Anomaly detection on chest x-rays with image-based deep learning. *IEEE Access*, *7*, 4466–4477.

Zhou, X., Wang, D., & Krahenbuhl, P. (2019). Objects as points.*arXiv preprint arXiv*:1904.07850

Index